Lecture Notes in Mathematics 2210

More information about this series at http://www.springer.com/series/304

Ta Quang Buu Library, Hanoi University of Science and Technology. Location of VIASM

VIASM – The Vietnamese Institute of Advanced Study in Mathematics, Hanoi

Mathematics in Vietnam goes back to ancient times. Over five hundred years ago in Hanoi the name of Luong Thê Vinh, an expert in geometry, was inscribed on a stele of honor in Vãn Miêú.

Over sixty years ago, the Viêt Minh published a geometry textbook written by Hoàng Tuy for schools in the liberated zones, a rare case of a guerrilla press publishing a mathematics book!

Founded in 2010 after the award of the Fields Medal to Ngo Baó Châu, the Vietnam Institute for Advanced Study in Mathematics VIASM officially opened in Hanoi in 2011, aiming to become a leading research center where Vietnamese mathematicians can develop projects and nurture young talent. Ngo Baó Châu, one of the initiators, became the scientific director in 2011.

VIASM engages in traditional research areas of pure and applied mathematics, as well as applying mathematics in other fields such as physics, computer science, biology and economics. The main activity of the Institute is the organization of research groups to conduct high quality research programs and projects. International and Vietnamese scientists in the same field gather and work together at the Institute. VIASM organizes conferences, workshops, seminars on topics associated with research groups working at the Institute, special schools for mathematics students, short-term training courses for mathematics teachers and common activities to disseminate scientific knowledge to the public and support the application of mathematics in socio-economic development.

The VIASM subseries of the Lecture Notes in Mathematics publishes high quality original articles or survey papers on topics of current interest. They are based on lectures delivered in special periods organized at the Vietnam Institute for Advanced Study in Mathematics (VIASM). With the agreement of the Editors of the LNM Series, and as a temporary arrangement, the first volumes are not subjected to the strict LNM rules of coherency for multi-author volumes.

VIASM

Nguyen Tu CUONG • Le Tuan HOA •
Ngo Viet TRUNG
Editors

Commutative Algebra and its Interactions to Algebraic Geometry

VIASM 2013–2014

Springer

Editors
Nguyen Tu CUONG
Institute of Mathematics
Vietnam Academy of Science
and Technology
Hanoi, Vietnam

Le Tuan HOA
Institute of Mathematics
Vietnam Academy of Science
and Technology
Hanoi, Vietnam

Ngo Viet TRUNG
Institute of Mathematics
Vietnam Academy of Science
and Technology
Hanoi, Vietnam

ISSN 0075-8434 ISSN 1617-9692 (electronic)
Lecture Notes in Mathematics
ISBN 978-3-319-75564-9 ISBN 978-3-319-75565-6 (eBook)
https://doi.org/10.1007/978-3-319-75565-6

Library of Congress Control Number: 2018942527

Mathematics Subject Classification (2010): 13-XX, 14-XX

Printed on acid-free paper

This Springer imprint is published by the registered company Springer International Publishing AG part of Springer Nature.
The registered company address is: Gewerbestrasse 11, 6330 Cham, Switzerland

Preface

This collection of notes is based on four lectures given during the programme *Commutative Algebra* at the Vietnam Institute of Advanced Study in Mathematics in the winter semester 2013–2014. The lectures provide introductions to recent research topics in Commutative Algebra, which are related to Algebraic Geometry and other fields. The topics were chosen to represent different aspects of the use of the basic tools of Commutative Algebra. The notes are mainly self-contained, with the hope that students with advanced backgrounds in algebra can get through and absorb different techniques and ideas in Commutative Algebra before settling on concrete research problems. They can also be used separately as courses for graduate students, depending on the level and interest of the students.

The first lecture, by M. Brodmann, offers an introduction to the theory of rings of differential operators and their modules, also known as Weyl algebras and D-modules. These concepts relate Non-commutative Algebra and Commutative Algebra with Algebraic Geometry and Analysis in a very appealing way. The lecture presents this theory from the viewpoint of Commutative Algebra and is aimed at an audience having only a basic background in Commutative Algebra. The main feature is therefore not to explain everything about Weyl algebras and D-modules, but only the relevant aspects which are directly related to Commutative Algebra, such as the characteristic variety via the theory of filtered algebras and modules. The last part also contains some recent results on the stability, deformation and defining equations of the characteristic variety. The material is developed systematically and is accompanied by examples and exercises. These notes are well suited for an undergraduate course.

The second lecture, by J. Elias, is a short introduction to the theory of inverse systems and its application in the classification of Artinian Gorenstein rings. The classification of Artinian rings (rings of finite length) up to analytic isomorphism is a basic problem in Commutative Algebra and Algebraic Geometry. This problem is even open for Artinian Gorenstein rings, when the ring is an injective module over itself. Inverse systems provide an important tool in Commutative Algebra, establishing a beautiful correspondence between Artinian Gorenstein quotient rings and certain polynomials via derivations. The notes give a thorough introduction to the theory of injective modules and inverse systems and show how to use these tools to classify Artinian Gorenstein rings and to compute their Betti numbers. The presented material combines several basic techniques of Commutative Algebra and could be used for a graduate course.

The third lecture, by R.M. Miró-Roig, is on the complexity of the structure of projective varieties. This complexity can be measured by the representation type, which is the dimension and the number of families of indecomposable arithmetically Cohen–Macaulay sheaves (i.e. sheaves without intermediate cohomology) on the underlying variety. This is a fascinating topic of Algebraic Geometry, which requires an advanced background in Commutative Algebra. The notes cover the basic facts on this and related subjects such as moduli spaces of sheaves, liaison theory, minimal resolutions and Hilbert schemes of points. Many interesting results are presented on arithmetically Cohen–Macaulay sheaves and bundles having natural extremal algebraic properties, and several examples of varieties of wild representation type are given. The exposition is self-contained and features numerous open problems and promising ideas for further investigation. It may serve as a graduate course in Algebraic Geometry.

The last lecture, by M. Morales, addresses a classical problem of both Commutative Algebra and Algebraic Geometry, namely, how many equations are needed to define an algebraic variety set-theoretically. This seemingly simple problem is wide open even for toric varieties, which are given parametrically by monomials. The notes provide an extensive survey on this problem in the case of simplicial toric varieties, which are defined by the property that the exponents of the parametrizing monomials span a simplicial complex. One can use arithmetical and combinatorial tools (semigroups, lattices) to obtain satisfactory results for large classes of simplicial toric varieties. The material is presented in a systematic way and can easily be followed by any reader with some basic background in Commutative Algebra. These notes are recommended as a first course for anyone who wants to see the interaction between algebra, combinatorics and geometry. They can be used as a starting point for graduate studies in Commutative Algebra.

Hanoi, Vietnam
14 October 2017

Nguyen Tu CUONG
Le Tuan HOA
Ngo Viet TRUNG

Contents

Contributors

Markus Brodmann Universität Zürich, Institut für Mathematik, Zürich, Switzerland

Juan Elias Department de Matemàtiques i Informàtica, Universitat de Barcelona, Barcelona, Spain

Rosa M. Miró-Roig Department de Matemàtiques i Informàtica, Universitat de Barcelona, Barcelona, Spain

Marcel Morales Université Grenoble Alpes, Institut Fourier, UMR 5582, Saint-Martin D'Hères Cedex, France

Chapter 1
Notes on Weyl Algebra and D-Modules

Markus Brodmann

Abstract Weyl algebras, sometimes called algebras of differential operators, are a fascinating and important subject, which relates Non-Commutative and Commutative Algebra, Algebraic Geometry and Analysis in very appealing way. The theory of modules over Weyl algebras, sometimes called D-modules, finds application in the theory of partial differential equations, and thus has a great impact to many fields of Mathematics. In our course, we shall give a short introduction to the subject, using only prerequisites from Linear Algebra, Basic Abstract Algebra, and Basic Commutative Algebra. In addition, in the last two sections, we present a few recent results.

1.1 Introduction

The present notes base on two short courses:

(1) *Introduction to Weyl Algebras*: five Twin Lessons, Thai Nguyen University of Science TNUS (Thai Nguyen, Vietnam), November 1–10, 2013.
(2) *Weyl Algebras, Universal Gröbner Bases, Filtration Deformations and Characteristic Varieties of D-Modules*: four Twin Lessons, Vietnam Institute for Advanced Study in Mathematics VIASM (Hanoi, Vietnam), November 12–26, 2013.

They were also the base for a third course:

(3) *Introduction to Weyl Algebras and D-Modules*: four Lessons and two Tutorial Sessions, "Workshop on Local Cohomology", St. Joseph's College Irinjalakuda, Kerala (India), June 20–July 2, 2016.

M. Brodmann (✉)
Universität Zürich, Institut für Mathematik, Zürich, Switzerland
e-mail: brodmann@math.uzh.ch

N. Tu CUONG et al. (eds.), *Commutative Algebra and its Interactions to Algebraic Geometry*, Lecture Notes in Mathematics 2210,
https://doi.org/10.1007/978-3-319-75565-6_1

These notes aim to give an approach to Boldini's stability and deformation results for characteristic varieties [11, 12] and to the bounding result [16] for the degrees of defining equations of characteristic varieties, including a self-contained introduction to the needed background on Weyl Algebras and D-modules. In particular, these notes should not be understood as an independent or complete introduction to the field of Weyl algebras and D-modules, which could replace one of the existing textbooks or monographs like [8, 9, 13, 24, 29, 37] or [38]: Too many core subjects are not treated at all or only marginally in these notes, as they are not needed on the way to our final results.

So, a few basic topics which are lacking in these notes—and which ought to be considered as indispensable in a complete introduction to the field—are:

- a systematic study of Bernstein's Inequality and holonomic D-modules (we treat these subjects only briefly in Exercise and Remark 1.14.3),
- Bernstein's result on singularities of generalized Γ-functions and Bernstein-Sato polynomials,
- weighted filtrations with negative weights,
- the sheaf theoretic and cohomological aspect,
- the analytic aspect.

Another subject which is not treated in our notes are Lyubeznik's finiteness results for local cohomology modules of regular local rings in characteristic 0 (see [33] and also [34]), which brought a break-through in Commutative Algebra, as they base on the use of (holonomic) D-modules—and hence present a very important link between these two fields.

These notes are divided up in 14 sections:

1. *Introduction*
2. *Filtered Algebras*
3. *Associated Graded Rings*
4. *Derivations*
5. *Weyl Algebras*
6. *Arithmetic in Weyl Algebras*
7. *The Standard Basis*
8. *Weighted Degrees and Filtrations*
9. *Weighted Associated Graded Rings*
10. *Filtered Modules*
11. *D-Modules*
12. *Gröbner Bases*
13. *Weighted Orderings*
14. *Standard Degree and Hilbert Polynomials*

Sections 1.1–1.9 were the subject of the introductory course (1) at TNUS. In our course (2) at the VIASM we gave an account on all 14 sections, and discussed a few applications (to the Gelfand-Kirillow dimension of D-modules for example), which are not contained in these notes. In the course (3) at St. Joseph's College we treated Sects. 1.1–1.9 and 1.14.

Our suggested basic reference is Coutinho's introduction [24], although we do not follow that introduction and we partly use our own terminology and notations. We start in a slightly more general setting, than Coutinho, and so also we recommend the references [4, 10, 11, 32] and [35]. Files of [10] and [11] are available on request at the author. For readers who have already some background in the subject, we recommend as possible references [8, 9, 13, 29, 37], or the first part of the PhD thesis [11].

Finally, we aim to fix a few notations and make a few conventions which we shall keep throughout these notes. We do this on a fairly elementary level, according to the original intention of the short course (1): To give a first introduction to the subject to an audience having only some background in Linear Algebra and basic Abstract Algebra. Only in Sect. 1.14 we will need some background from Commutative Algebra, notably Hilbert functions and -polynomials, Local Cohomology and Castelnuovo-Mumford regularity. We shall give brief reminders on these more advanced preliminaries in Sect. 1.14.

Conventions, Reminders and Notations 1.1.1 (A) *(General Notations)* By $\mathbb{Z}$, $\mathbb{Q}$ and $\mathbb{R}$ we respectively denote the set of integers, of rationals and of real numbers. We also write

$$\mathbb{R}_{\geq 0} := \{x \in \mathbb{R} \mid x \geq 0\} \text{ and } \mathbb{R}_{>0} := \{x \in \mathbb{R} \mid x > 0\}$$

for the set of non-negative respectively of positive real numbers. Moreover, we use the following notations for the set of non-negative respectively the set of positive integers:

$$\mathbb{N}_0 := \mathbb{Z} \cap \mathbb{R}_{\geq 0} \text{ and } \mathbb{N} := \mathbb{Z} \cap \mathbb{R}_{>0} = \mathbb{N}_0 \setminus \{0\}.$$

If $\mathbb{S} \subset \mathbb{R}$ we form the supremum and infimum $\sup(\mathbb{S})$ resp. $\inf(\mathbb{S})$ within the set $\mathbb{R} \cup \{-\infty, \infty\}$, using the convention that $\sup(\emptyset) = -\infty$ and $\inf(\emptyset) = \infty$.
Empty sums and empty products are respectively understood to be 0 or 1. We thus set

$$\sum_{i=0}^{-1} x_i := 0 \text{ and } \prod_{i=0}^{-1} x_i := 1 \text{ with } x_1, x_2, \cdots \in \mathbb{R}.$$

(B) *(Rings)* All rings R are understood to be associative, non-trivial and unital, so that they have a unit-element $1 = 1_R \in R \setminus \{0\}$ and the following properties hold

(a) $0x = x0 = 0$ and $1x = x1 = x$ for all $x \in R$;
(b) $x(yz) = (xy)z$, $x(y+z) = xy+xz$ and $(x+y)z = xz+yz$ for all $x, y, z \in R$.

Rings need not be commutative.
If R is a ring, a *subring* of R is a subset $R_0 \subseteq R$, such that $1_R \in R_0$ and $x+y, xy \in R_0$ whenever $x, y \in R_0$. If $R_0 \subseteq R$ is a subring and $S \subseteq R$ is an arbitrary subset, we write $R_0[S]$ for the *subring of* R *generated by* R_0 *and* S, hence for the smallest

subring of R which contains R_0 and S. Thus, $R_0[S]$ is the intersection of all subrings of R which contain R_0 and S, and may be written in the form

$$R_0[S] = \{\sum_{i=1}^{r}\prod_{j=1}^{k_r} a_{i,j} \mid r, k_1, \ldots, k_r \in \mathbb{N}, a_{i,j} \in R_0 \cup S, \forall i \leq r, \forall j \leq k_i\}.$$

If $a_1, a_2, \ldots, a_r$ is a finite collection of elements of R, we set

$$R_0[a_1, a_2, \ldots, a_r] := R_0[\{a_1, a_2, \ldots, a_r\}].$$

(C) *(Homomorphisms of Rings)* All *homomorphisms of rings* are understood to be unital, and hence are maps $h : R \longrightarrow S$, with R and S rings, such that

(a) $h(x + y) = h(x) + h(y)$ and $h(xy) = h(x)h(y)$ for all $x, y \in R$;
(b) $h(1_R) = 1_S$.

Clearly, the identity map $\mathrm{Id}_R : R \longrightarrow R$ is a homomorphism of rings, and the composition of homomorphisms of rings is again a homomorphism of rings. An *isomorphism of rings* is a homomorphism of rings admitting an inverse homomorphism. A homomorphism of rings is an isomorphism, if and only if it is bijective.

(D) (K*-Algebras*) All fields are considered as commutative. If K is a field, a K-algebra is understood to be a ring A together with a homomorphism of rings $\varepsilon : K \longrightarrow A$ such that

$$\varepsilon(c)a = a\varepsilon(c) \text{ for all } c \in K \text{ and all } a \in A.$$

As the ring A is non-trivial, the homomorphism $\varepsilon : K \longrightarrow A$ is injective. So, we can and do always embed K into A by means of ε and thus identify c with $\varepsilon(c)$ for all $c \in K$. Hence we have

$$c := \varepsilon(c) = c1_A = 1_A c \text{ and } ca = ac \text{ for all } c \in K \text{ and all } a \in A.$$

Keep in mind, that a K-algebra A is a K-vector space in a natural way.

(E) (*Homomorphisms of* K*-Algebras*) Let K be a field. A homomorphism of K-algebras $h : A \longrightarrow B$ is a map with K-algebras A and B such that:

(a) $h : A \longrightarrow B$ is a homomorphism of rings;
(b) $h(c) = c$ for all $c \in K$.

Observe, that a homomorphism of K-algebras is also a homomorphism of K-vector spaces.

(F) *(Modules)* We usually shall consider unital left-modules, hence modules M over a ring R, such that

$$x(m + n) = xm + xn, \quad (x + y)m = xm + ym, \quad (xy)m = x(ym) \text{ and } 1m = m$$

for all $x, y \in R$ and all $m, n \in M$. We shall refer to left-modules just as modules. By a *homomorphism of R-modules* we mean a map $h : M \longrightarrow N$, with M and N both R-modules, such that

(a) $h(m+n) = h(m) + h(n)$ for all $m, n \in M$.
(b) $h(xm) = xh(m)$ for all $x \in R$ and all $m \in M$.

A *submodule* of a R-module M is a subset $N \subseteq M$, such that $m + n \in N$ and $xm \in N$ whenever $m, n \in N$ and $x \in R$. Clearly $0 := \{0\}$ and M are submodules of M.
If $h : M \longrightarrow N$ is a homomorphism of R-modules, the *kernel* $\mathrm{Ker}(h) := \{m \in M \mid h(m) = 0\}$ of h is a submodule of M and the *image* $\mathrm{Im}(h) := h(M)$ of h is a submodule of N.
A sequence of (homomorphisms of) R-modules

$$M_0 \xrightarrow{h_0} M_1 \xrightarrow{h_1} M_2 \cdots M_{i-1} \xrightarrow{h_{i-1}} M_i \xrightarrow{h_i} M_{i+1} \cdots M_{r-1} \xrightarrow{h_{r-1}} M_r$$

is said to be *exact* if $\mathrm{Ker}(h_i) = \mathrm{Im}(h_{i-1})$ for all $i = 1, 2, \ldots, r-1$. A *short exact sequence* of R-modules is an exact sequence of the form $0 \longrightarrow M \xrightarrow{h} N \xrightarrow{l} P \longrightarrow 0$, meaning that h is injective, l is surjective and $\mathrm{Ker}(l) = \mathrm{Im}(h)$.
The *annihilator* of an R-module M is defined as the left ideal of R consisting of all elements which annihilate M, thus:

$$\mathrm{Ann}_R(M) := \{x \in R \mid xM = 0\}.$$

(G) *(Noetherian Modules and Rings)* Let R be a ring. A left R-module is said to be *Noetherian*, if it satisfies the following equivalent conditions

(i) Each left submodule $N \subseteq M$ if finitely generated, and hence of the form $N = \sum_{i=1}^{r} Rn_i$ with $r \in \mathbb{N}_0$ and $n_1, n_2, \ldots, n_r \in N$.
(ii) Each ascending sequence $N_0 \subseteq N_1 \subseteq \cdots N_i \subseteq N_{i+1} \subseteq \cdots$ of left submodules $N_i \subseteq M$ ultimately becomes stationary and thus satisfies $N_{i_0} = N_{i_0+1} = N_{i_0+2} = \ldots$ for some $i_0 \in \mathbb{N}_0$.

We say that the ring R is *left Noetherian* if it is Noetherian as a left module.
Keep in mind the following facts:

(a) If $0 \longrightarrow N \longrightarrow M \longrightarrow P \longrightarrow 0$ is an exact sequence of left R-modules then M is Noetherian if and only N and P are both Noetherian.
(b) If M and N are two Noetherian left R-modules, then their direct sum $M \oplus N$ is Noetherian, too.
(c) If R is left Noetherian, a left R-module M is Noetherian if and only if it is finitely generated.

(H) *(Modules of Finite Presentation)* Let R be a ring. A left R-module M is said to be of *finite presentation* if there is an exact sequence of left R-modules

$$R^s \xrightarrow{h} R^r \longrightarrow M \longrightarrow 0 \quad \text{with } r, s \in \mathbb{N}_0.$$

In this situation, the above exact sequence is called a *(finite) presentation* of M and $R^s \xrightarrow{h} R^r$ is called a *presenting homomorphism* for M.
Keep in mind, that the presenting homomorphism is given by a matrix with entries in R, more precisely: There is a matrix

$$A = \begin{pmatrix} a_{11} & a_{12} & \dots & a_{1r} \\ a_{21} & a_{22} & \dots & a_{2r} \\ \vdots & \vdots & & \vdots \\ a_{s1} & a_{s2} & \dots & a_{sr} \end{pmatrix} \in R^{s \times r} \text{ such that}$$

$$h(x_1, x_2, \dots, x_s) = (x_1, x_2, \dots, x_s)A = \Big(\sum_{i=1}^{s} x_i a_{i1}, \sum_{i=1}^{s} x_i a_{i2}, \dots, \sum_{i=1}^{s} x_i a_{ir}\Big)$$

for all $(x_1, x_2, \dots, x_s) \in R^s$. This matrix A is called a *presentation matrix* for M.
Note the following facts:

(a) A left R-module M of finite presentation is finitely generated.
(b) If R is left Noetherian, then each finitely generated left R-module is of finite presentation.

(I) *(Graded Rings and Modules)* A *(positively) graded ring* is a ring R together with a family $(R_i)_{i \in \mathbb{N}_0}$ of additive subgroups $R_i \subseteq R$ such that

(1) $R = \bigoplus_{i \in \mathbb{N}_0} R_i$;
(2) $1 \in R_0$;
(3) for all $i, j \in \mathbb{N}_0$ and all $a \in R_i$ and all $b \in R_j$ it holds $ab \in R_{i+j}$.

In this situation we also refer to $R = \bigoplus_{i \in \mathbb{N}_0} R_i$ as (positively) *graded* R_0*-algebra*. If $a \in R_i \setminus \{0\}$, we call a a *homogeneous element of degree* i.
Let $R' = \bigoplus_{i \in \mathbb{N}_0} R'_i$ be a second graded ring. A *homomorphism of graded rings* is a homomorphism $f : R \longrightarrow R'$ of rings which *respects gradings*, hence such that $f(R_i) \subseteq R'_i$ for all $i \in \mathbb{N}_0$. Clearly, the identity map $\mathrm{Id}_R : R \longrightarrow R$ of a graded ring as well as the composition of two homomorphisms of graded rings is a homomorphism of graded rings. An *isomorphism of graded rings* is a homomorphism of graded rings which admits an inverse which is a homomorphism of graded rings—or, equivalently—a bijective homomorphism of graded rings.
The (positively) graded ring $R = \bigoplus_{i \in \mathbb{N}_0} R_i$ is called a *homogeneous ring* if it is generated over R_0 by homogeneous elements of degree 1, hence if (in the notation introduced in part (B)) we have $R = R_0[R_1]$.

A *graded (left) module* over the graded ring R is a left R-module together with a family $(M_j)_{j\in\mathbb{Z}}$ of additive subgroups $M_j \subseteq M$ such that

(1) $M = \bigoplus_{j\in\mathbb{Z}} M_j$;
(2) For all $i \in \mathbb{N}_0$, all $j \in \mathbb{Z}$, all $a \in R_i$ and all $m \in M_j$ it holds $am \in M_{i+j}$.

A *homomorphism of graded (left) modules* is a homomorphism $h : M \longrightarrow N$ of R-modules, in which $M = \bigoplus_{j\in\mathbb{Z}} M_j$ and $N = \bigoplus_{j\in\mathbb{Z}} N_j$ are both graded and $h(M_j) \subseteq N_j$ for all $j \in \mathbb{Z}$. Clearly, the identity map of a graded R-module and the composition of two homomorphisms of graded R-modules are again homomorphisms of graded R-modules. An *isomorphism of graded R-modules* is a homomorphism of graded R-modules which admits an inverse which is a homomorphism of graded R-modules—or, equivalently—a bijective homomorphism of graded R-modules.

(K) *(Prime Varieties)* Let R be a commutative ring. We denote the *prime spectrum* of R, hence the set of all prime ideals in R, by $\mathrm{Spec}(R)$. If $\mathfrak{a} \subseteq R$ is an ideal, we denote by $\mathrm{Var}(\mathfrak{a})$ the *prime variety* of $\mathfrak{a}$, thus

$$\mathrm{Var}(\mathfrak{a}) := \{\mathfrak{p} \in \mathrm{Spec}(R) \mid \mathfrak{a} \subseteq \mathfrak{p}\}.$$

Let

$$\sqrt{\mathfrak{a}} := \{a \in R \mid \exists n \in \mathbb{N} : a^n \in \mathfrak{a}\}.$$

denote the *radical ideal* of $\mathfrak{a}$. Keep in mind the following facts:

(a) $\mathrm{Var}(\mathfrak{a}) = \mathrm{Var}(\sqrt{\mathfrak{a}})$.
(b) If $\mathfrak{a}, \mathfrak{b} \subseteq R$ are ideals, then $\mathrm{Var}(\mathfrak{a}) = \mathrm{Var}(\mathfrak{b})$ if and only if $\sqrt{\mathfrak{a}} = \sqrt{\mathfrak{b}}$.

(L) *(Krull Dimension)* Let R be as in part (K) and let M be a finitely generated R-module. Then, the *(Krull) dimension* $\dim_R(M)$ of M is defined as the supremum of the lengths of chains of prime ideals which can be found in the prime variety of the annihilator of M:

$$\dim_R(M) := \sup\{r \in \mathbb{N}_0 \mid \exists \mathfrak{p}_0, \ldots, \mathfrak{p}_r \in \mathrm{Var}\big(\mathrm{Ann}_R(M)\big) \text{ with } \mathfrak{p}_{i-1} \subsetneq \mathfrak{p}_i \text{ for } i = 1, \ldots, r\}.$$

In particular, the (Krull) dimension $\dim(R)$ of R is the dimension of the R-module R:

$$\dim(R) = \sup\{r \in \mathbb{N}_0 \mid \exists \mathfrak{p}_0, \ldots, \mathfrak{p}_r \in \mathrm{Spec}(R) \text{ with } \mathfrak{p}_{i-1} \subsetneq \mathfrak{p}_i \text{ for } i = 1, \ldots, r\}.$$

Before giving a formal acknowledgement, the author likes very much to express his gratitude toward his Vietnamese fellow mathematicians, who gave him the chance to visit the Country so many times, to teach several invited short courses, to present talks and to discuss on Mathematics at various Universities since his first visit of Vietnam in 1996. He also looks back with pleasure to the many visits of Vietnamese mathematicians in Zürich as well as the numerous mathematical cooperations and the many personal friendships which resulted from them.

Acknowledgements

(1) The author expresses his gratitude toward the VIASM and the Mathematical Institute of the Vietnam Academy of Science and Technology MIVAST in Hanoi for the invitation and generous financial and institutional support during his stay in Vietnam in October–December 2013, but also toward the Universities of Thai Nguyen, of Hué and toward the Ho Chi Minh City University of Education for their intermediate invitations and financial support.
(2) The author expresses his gratitude toward the Indian National Centre for Mathematics of IIT-B and TIFR-Mumbai and toward St. Joseph's College in Irinjalakude, Kerala, India for their financial and institutional support during his visit in June-July 2016.
(3) In particular, the author to thanks to the referee, for his very careful and critical study of the manuscript, for his valuable historical and bibliographical hints—in particular to the paper [2] (see Conclusive Remark 1.14.10 (C))—and for his suggestions which helped to improve and clarify a number of arguments, and also for his long list of misprints.

1.2 Filtered Algebras

We begin with a few general preliminaries, which will pave our way to introduce and to treat Weyl algebras and D-modules. Our first preliminary theme are filtered algebras over a field. It will turn out later, that this concept is of basic significance for the theory of Weyl algebras.

Definition and Remark 1.2.1 (A) Let K be a field and let A be K-algebra (see Conventions, Reminders and Notations 1.1.1 (D)). By a *filtration* of A we mean a family

$$A_\bullet = (A_i)_{i \in \mathbb{N}_0}$$

such that the following conditions hold:

(a) Each A_i is a K-vector subspace of A;
(b) $A_i \subseteq A_{i+1}$ for all $i \in \mathbb{N}_0$;
(c) $1 \in A_0$;
(d) $A = \bigcup_{i \in \mathbb{N}_0} A_i$;
(e) $A_i A_j \subseteq A_{i+j}$ for all $i, j \in \mathbb{N}_0$.

In requirement (e) we have used the standard notation

$$A_i A_j := \sum_{(f,g) \in A_i \times A_j} Kfg \text{ for all } i, j \in \mathbb{N}_0,$$

which we shall use from now on without further mention. To simplify notation, we also often set

$$A_i = 0 \text{ for all } i < 0$$

and then write our filtration in the form

$$A_\bullet = (A_i)_{i\in\mathbb{Z}}.$$

If a filtration of A is given, we say that $(A, A_\bullet)$ or—by abuse of language—that A is a *filtered K-algebra.*

(B) Keep the notations and hypotheses of part (A) and let $A_\bullet = (A_i)_{i\in\mathbb{Z}}$ be a filtered K-algebra. Observe that we have the following statements:

(a) A_0 is a K-subalgebra of A.
(b) For all $i \in \mathbb{Z}$ the K-vector space A_i is a left- and a right- A_0-submodule of A.

Example 1.2.2 (The Degree Filtration of a Commutative Polynomial Ring) Let $n \in \mathbb{N}$ and let $A = K[X_1, X_2, \ldots, X_n]$ be the commutative polynomial algebra over the field K in the indeterminates $X_1, X_2, \ldots, X_n$. Then clearly A is a K-space over its *monomial basis*:

$$A = K[X_1, X_2, \ldots, X_n] = \bigoplus_{\nu_1,\nu_2,\ldots,\nu_n\in\mathbb{N}_0} KX_1^{\nu_1}X_2^{\nu_2}\ldots X_n^{\nu_n} = \bigoplus_{\underline{\nu}\in\mathbb{N}_0^n} K\underline{X}^{\underline{\nu}},$$

where we have used use the standard notation

$$\underline{X}^{\underline{\nu}} := X_1^{\nu_1}X_2^{\nu_2}\ldots X_n^{\nu_n}, \text{ for } \underline{\nu} := (\nu_1, \nu_2 \ldots \nu_n) \in \mathbb{N}_0^n.$$

So, each $f \in A$ can be written as

$$f = \sum_{\underline{\nu}\in\mathbb{N}_0^n} c_{\underline{\nu}}^{(f)}\underline{X}^{\underline{\nu}}$$

with a unique family

$$\left(c_{\underline{\nu}}^{(f)}\right)_{\underline{\nu}\in\mathbb{N}_0^n} \in \prod_{\underline{\nu}\in\mathbb{N}_0^n} K = K^{\mathbb{N}_0^n},$$

whose *support*

$$\operatorname{supp}(f) = \operatorname{supp}\left(\left(c_{\underline{\nu}}^{(f)}\right)_{\underline{\nu}\in\mathbb{N}_0^n}\right) := \{\underline{\nu} \in \mathbb{N}_0^n \mid c_{\underline{\nu}}^{(f)} \neq 0\}$$

is finite. We also introduce the notation

$$|\underline{\nu}| = \sum_{i=1}^{n} \nu_i, \text{ for } \underline{\nu} = (\nu_1, \nu_2, \ldots, \nu_n) \in \mathbb{N}_0^n.$$

Then, with the usual convention of Conventions, Reminders and Notations 1.1.1 (A) we may describe the *degree* of the polynomial $f \in A$ by

$$\deg(f) := \sup\{|\underline{\nu}| \mid c_{\underline{\nu}}^{(f)} \neq 0\} = \sup\{|\underline{\nu}| \mid \underline{\nu} \in \operatorname{supp}(f)\}.$$

Now, for each $i \in \mathbb{N}_0$ we introduce the K-subspace A_i of A which is given by

$$A_i := \{f \in A \mid \deg(f) \leq i\} = \bigoplus_{\underline{\nu} \in \mathbb{N}_0^n \text{ with } |\underline{\nu}| \leq i} K\underline{X}^{\underline{\nu}}.$$

With the usual convention that $u + (-\infty) = -\infty$ for all $u \in \mathbb{Z} \cup \{-\infty\}$, we have the obvious relation

$$\deg(fg) = \deg(f) + \deg(g) \text{ for all } f, g \in A = K[X_1, X_2, \ldots, X_n].$$

From this it follows easily:

$$\text{The family } A_\bullet = \big(A_i := \{f \in A \mid \deg(f) \leq i\}\big)_{i \in \mathbb{N}_0}$$

is a filtration of A. This filtration is called the *degree filtration* of the polynomial algebra $A = K[X_1, X_2, \ldots, X_n]$.

Clearly filtrations also may occur in non-commutative algebras. The next example presents somehow the "generic occurrence" of this.

Example 1.2.3 (The Degree Filtration of a Free Associative Algebra) Let $n \in \mathbb{N}$, let K be a field and let $A = K\langle X_1, X_2, \ldots, X_n\rangle$ be the free associative algebra over K in the indeterminates $X_1, X_2, \ldots, X_n$. We suppose in particular that (see Conventions, Reminders and Notations 1.1.1 (D))

$$cX_i = X_i c \text{ for all } c \in K \text{ and all } i = 1, 2, \ldots, n,$$

and hence

$$cf = fc \text{ for all } c \in K \text{ and all } f \in A.$$

Let $i \in \mathbb{N}_0$. If

$$\underline{\sigma} = (\sigma_1, \sigma_2, \ldots, \sigma_i) \in \{1, 2, \ldots, n\}^i$$

is a sequence of length i with values in the set $\{1, 2, \ldots, n\}$ we write

$$\underline{X_{\sigma}} := \prod_{j=1}^{i} X_{\sigma_j} = X_{\sigma_1} X_{\sigma_2} \ldots X_{\sigma_i}.$$

Then, with the usual convention that the product $\prod_{j\in\emptyset} X_j$ of an empty family of factors equals 1 and using the notation

$$\mathbb{S}_n := \left\{\{1, 2, \ldots, n\}^i \mid i \in \mathbb{N}_0\right\}$$

we can write A as a K-space over its *monomial basis* as follows:

$$\begin{aligned} A &= K\langle X_1, X_2, \ldots, X_n\rangle \\ &= \bigoplus_{i\in\mathbb{N}_0} \bigoplus_{(\sigma_1,\sigma_2\ldots\sigma_i)\in\{1,2,\ldots,n\}^i} K X_{\sigma_1} X_{\sigma_2} \ldots X_{\sigma_i} \\ &= \bigoplus_{i\in\mathbb{N}_0} \bigoplus_{\underline{\sigma}\in\{1,2,\ldots,n\}^i} K \underline{X_{\sigma}} \\ &= \bigoplus_{\underline{\sigma}\in\mathbb{S}_n} K \underline{X_{\sigma}}. \end{aligned}$$

Clearly, as in the case of a commutative polynomial ring, each $f \in A$ may be written in the form

$$f = \sum_{\underline{\sigma}\in\mathbb{S}_n} c_{\underline{\sigma}}^{(f)} \underline{X_{\sigma}}$$

with a unique family

$$\left(c_{\underline{\sigma}}^{(f)}\right)_{\underline{\sigma}\in\mathbb{S}_n} \in \prod_{\underline{\sigma}\in\mathbb{S}_n} K = K^{\mathbb{S}_n},$$

whose *support*

$$\operatorname{supp}(f) = \operatorname{supp}\left((c_{\underline{\sigma}}^{(f)})_{\underline{\sigma}\in\mathbb{S}_n}\right) := \{\underline{\sigma} \in \mathbb{S}_n \mid c_{\underline{\sigma}}^{(f)} \neq 0\}$$

is finite. We also introduce the notion of *length* of a sequence $\underline{\sigma} \in \mathbb{S}_n$ by setting

$$\lambda(\underline{\sigma}) := i, \text{ if } \underline{\sigma} \in \{1, 2, \ldots, n\}^i.$$

Now, we may define the *degree* of an element $f \in A$ by

$$\deg(f) := \sup\{\lambda(\underline{\sigma}) \mid c_{\underline{\sigma}}^{(f)} \neq 0\} = \sup\{\lambda(\underline{\sigma}) \mid \underline{\sigma} \in \operatorname{supp}(f)\}.$$

For each $i \in \mathbb{N}_0$ we introduce a K-subspace A_i of A, by setting

$$A_i := \{f \in A \mid \deg(f) \leq i\} = \bigoplus_{\underline{\sigma} \in \mathbb{S}_n \text{ with } \lambda(\underline{\sigma}) \leq i} K\underline{X}_{\underline{\sigma}}.$$

We obviously have the relation

$$\deg(fg) \leq \deg(f) + \deg(g) \text{ for all } f, g \in A = K\langle X_1, X_2, \ldots, X_n\rangle.$$

Moreover, it is easy to see:

$$\text{The family } A_\bullet = \big(A_i = \{f \in A \mid \deg(f) \leq i\}\big)_{i \in \mathbb{N}_0}$$

is a filtration of A. This filtration is called the *degree filtration* of the free associative K-algebra $A = K\langle X_1, X_2, \ldots, X_n\rangle$.

Later, our basic filtered algebras will be Weyl algebras. These are non-commutative too, but they also admit the notion of degree and of degree filtration. From the point of view of filtrations, these algebras will turn out to be "close to commutative", as we shall see later. To make this more precise, we will introduce the notion of associated graded ring with respect to a filtration in the next section.

1.3 Associated Graded Rings

Remark and Definition 1.3.1 (A) Let K be a field and let $A = (A, A_\bullet)$ be a filtered K-algebra. We consider the K-vector space

$$\mathrm{Gr}(A) = \mathrm{Gr}_{A_\bullet}(A) = \bigoplus_{i \in \mathbb{N}_0} A_i/A_{i-1}.$$

For all $i \in \mathbb{N}_0$ we also use the notation

$$\mathrm{Gr}(A)_i = \mathrm{Gr}_{A_\bullet}(A)_i := A_i/A_{i-1},$$

so that we may write

$$\mathrm{Gr}(A) = \mathrm{Gr}_{A_\bullet}(A) = \bigoplus_{i \in \mathbb{N}_0} \mathrm{Gr}_{A_\bullet}(A)_i.$$

(B) Let $i, j \in \mathbb{N}_0$, let $f, f' \in A_i$ and let $g, g' \in A_j$ such that

$$h := f - f' \in A_{i-1} \text{ and } k := g - g' \in A_{j-1}.$$

It follows that

$$\begin{aligned} fg - f'g' = fg - (f-h)(g-k) &= fk + hg - hk \\ &\in A_iA_{j-1} + A_{i-1}A_j + A_{i-1}A_{j-1} \subseteq \\ &\subseteq A_{i+(j-1)} + A_{j+(i-1)} + A_{(i-1)+(j-1)} \subseteq A_{i+j-1}. \end{aligned}$$

So in $A_{i+j}/A_{i+j-1} = \mathrm{Gr}_{A_\bullet}(A)_{i+j} \subset \mathrm{Gr}_{A_\bullet}(A)$ we get the relation

$$fg + A_{i+j-1} = f'g' + A_{i+j-1}.$$

This allows to define a *multiplication* on the K-space $\mathrm{Gr}_{A_\bullet}(A)$ which is induced by

$$(f+A_{i-1})(g+A_{j-1}) := fg+A_{i+j-1} \text{ for all } i, j \in \mathbb{N}_0, \text{ all } f \in A_i \text{ and all } g \in A_j.$$

With respect to this multiplication, the K-vector space $\mathrm{Gr}_{A_\bullet}(A)$ acquires a structure of K-algebra.
Observe that, if $r, s \in \mathbb{N}_0$ and

$$\overline{f} = \sum_{i=0}^{r} \overline{f_i}, \text{ with } f_i \in A_i \text{ and } \overline{f_i} = (f_i + A_{i-1}) \in \mathrm{Gr}_{A_\bullet}(A)_i \text{ for all } i = 0, 1, \ldots, r,$$

and, moreover

$$\overline{g} = \sum_{j=0}^{s} \overline{g_j}, \text{ with } g_j \in A_j \text{ and } \overline{g_j} = (g_j + A_{j-1}) \in \mathrm{Gr}_{A_\bullet}(A)_j \text{ for all } j = 0, 1, \ldots, s,$$

then

$$\overline{f}\overline{g} = \sum_{k=0}^{r+s} \sum_{i+j=k} \overline{f_i}\overline{g_j} = \sum_{k=0}^{r+s} \sum_{i+j=k} (f_ig_j + A_{i+j-1}).$$

(C) Keep the above notations and hypotheses. Observe in particular, that $\mathrm{Gr}_{A_\bullet}(A)_0$ is a K-subalgebra of $\mathrm{Gr}_{A_\bullet}(A)$, and that there is an isomorphism of K-algebras

$$\mathrm{Gr}_{A_\bullet}(A)_0 \cong A_0.$$

Moreover, with respect to our multiplication on $\mathrm{Gr}_{A_\bullet}(A)$ we have the relations

$$\mathrm{Gr}_{A_\bullet}(A)_i \mathrm{Gr}_{A_\bullet}(A)_j \subseteq \mathrm{Gr}_{A_\bullet}(A)_{i+j} \text{ for all } i, j \in \mathbb{N}_0.$$

So, the K-vector space $\mathrm{Gr}_{A_\bullet}(A)$ is turned into a (positively) graded ring

$$\mathrm{Gr}_{A_\bullet}(A) = \big(\mathrm{Gr}_{A_\bullet}(A), (\mathrm{Gr}_{A_\bullet}(A)_i)_{i\in\mathbb{N}_0}\big) = \bigoplus_{i\in\mathbb{N}_0} \mathrm{Gr}_{A_\bullet}(A)_i$$

by means of the above multiplication. We call this ring the *associated graded ring* of A with respect to the filtration $A_\bullet$. From now on, we always furnish $\mathrm{Gr}_{A_\bullet}(A)$ with this multiplication.

Example and Exercise 1.3.2 (A) Let $n \in \mathbb{N}$, let K be a field and consider the commutative polynomial ring $A = K[X_1, X_2, \ldots, X_n]$. Show that A has the following *universal property* within the category of all commutative K-algebras:

If B is a commutative K-algebra and $\phi : \{X_1, X_2, \ldots, X_n\} \longrightarrow B$ is a map, then there is a unique homomorphism of K-algebras $\widetilde{\phi} : A \longrightarrow B$ such that $\widetilde{\phi}(X_i) = \phi(X_i)$ for all $i = 1, 2, \ldots, n$.

Show also, that A has the following *relational universal property* within the category of all associative K-algebras:

If B is an associative K-algebra and $\phi : \{X_1, X_2, \ldots, X_n\} \longrightarrow B$ is a map such that $\phi(X_i)\phi(X_j) = \phi(X_j)\phi(X_i)$ for all $i, j \in \{1, 2, \ldots, n\}$, then there is a unique homomorphism of K-algebras $\widetilde{\phi} : A \longrightarrow B$ such that $\widetilde{\phi}(X_i) = \phi(X_i)$ for all $i = 1, 2, \ldots, n$.

(B) Now, furnish $A = K[X_1, X_2, \ldots, X_n]$ with its degree filtration (see Example 1.2.2). Then, on use of the above universal property of A it is not hard to show that there is an isomorphism of K-algebras

$$K[X_1, X_2, \ldots, X_n] \xrightarrow{\cong} \mathrm{Gr}_{A_\bullet}(A),$$

given by $X_i \mapsto (X_i + A_0) \in A_1/A_0 = \mathrm{Gr}_{A_\bullet}(A)_1 \subset \mathrm{Gr}_{A_\bullet}(A)$ for all $i = 1, 2 \ldots, n$.

We now introduce a class of filtrations, which will be of particular interest for our lectures.

Definition 1.3.3 Let K be a field and let $A = (A, A_\bullet)$ be a filtered K-algebra. The filtration $A_\bullet$ is said to be *commutative* if

$$fg - gf \in A_{i+j-1} \text{ for all } i, j \in \mathbb{N}_0 \text{ and for all } f \in A_i \text{ and all } g \in A_j.$$

It is equivalent to say that the associated graded ring $\mathrm{Gr}_{A_\bullet}(A)$ is commutative. In this situation, we also say that $(A, A_\bullet)$ is a *commutatively filtered K-algebra*.

Later, in the case of Weyl algebras, we shall meet various interesting commutative filtrations—and precisely this makes these algebras to a subject which is intimately tied to Commutative Algebra. We now shall define three special types of commutative filtrations, which will play a particularly important rôle in Weyl algebras.

Definition and Remark 1.3.4 (A) Let $(A, A_\bullet)$ be a filtered K-algebra. The filtration $A_\bullet$ is said to be *very good* if it satisfies the following conditions:

(a) The filtration $A_\bullet$ is commutative;
(b) $A_0 = K$;

(c) $\dim_K(A_1) < \infty$;
(d) $A_i = A_1 A_{i-1}$ for all $i \in \mathbb{N}$.

Under these circumstances and on use of the notation introduced in Conventions, Reminders and Notations 1.1.1 (B) we clearly have

$$\dim_K(A_1/A_0) = \dim_K\big(\mathrm{Gr}_{A_\bullet}(A)_1\big) = \dim_K(A_1) - 1 < \infty \text{ and } \mathrm{Gr}_{A_\bullet}(A) = K[\mathrm{Gr}_{A_\bullet}(A)_1].$$

So, in this situation, the associated graded ring $\mathrm{Gr}_{A_\bullet}(A)$ is a commutative homogeneous (thus standard graded) Noetherian K-algebra (see Conventions, Reminders and Notations 1.1.1 (I)). If $A_\bullet$ is a very good filtration of A, we say that $(A, A_\bullet)$—or briefly A—is a *very well-filtered* K-algebra.

(B) Let $(A, A_\bullet)$ be a filtered K-algebra. The filtration $A_\bullet$ is said to be *good* if it satisfies the following conditions:

(a) The filtration $A_\bullet$ is commutative;
(b) A_0 is a K-algebra of finite type;
(c) A_1 is finitely generated as a (left-)module over A_0;
(d) $A_i = A_1 A_{i-1}$ for all $i \in \mathbb{N}$.

Under these circumstances we clearly have

$$\begin{aligned} A_0 &\cong \mathrm{Gr}_{A_\bullet}(A)_0 \text{ is commutative and Noetherian} \\ A_1/A_0 &= \mathrm{Gr}_{A_\bullet}(A)_1 \text{ is a finitely generated } A_0\text{-module, and} \\ \mathrm{Gr}_{A_\bullet}(A) &= \mathrm{Gr}_{A_\bullet}(A)_0[\mathrm{Gr}_{A_\bullet}(A)_1]. \end{aligned}$$

So, in this situation, the associated graded ring $\mathrm{Gr}_{A_\bullet}(A)$ is a commutative homogeneous Noetherian A_0-algebra (see Conventions, Reminders and Notations 1.1.1 (I)). If $A_\bullet$ is a good filtration of A, we say that $(A, A_\bullet)$—or briefly A—is a *well-filtered* K-algebra.

Clearly, a very well-filtered K-algebra is also well-filtered. (C) Let $(A, A_\bullet)$ be a filtered K-algebra. The filtration $A_\bullet$ is said to be *of finite type* if it satisfies the following conditions:

(a) The filtration $A_\bullet$ is commutative;
(b) A_0 is a K-algebra of finite type;
(c) There is an integer $\delta \in \mathbb{N}$ such that A_j is finitely generated as a (left-)module over A_0 for all $j \leq \delta$ and
(d) $A_i = \sum_{j=1}^{\delta} A_j A_{i-j}$ for all $i > \delta$.

In this situation, we call the number δ a *generating degree* of the filtration $A_\bullet$. Under these circumstances we clearly have

$$A_i = \sum_{1 \leq j_1, \ldots, j_s \leq \delta : j_1 + \cdots + j_s = i} A_{j_1} \cdots A_{j_s}, \quad (\forall i \in \mathbb{N})$$

$$A_0 \cong \mathrm{Gr}_{A_\bullet}(A)_0 \text{ is commutative and Noetherian}$$

$$A_1/A_0 = \mathrm{Gr}_{A_\bullet}(A)_1 \text{ is a finitely generated } A_0\text{-module, and}$$

$$\mathrm{Gr}_{A_\bullet}(A) = \mathrm{Gr}_{A_\bullet}(A)_0[\sum_{i=1}^{\delta} \mathrm{Gr}_{A_\bullet}(A)_i].$$

So, in this situation, the associated graded ring $\mathrm{Gr}_{A_\bullet}(A)$ is a commutative Noetherian graded A_0-algebra, which is generated by finitely many homogeneous elements of degree $\leq \delta$. If $A_\bullet$ is a filtration of A, which is of finite type, we say that $(A, A_\bullet)$ is a *filtered algebra of finite type*.
Clearly, a well-filtered K-algebra is also finitely filtered. Moreover, if $A_\bullet$ is of finite type and $\delta = 1$, the filtration $A_\bullet$ is good.

Example and Exercise 1.3.5 (A) Let $n \in \mathbb{N}$, let K be a field and consider the commutative polynomial ring $A = K[X_1, X_2, \ldots, X_n]$, furnished with its degree filtration. Then, it is easy to see, that $A = K[X_1, X_2, \ldots, X_n]$ is a very well filtered K-algebra.

(B) Let $n \in \mathbb{N}$, let K be a field and consider the commutative polynomial ring $A = K[X_1, X_2, \ldots, X_n]$. Let $m \in \{0, 1, \ldots, n-1\}$ and consider the subring $B := K[X_1, X_2, \ldots, X_m] \subset A$, so that $A = B[X_{m+1}, X_{m+2}, \ldots, X_n]$. For each polynomial $f = \sum_{\underline{\nu}} c_{\underline{\nu}}^{(f)} \underline{X}^{\underline{\nu}} \in A$ we denote by $\deg_B(f)$ the degree of f with respect to the indeterminates $X_{m+1}, X_{m+2}, \ldots, X_n$, hence the degree of f considered as a polynomial in these indeterminates with coefficients in B. Thus we may write

$$\deg_B(f) = \sup\{\sum_{i=1}^{n} w_i \nu_i \mid (\nu_1, \nu_2, \ldots, \nu_n) \in \mathrm{supp}(f)\}$$

where

$$w_1 = w_2 = \cdots = w_m = 0 \text{ and } w_{m+1} = w_{m+2} = \cdots = w_n = 1.$$

Show, that by

$$A_i := \{f \in A \mid \deg_B(f) \leq i\} \text{ for all } i \in \mathbb{N}_0$$

a good filtration $A_\bullet$ on A is defined and that there is a canonical isomorphism of graded B-algebras

$$A = B[X_{m+1}, X_{m+2}, \ldots, X_n] \cong \mathrm{Gr}_{A_\bullet}(A),$$

where $A = B[X_{m+1}, X_{m+2}, \ldots, X_n]$ is endowed with the standard grading with respect to the indeterminates $X_{m+1}, \ldots, X_n$, hence with the grading given by $\deg(X_i) = 0$ if $1 \leq i \leq m$ and $\deg(X_i) = 1$ for $m < i \leq n$.

(C) Let $n \in \mathbb{N}$, with $n > 1$, let K be a field and consider the free associative K-algebra $A = K\langle X_1, X_2, \ldots, X_n\rangle$, furnished with its degree filtration $A_\bullet$. For each $i \in \{1, 2, \ldots, n\}$, let

$$\overline{X}_i := (X_i + A_0) \in A_1/A_0 = \mathrm{Gr}_{A_\bullet}(A)_1 \subset \mathrm{Gr}_{A_\bullet}(A).$$

Show that

$$\overline{X}_i\overline{X}_j = \overline{X}_j\overline{X}_i \text{ if and only if } i = j.$$

(D) Let the notations and hypotheses be as in part (C). Show that $A = K\langle X_1, X_2, \ldots, X_n\rangle$ has the following universal property in the category of K-algebras:

If B is a K-algebra and $\phi : \{X_1, X_2, \ldots, X_n\} \longrightarrow B$ is a map, there is a unique homomorphism of K-algebras $\widetilde{\phi} : A \longrightarrow B$ such that $\widetilde{\phi}(X_i) = \phi(X_i)$ for all $i = 1, 2, \ldots, n$.

Use this to show, that there is a unique homomorphism of (graded) K-algebras (which must be in addition surjective)

$$\widetilde{\phi} : A \twoheadrightarrow \mathrm{Gr}_{A_\bullet}(A), \text{ such that } X_i \mapsto \overline{X}_i := (X_i + A_0) \in A_1/A_0 = \mathrm{Gr}_{A_\bullet}(A)_1.$$

(E) Let $(A, A_\bullet)$ be a filtered K-algebra, let $r \in \mathbb{N}$ and let $i_1, i_2, \ldots, i_r \in \mathbb{N}_0$. We define inductively

$$A_{i_1}A_{i_2}\ldots A_{i_r} = \prod_{j=1}^{r} A_{i_j} := \begin{cases} A_{i_1}, & \text{if } r = 1, \\ \left(\prod_{j=1}^{r-1} A_{i_j}\right)A_{i_r}, & \text{if } r > 1. \end{cases}$$

In particular, if $i \in \mathbb{N}_0$ we set

$$(A_i)^r := \prod_{j=1}^{r} A_i.$$

Assume now, that the filtration $A_\bullet$ is good and prove that

$$A_r = (A_1)^r \text{ and } A_iA_j = A_{i+j} \text{ for all } r \in \mathbb{N} \text{ and all } i, j \in \mathbb{N}_0.$$

Assume that the filtration $A_\bullet$ is of finite type and has generating degree δ. Prove that

$$A_i = \sum_{\nu_0,\nu_1,\ldots,\nu_\delta \in \mathbb{N}_0 : i = \sum_{j=0}^{\delta} j\nu_j} \prod_{j=0}^{\delta} A_j^{\nu_j} \text{ for all } i \in \mathbb{N}_0.$$

1.4 Derivations

Filtered K-algebras and their associated graded rings are one basic ingredient of the theory of Weyl algebras. Another basic ingredient are derivations (or derivatives). The present section is devoted to this subject.

Definition and Remark 1.4.1 (A) Let K be a field, let A be a commutative K-algebra and let M be an A-module. A *K-derivation* (or *K-derivative*) on A with values in M is a map $d : A \longrightarrow M$ such that:

(a) d is K-linear: $d(\alpha a + \beta b) = \alpha d(a) + \beta d(b)$ for all $\alpha, \beta \in K$ and all $a, b \in A$.
(b) d satisfies the *Leibniz Product Rule*: $d(ab) = ad(b) + bd(a)$ for all $a, b \in A$.

We denote the set of all K-derivations on A with values in M by $\mathrm{Der}_K(A, M)$, thus:

$$\mathrm{Der}_K(A, M) := \{d \in \mathrm{Hom}_K(A, M) \mid d(ab) = ad(b) + bd(a) \text{ for all } a, b \in A\}.$$

To simplify notations, we also write

$$\mathrm{Der}_K(A, A) =: \mathrm{Der}_K(A).$$

(B) Keep in mind, that $\mathrm{Hom}_K(A, M)$ carries a natural structure of A-module, with scalar multiplication given by

$$(ah)(x) := a(h(x)) \text{ for all } a \in A, \text{ all } h \in \mathrm{Hom}_K(A, M) \text{ and all } x \in A.$$

It is easy to verify:

$$\mathrm{Der}_K(A, M) \text{ is a submodule of the } A\text{-module } \mathrm{Hom}_K(A, M).$$

With our usual convention (suggested in Conventions, Reminders and Notations 1.1.1 (D)) that we identify $c \in K$ with $c1_A \in A$, the rules (a) and (b) of part (A) imply $d(c) = d(c1) = cd(1)$ and $d(c1) = 1d(c) + cd(1)$, hence $d(c) = d(c) + cd(1) = d(c) + d(c)$, thus

$$d(c) = 0 \text{ for all } c \in K \text{ and all } d \in \mathrm{Der}_K(A, M) : \text{ “Derivations vanish on constants.”}$$

Next, we shall look at the arithmetic properties of derivations and gain an important embedding procedure for modules of derivations of K-algebras of finite type.

Exercise and Definition 1.4.2 (A) Let K be a field, let A be a commutative K-algebra and let M be an A-module. Let $d \in \mathrm{Der}_K(A, M)$, let $r \in \mathbb{N}$, let $\nu_1, \nu_2, \ldots, \nu_r \in \mathbb{N}$ and let $a_1, a_2, \ldots, a_r \in A$. Use induction on r to prove the *Generalized Product Rule*

$$d\Big(\prod_{j=1}^{r} a_j^{\nu_j}\Big) = \sum_{i=1}^{r} \nu_i a_i^{\nu_i - 1}\Big(\prod_{j \neq i} a_j^{\nu_j}\Big) d(a_i)$$

and the resulting *Power Rule*

$$d(a^r) = ra^{r-1}d(a) \text{ for all } a \in A.$$

(B) Let the notations and hypotheses be as in part (A). Assume in addition that $A = K[a_1, a_2, \ldots, a_r]$. Let $e \in \mathrm{Der}_K(A, M)$. Use what you have shown in part (A) together with the fact that e and d are K-linear to prove that the following uniqueness statement holds:

$$e = d \text{ if and only if } e(a_i) = d(a_i) \text{ for all } i = 1, 2, \ldots, r.$$

(C) Yet assume that $A = K[a_1, a_2, \ldots, a_r]$. Prove that there is a monomorphism (thus an injective homomorphism) of A-modules

$$\Theta^M_{\underline{a}} = \Theta^M_{(a_1,a_2,\ldots,a_r)} : \mathrm{Der}_K(A, M) \longrightarrow M^r, \text{ given by } d \mapsto \big(d(a_1), d(a_2), \ldots, d(a_r)\big).$$

This monomorphism $\Theta^M_{\underline{a}}$ is called the *embedding* of $\mathrm{Der}_K(A, M)$ in M^r with respect to $\underline{a} := (a_1, a_2, \ldots, a_r)$.

(D) Let the notations and hypotheses be as in part (C). Assume that M is finitely generated. Prove, that the A-module $\mathrm{Der}_K(A, M)$ is finitely generated.

Now, we turn to derivatives in polynomial algebras, a basic ingredient of Weyl algebras.

Exercise and Definition 1.4.3 (Partial Derivatives in Polynomial Rings) (A) Let $n \in \mathbb{N}$, let K be a field and consider the polynomial algebra $K[X_1, X_2, \ldots, X_n]$. Fix $i \in \{1, 2, \ldots, n\}$. Then, using the monomial basis of $K[X_1, X_2, \ldots, X_n]$ we see that there is a unique K-linear map

$$\partial_i = \frac{\partial}{\partial X_i} : K[X_1, X_2, \ldots, X_n] \longrightarrow K[X_1, X_2, \ldots, X_n]$$

such that for all $\underline{\nu} = (\nu_1, \nu_2, \ldots, \nu_n) \in \mathbb{N}_0^n$ we have

$$\partial_i(\underline{X}^{\underline{\nu}}) = \frac{\partial}{\partial X_i}\Big(\prod_{j=1}^{n} X_j^{\nu_j}\Big) = \begin{cases} \nu_i X_i^{\nu_i - 1} \prod_{j \neq i} X_j^{\nu_j}, & \text{if } \nu_i > 0 \\ 0, & \text{if } \nu_i = 0. \end{cases}$$

(B) Keep the notations and hypotheses of part (A). Let

$$\underline{\mu} = (\mu_1, \mu_2, \ldots, \mu_n), \quad \underline{\nu} = (\nu_1, \nu_2, \ldots, \nu_n) \in \mathbb{N}_0^n$$

and prove that

$$\partial_i\big(\underline{X}^{\underline{\mu}}\underline{X}^{\underline{\nu}}\big) = \underline{X}^{\underline{\mu}}\partial_i\big(\underline{X}^{\underline{\nu}}\big) + \underline{X}^{\underline{\nu}}\partial_i\big(\underline{X}^{\underline{\mu}}\big).$$

Use the K-linearity of ∂_i to conclude that

$$\partial_i = \frac{\partial}{\partial X_i} \in \mathrm{Der}_K\big(K[X_1, X_2, \ldots, X_n]\big) \text{ for all } i = 1, 2 \ldots, n.$$

The derivation $\partial_i = \frac{\partial}{\partial X_i}$ is called the *i-th partial derivative* in $K[X_1, X_2, \ldots, X_n]$.

As we shall see in the proposition below, the embedding introduced in Exercise and Definition 1.4.2 (C) takes a particularly favorable shape in the case of polynomial algebras. The exercise to come is aimed to prepare the proof for this.

Exercise 1.4.4 (A) Let the notations and hypotheses be as in Exercise and Definition 1.4.3. For all $i, j \in \mathbb{Z}$ let $\delta_{i,j}$ denote the *Kronecker symbol*, so that

$$\delta_{i,j} = \begin{cases} 1, & \text{if } i = j, \\ 0, & \text{if } i \neq j. \end{cases}$$

Check that

$$\partial_i(X_j) = \delta_{i,j}, \text{ for all } i, j \in \{1, 2 \ldots, n\}.$$

(B) Keep the above notations and hypotheses. Show that

(a) For each $i \in \{1, 2, \ldots, n\}$ it holds $K[X_1, X_2, \ldots, X_{i-1}, X_{i+1}, \ldots, X_n] \subseteq \mathrm{Ker}(\partial_i)$ with equality if and only if $\mathrm{Char}(K) = 0$.
(b) $K \subseteq \bigcap_{i=1}^n \mathrm{Ker}(\partial_i)$ with equality if and only if $\mathrm{Char}(K) = 0$.

Proposition 1.4.5 (The Canonical Basis for the Derivations of a Polynomial Ring) *Let $n \in \mathbb{N}$, let K be a field and consider the polynomial algebra $K[X_1, X_2, \ldots, X_n]$. Then the canonical embedding of* $\mathrm{Der}_K\big(K[X_1, X_2, \ldots, X_n]\big)$ *into $K[X_1, X_2, \ldots, X_n]^n$ with respect to $X_1, X_2, \ldots, X_n$ (see Exercise and Definition 1.4.2 (C)) yields an isomorphism of $K[X_1, X_2, \ldots, X_n]$-modules*

$$\Theta := \Theta_{X_1, X_2, \ldots, X_n} : \mathrm{Der}_K\big(K[X_1, X_2, \ldots, X_n]\big) \xrightarrow{\cong} K[X_1, X_2, \ldots, X_n]^n,$$

given by

$$d \mapsto \Theta(d) := \Theta_{X_1, X_2, \ldots, X_n}(d) = \big(d(X_1), d(X_2), \ldots, d(X_n)\big),$$
$$\textit{for all } d \in \mathrm{Der}_K\big(K[X_1, X_2, \ldots, X_n]\big).$$

In particular, the n partial derivatives $\partial_1, \partial_2, \ldots, \partial_n$ form a free basis of the $K[X_1, X_2, \ldots, X_n]$-module $\mathrm{Der}_K\big(K[X_1, X_2, \ldots, X_n]\big)$, *hence*

$$\mathrm{Der}_K\big(K[X_1, X_2, \ldots, X_n]\big) = \bigoplus_{i=1}^n K[X_1, X_2, \ldots, X_n]\partial_i.$$

Proof According to Exercise and Definition 1.4.2 (C), the map Θ is a monomorphism of $K[X_1, X_2, \ldots, X_n]$-modules. By what we have seen in Exercise 1.4.4 (A) we have

$$\Theta(\partial_i) = \left(\delta_{i,1}, \delta_{i,2}, \ldots, \delta_{i-1,i}, \delta_{i,i}, \delta_{i,i+1}, \ldots, \delta_{i,n}\right) = \left(\delta_{i,j}\right)_{j=1}^{n} =: \mathrm{e}_i$$

for all $i = 1, 2, \ldots, n$. As the n elements

$$\mathrm{e}_i = \left(\delta_{i,j}\right)_{j=1}^{n} \in K[X_1, X_2, \ldots, X_n]^n \text{ with } i = 1, 2, \ldots, n$$

form the canonical free basis of the $K[X_1, X_2, \ldots, X_n]$-module $K[X_1, X_2, \ldots, X_n]^n$ our claims follow immediately.

1.5 Weyl Algebras

Now, we are ready to introduce Weyl algebras. We first remind a few facts on endomorphism rings of commutative K-algebras and relate these to modules of derivations.

Reminder and Remark 1.5.1 (A) Let K be a field and let A be a commutative K-algebra and let M be an A-module. Keep in mind, that the A-module

$$\mathrm{End}_K(M) := \mathrm{Hom}_K(M, M)$$

carries a natural structure of K-algebra, whose multiplication is given by composition of maps, thus:

$$fg := f \circ g, \text{ hence } (fg)(m) := f(g(m)) \text{ for all } f, g \in \mathrm{End}_K(M) \text{ and all } m \in M.$$

The module $\mathrm{End}_K(M)$ endowed with this multiplication is called the *K-endomorphism ring* of M. Observe, that this endomorphism ring is not commutative in general.

(B) Keep the above notations and hypothesis. Then, we have a *canonical homomorphism* of rings

$$\varepsilon_M : A \longrightarrow \mathrm{End}_K(M) \text{ given by } a \mapsto \varepsilon_M(a) := a\mathrm{id}_M \text{ for all } a \in A,$$

where $\mathrm{id}_M : M \longrightarrow M$ is the *identity map* on M, so that

$$\varepsilon_M(a)(m) = am \text{ for all } a \in A \text{ and all } m \in M.$$

It is immediate to verify that this canonical homomorphism is injective if $M = A$:

$$\text{The canonical homomorphism } \varepsilon_A : A \longrightarrow \mathrm{End}_K(A) \text{ is injective.}$$

We therefore call the map $\varepsilon_A : A \longrightarrow \mathrm{End}_K(A)$ the *canonical embedding* of A into its K-endomorphism ring and we consider A as a subalgebra of $\mathrm{End}_K(A)$ by means of this canonical embedding. So, for all $a \in A$ we identify a with $\varepsilon_A(a)$.

Remark and Definition 1.5.2 (A) Let K be a field and let A be a commutative K-algebra. By the convention made in Reminder and Remark 1.5.1 we may consider A as a subalgebra of the endomorphism ring $\mathrm{End}_K(A)$. We obviously also have $\mathrm{Der}_K(A) \subseteq \mathrm{End}_K(A)$. So using the notation introduced in Conventions, Reminders and Notations 1.1.1 (B), we have may consider the K-subalgebra

$$W_K(A) := K[A \cup \mathrm{Der}_K(A)] = A[\mathrm{Der}_K(A)] \subseteq \mathrm{End}_K(A).$$

of the K-endomorphism ring of A which is generated by A and all derivations on A with values in A. We call $W_K(A)$ the *Weyl algebra of the K-algebra A*.

(B) Keep the hypotheses and notations of part (A). Assume in addition, that the commutative K-algebra A is of finite type, so that we find some $r \in \mathbb{N}_0$ and elements $a_1, a_2, \ldots, a_r \in A$ such that

$$A = K[a_1, a_2, \ldots, a_r].$$

Then according to Exercise and Definition 1.4.2 (D), the A-module $\mathrm{Der}_K(A)$ is finitely generated. We thus find some $s \in \mathbb{N}_0$ and derivations $d_1, d_2, \ldots, d_s \in \mathrm{Der}_K(A)$ such that

$$\mathrm{Der}_K(A) = \sum_{i=1}^{s} A d_i.$$

A straight forward computation now allows to see, that

$$W_K(A) = K[a_1, a_2 \ldots, a_r, d_1, d_2, \ldots, d_s] \subseteq \mathrm{End}_K(A).$$

In particular we may conclude, that the K-algebra $W_K(A)$ is finitely generated.

(C) Keep the above notations and let $n \in \mathbb{N}$. The *n-th standard Weyl algebra* $\mathbb{W}(K, n)$ over the field K is defined as the Weyl algebra of the polynomial ring $K[X_1, X_2, \ldots, X_n]$, thus

$$\mathbb{W}(K, n) := W_K\big(K[X_1, X_2, \ldots, X_n]\big) \subseteq \mathrm{End}_K\big(K[X_1, X_2, \ldots, X_n]\big).$$

Observe, that by Proposition 1.4.5 and according to the observations made in part (B) we may write

$$\mathbb{W}(K, n) = K[X_1, X_2, \ldots, X_n, \partial_1, \partial_1, \partial_2, \ldots, \partial_n] \subseteq \mathrm{End}_K\big(K[X_1, X_2, \ldots, X_n]\big).$$

The elements of $\mathbb{W}(K,n)$ are called *polynomial differential operators* in the indeterminates $X_1, X_2, \ldots, X_n$ over the field K. They are all K-linear combinations of products of indeterminates X_i and partial derivatives ∂_j.
The differential operators of the form

$$\underline{X}^{\underline{\nu}}\underline{\partial}^{\underline{\mu}} := X_1^{\nu_1} \ldots X_n^{\nu_n} \partial_1^{\mu_1} \ldots \partial_n^{\mu_n} = \prod_{i=1}^{n} X^{\nu_i} \prod_{j=1}^{n} \partial^{\mu_j} \in \mathbb{W}(K,n)$$

with

$$\underline{\nu} := (\nu_1, \ldots, \nu_n), \quad \underline{\mu} := (\mu_1, \ldots, \mu_n) \in \mathbb{N}_0^n$$

are called *elementary differential operators* in the indeterminates $X_1, X_2, \ldots, X_n$ over the field K.

We now aim to study the structure of standard Weyl algebras. One of the main goals we are heading for is to find an appropriate "monomial basis" in each of these algebras. We namely shall see later that the previously introduced elementary differential operators form a K-basis of the standard Weyl algebra $\mathbb{W}(K,n)$, provided K is of characteristic 0. To pave our way to this fundamental result, we first of all have to prove that in standard Weyl algebras certain commutation relations hold: the so-called Heisenberg relations. To establish these relations, we begin with the following preparations.

Remark and Exercise 1.5.3 (A) If K is a field and B is a K-algebra, we introduce the *Poisson operation*, that is the map

$$[\bullet,\bullet] : B \times B \longrightarrow B, \text{ defined by } [a,b] := ab - ba \text{ for all } a, b \in B.$$

Show, that the Poisson operation has the following properties:

(a) $[a,b] = -[b,a]$ for all $a, b \in B$.
(b) $[[a,b],c] + [[b,c],a] + [[c,a],b] = 0$ for all $a, b, c \in B$.
(c) $[\alpha a + \alpha' a', \beta b + \beta' b'] = \alpha\beta[a,b] + \alpha\beta'[a,b'] + \alpha'\beta[a',b] + \alpha'\beta'[a',b']$ for all $\alpha, \alpha', \beta, \beta' \in K$ and all $a, a', b, b' \in B$.

Observe in particular, that statement (a) says that the Poisson operation is *anti-commutative*, whereas statement (c) says that this operation is *K-bilinear*. We call $[a,b]$ the *commutator* of a and b.

(B) Now, let K be a field, let A be a commutative K-algebra and consider the Weyl algebra $W_K(A) := A[\mathrm{Der}_K(A)]$. Show that the following relations hold:

(a) $[a,b] = 0$ for all $a, b \in A$.
(b) $[a,d] = -d(a)$ for all $a \in A$ and all $d \in \mathrm{Der}_K(A)$.
(c) $[d,e] \in \mathrm{Der}_K(A)$ for all $d, e \in \mathrm{Der}_K(A)$.

(C) Let the notations and hypotheses be as in part (B). Let $d, e \in \mathrm{Der}_K(A)$, let $r \in \mathbb{N}$, let $\nu_1, \nu_2, \ldots, \nu_r \in \mathbb{N}$ and let $a_1, a_2, \ldots, a_r \in A$. Use statement (c) of part (B) and the Generalized Product Rule of Exercise and Definition 1.4.2 (A) to prove that

$$[d, e]\Big(\prod_{j=1}^{r} a_j^{\nu_j}\Big) = \sum_{i=1}^{r} \nu_i a_i^{\nu_i - 1}\Big(\prod_{j \neq i} a_j^{\nu_j}\Big)[d, e](a_i).$$

Proposition 1.5.4 (The Heisenberg Relations) *Let $n \in \mathbb{N}$, and let K be a field. Then, in the standard Weyl algebra*

$$\mathbb{W}(K, n) = K[X_1, X_2, \ldots, X_n, \partial_1, \partial_2, \ldots, \partial_n]$$

the following relations hold:

(a) $[X_i, X_j] = 0$, *for all $i, j \in \{1, 2, \ldots, n\}$;*

(b) $[X_i, \partial_j] = -\delta_{i,j}$, *for all $i, j \in \{1, 2, \ldots, n\}$;*

(c) $[\partial_i, \partial_j] = 0$, *for all $i, j \in \{1, 2, \ldots, n\}$.*

Proof

(a) This is clear on application of Remark and Exercise 1.5.3 (B)(a) with $a = X_i$ and $b = X_j$.
(b) If we apply Remark and Exercise 1.5.3 (B)(b) with $a = X_i$ and $d = \partial_j$, and observe that $\partial_j(X_i) = \delta_{j,i} = \delta_{i,j}$ we get our claim.
(c) Observe that for all $i, k \in \{1, 2, \ldots, n\}$ we have $\partial_i(X_k) \in \{0, 1\} \subseteq K$. So for all $i, j, k \in \{1, 2, \ldots, n\}$ we obtain (see Definition and Remark 1.4.1 (B)):

$$[\partial_i, \partial_j](X_k) = \partial_i\big(\partial_j(X_k)\big) - \partial_j\big(\partial_i(X_k)\big) \in \partial_i(K) + \partial_j(K) = \{0\} + \{0\} = \{0\}.$$

Now, we get our claim by Exercise and Definition 1.4.2 (B) and Remark and Exercise 1.5.3 (B) (c) and (C).

The Heisenberg relations are of basic significance for the arithmetic in standard Weyl algebras. Before we show that the elementary differential operators provide a basis for a standard Weyl algebra we shall study the arithmetic of these algebras. In particular, in the next section we shall prove a product formula for elementary differential operators, which will be of basic significance. We shall do this in a slightly more general setting, namely just for K-algebras "mimicking" the Heisenberg relations. The next exercise is aimed to prepare this.

Exercise 1.5.5 (A) Let $n \in \mathbb{N}$, let K be a field, let B be a K-algebra and let

$$a_1, a_1, \ldots, a_n, d_1, d_2, \ldots, d_n \in B$$

be elements *mimicking the Heisenberg relations*, which means:

(1) $[a_i, a_j] = 0$, for all $i, j \in \{1, 2, \ldots, n\}$;
(2) $[a_i, d_j] = -\delta_{i,j}$, for all $i, j \in \{1, 2, \ldots, n\}$;
(3) $[d_i, d_j] = 0$, for all $i, j \in \{1, 2, \ldots, n\}$.

Let $\mu, \nu \in \mathbb{N}_0$. To simplify notations, we set

$$0b^k := 0 \text{ for all } b \in B \text{ and all } k \in \mathbb{Z}.$$

prove the following statements (using induction on μ and ν):

(a) $a_i^\mu a_j^\nu = a_j^\nu a_i^\mu$;
(b) $d_i^\mu d_j^\nu = d_j^\nu d_i^\mu$;
(c) $d_i^\mu a_j^\nu = a_j^\nu d_i^\mu$ for all $i, j \in \{1, 2, \ldots, n\}$ with $i \neq j$.
(d) $d_i a_i^\nu = a_i^\nu d_i + \nu a_i^{\nu-1}$ for all $i \in \{1, 2, \ldots, n\}$.

(B) Keep the notations and hypotheses of part (A). For all $(\lambda_1, \lambda_2, \ldots, \lambda_n) \in \mathbb{N}_0^n$ and each sequence $(b_1, b_2, \ldots, b_n) \in B^n$ we use again our earlier standard notation

$$\underline{\lambda} := (\lambda_1, \lambda_2, \ldots, \lambda_n) \text{ and } \underline{b}^{\underline{\lambda}} := b_1^{\lambda_1} b_2^{\lambda_2} \ldots b_n^{\lambda_n} = \prod_{i=1}^n b_i^{\lambda_i}.$$

Now, let

$$\underline{\mu} := (\mu_1, \mu_1, \ldots, \mu_n), \quad \underline{\nu} := (\nu_1, \nu_2, \ldots, \nu_n), \text{ and}$$

$$\underline{\mu}' := (\mu_1', \mu_1', \ldots, \mu_n'), \quad \underline{\nu}' := (\nu_1', \nu_2', \ldots, \nu_n') \in \mathbb{N}_0^n.$$

Prove that the following relations hold

(a) $\underline{a}^{\underline{\nu}} \underline{d}^{\underline{\mu}} = \prod_{i=1}^n a_i^{\nu_i} \prod_{j=1}^n d_j^{\mu_j} = \prod_{i=1}^n a_i^{\nu_i} d_i^{\mu_i}$.
(b) $(\underline{a}^{\underline{\nu}} \underline{d}^{\underline{\mu}})(\underline{a}^{\underline{\nu}'} \underline{d}^{\underline{\mu}'}) = \left(\prod_{i=1}^n a_i^{\nu_i} \prod_{j=1}^n d_j^{\mu_j}\right)\left(\prod_{i=1}^n a_i^{\nu_i'} \prod_{j=1}^n d_j^{\mu_j'}\right) = \prod_{i=1}^n a_i^{\nu_i} d_i^{\mu_i} a_i^{\nu_i'} d_i^{\mu_i'}$.

1.6 Arithmetic in Weyl Algebras

As announced above, we now aim to do some Arithmetic in standard Weyl algebras. This means in particular, that we make explicit a number of computations in the hope that readers who up to now were mainly faced with commutative rings, get fascinated by the complexity of the arithmetic in Weyl algebras.
The following arithmetical Lemma is formulated in a more general framework, namely in a situation, which “mimicks” the Heisenberg relation. If we specialize

this Lemma to standard Weyl algebras, we get a most important formula, which expresses the product of two elementary differential operators as a $\mathbb{Z}$-linear combination of elementary differential operators. This will also give us an explicit presentation of the commutator $[d, e]$ (see Remark and Exercise 5.3 (A)) of two elementary differential operators d and e. As a further application we will get the Reduction Principle for arbitrary products of elementary differential operators and thus pave our way to the standard basis presentation of Weyl algebras, which we shall introduce in the next section.
We prove the announced Lemma in a setting which is more general than just the framework of Weyl algebras, because in this form it will help us to prove the universal property of Weyl algebras formulated in Corollary 1.7.5. This property is an analogue of the (relational) universal property of commutative polynomial algebras (see Example and Exercise 1.3.2 (A)) or of free associative algebras (see Example and Exercise 1.3.5 (D)).

Lemma 1.6.1 *Let $n \in \mathbb{N}$, let K be a field, let B be a K-algebra and let*

$$a_1, a_2, \ldots, a_n, d_1, d_2, \ldots, d_n \in B$$

such that:

(1) $[a_i, a_j] = 0,$ *for all* $i, j \in \{1, 2, \ldots, n\}$;
(2) $[a_i, d_j] = -\delta_{i,j},$ *for all* $i, j \in \{1, 2, \ldots, n\}$;
(3) $[d_i, d_j] = 0,$ *for all* $i, j \in \{1, 2, \ldots, n\}$.

Then, the following statements hold:

(a) *For all* $\mu, \nu \in \mathbb{N}_0$ *and all* $i \in \{1, 2, \ldots, n\}$ *we have*

$$d_i^{\mu} a_i^{\nu} = \sum_{k=0}^{\min\{\mu,\nu\}} \binom{\mu}{k} \prod_{p=0}^{k-1} (\nu - p) a_i^{\nu-k} d_i^{\mu-k}.$$

(b) *Let*

$$\underline{\mu} := (\mu_1, \mu_1, \ldots, \mu_n), \quad \underline{\nu} := (\nu_1, \nu_2, \ldots, \nu_n), \text{ and}$$

$$\underline{\mu'} := (\mu'_1, \mu'_1, \ldots, \mu'_n), \quad \underline{\nu'} := (\nu'_1, \nu'_2, \ldots, \nu'_n) \in \mathbb{N}_0^n.$$

Set

$$\mathbb{I} := \{\underline{k} := (k_1, k_2, \ldots, k_n) \in \mathbb{N}_0^n \mid k_i \leq \min\{\mu_i, \nu'_i\} \text{ for } i = 1, 2, \ldots, n\},$$

and let

$$\lambda_{\underline{k}} := [\prod_{i=1}^{n} \binom{\mu_i}{k_i}] \times [\prod_{i=1}^{n} \prod_{p=0}^{k_i-1} (\nu'_i - p)].$$

Then, we have the relation

$$(\underline{a}^{\underline{\nu}}\underline{d}^{\underline{\mu}})(\underline{a}^{\underline{\nu}'}\underline{d}^{\underline{\mu}'}) := (\prod_{i=1}^{n} a_i^{\nu_i} \prod_{j=1}^{n} d_j^{\mu_j}) \prod_{i=1}^{n} a_i^{\nu'_j} \prod_{j=1}^{n} d_i^{\mu'_j})$$

$$= \prod_{i=1}^{n} a_i^{\nu_i+\nu'_i} \prod_{i=1}^{n} d_i^{\mu_i+\mu'_i} + \sum_{\underline{k}\in\mathbb{I}\setminus\{\underline{0}\}} \lambda_{\underline{k}} \prod_{i=1}^{n} a_i^{\nu_i+\nu'_i-k_i} \prod_{i=1}^{n} d_i^{\mu_i+\mu'_i-k_i}$$

$$= \underline{a}^{\underline{\nu}+\underline{\nu}'}\underline{d}^{\underline{\mu}+\underline{\mu}'} + \sum_{\underline{k}\in\mathbb{I}\setminus\{\underline{0}\}} \lambda_{\underline{k}} \underline{a}^{\underline{\nu}+\underline{\nu}'-\underline{k}}\underline{d}^{\underline{\mu}+\underline{\mu}'-\underline{k}}.$$

Proof (a) To simplify matters we use the notation

$$0b^k := 0 \text{ for all } b \in B \text{ and all } k \in \mathbb{Z}$$

already introduced in the previous Exercise 1.5.5 (A). Then, it suffices to show that

$$d_i^{\mu} a_i^{\nu} = \sum_{k=0}^{\mu} \binom{\mu}{k} \prod_{p=0}^{k-1} (\nu - p) a_i^{\nu-k} d_i^{\mu-k}.$$

We proceed by induction on μ. The case $\mu = 0$ is obvious. The case $\mu = 1$ is clear by Exercise 1.5.5 (A)(d). So, let $\mu > 1$. By induction we have

$$d_i^{\mu-1} a_i^{\nu} = \sum_{k=0}^{\mu-1} \binom{\mu-1}{k} \prod_{p=0}^{k-1} (\nu - p) a_i^{\nu-k} d_i^{\mu-1-k}.$$

It follows on use of Exercise 1.5.5 (A)(d) and the Pascal formulas for the sum of binomial coefficients, that

$$d_i^{\mu} a_i^{\nu} = d_i(d_i^{\mu-1} a_i^{\nu}) = d_i(\sum_{k=0}^{\mu-1} \binom{\mu-1}{k} \prod_{p=0}^{k-1} (\nu - p) a_i^{\nu-k} d_i^{\mu-1-k})$$

$$= \sum_{k=0}^{\mu-1} \binom{\mu-1}{k} \prod_{p=0}^{k-1} (\nu - p)(d_i a_i^{\nu-k}) d_i^{\mu-1-k}$$

$$= \sum_{k=0}^{\mu-1} \binom{\mu-1}{k} \prod_{p=0}^{k-1} (\nu - p)\big(a_i^{\nu-k} d_i + (\nu - k) a_i^{\nu-k-1}\big) d_i^{\mu-1-k}$$

$$= \sum_{k=0}^{\mu-1} [\binom{\mu-1}{k} \prod_{p=0}^{k-1} (\nu - p) a_i^{\nu-k} d_i^{\mu-k}$$

$$+ \binom{\mu-1}{k} \prod_{p=0}^{k-1} (\nu - p)(\nu - k) a_i^{\nu-k-1} d_i^{\mu-1-k}]$$

$$= \sum_{k=0}^{\mu-1} \binom{\mu-1}{k} \prod_{p=0}^{k-1} (\nu - p) a_i^{\nu-k} d_i^{\mu-k}$$

$$+ \sum_{k=0}^{\mu-1} \binom{\mu-1}{k} \prod_{p=0}^{k} (\nu - p) a_i^{\nu-k-1} d_i^{\mu-1-k}$$

$$= \sum_{k=0}^{\mu-1} \binom{\mu-1}{k} \prod_{p=0}^{k-1} (\nu - p) a_i^{\nu-k} d_i^{\mu-k} + \sum_{k=1}^{\mu} \binom{\mu-1}{k-1} \prod_{p=0}^{k-1} (\nu - p) a_i^{\nu-k} d_i^{\mu-k}$$

$$= a_i^{\nu} d_i^{\mu} + \sum_{k=1}^{\mu-1} \binom{\mu-1}{k} \prod_{p=0}^{k-1} (\nu - p) a_i^{\nu-k} d_i^{\mu-k} +$$

$$+ \sum_{k=1}^{\mu-1} \binom{\mu-1}{k-1} \prod_{p=0}^{k-1} (\nu - p) a_i^{\nu-k} d_i^{\mu-k} + \prod_{p=0}^{\mu-1} (\nu - p) a_i^{\nu-\mu}$$

$$= a_i^{\nu} d_i^{\mu} + \sum_{k=1}^{\mu-1} [\binom{\mu-1}{k} + \binom{\mu-1}{k-1}] \prod_{p=0}^{k-1} (\nu - p) a_i^{\nu-k} d_i^{\mu-k} + \prod_{p=0}^{\mu-1} (\nu - p) a_i^{\nu-\mu}$$

$$= a_i^{\nu} d_i^{\mu} + \sum_{k=1}^{\mu-1} \binom{\mu}{k} \prod_{p=0}^{k-1} (\nu - p) a_i^{\nu-k} d_i^{\mu-k} + \prod_{p=0}^{\mu-1} (\nu - p) a_i^{\nu-\mu}$$

$$= \sum_{k=0}^{\mu} \binom{\mu}{k} \prod_{p=0}^{k-1} (\nu - p) a_i^{\nu-k} d_i^{\mu-k}.$$

(b) According to Exercise 1.5.5 (B)(a),(b), the previous statement (a) and Exercise 1.5.5 (A)(a),(b) and (c) we may write

$$(\underline{a}^{\underline{\nu}} \underline{d}^{\underline{\mu}})(\underline{a}^{\underline{\nu}'} \underline{d}^{\underline{\mu}'}) := (\prod_{i=1}^{n} a_i^{\nu_i} \prod_{j=1}^{n} d_j^{\mu_j})(\prod_{i=1}^{n} a_i^{\nu_i'} \prod_{j=1}^{n} d_j^{\mu_j'}) = \prod_{i=1}^{n} a_i^{\nu_i} d_i^{\mu_i} a_i^{\nu_i'} d_i^{\mu_i'}$$

$$= \prod_{i=1}^{n} a_i^{\nu_i} (d_i^{\mu_i} a_i^{\nu_i'}) d_i^{\mu_i'}$$

$$= \prod_{i=1}^{n} a_i^{\nu_i} [\sum_{k=0}^{\min\{\mu_i, \nu_i'\}} \binom{\mu_i}{k} \prod_{p=0}^{k-1} (\nu_i' - p) a_i^{\nu_i'-k} d_i^{\mu_i-k}] d_i^{\mu_i'}$$

$$= \prod_{i=1}^{n} (\sum_{k=0}^{\min\{\mu_i, \nu_i'\}} \binom{\mu_i}{k} \prod_{p=0}^{k-1} (\nu_i' - p) a_i^{\nu_i+\nu_i'-k} d_i^{\mu_i+\mu_i'-k}]$$

$$= \sum_{\underline{k}:=(k_1,k_2,\ldots,k_n)\in\mathbb{I}} \prod_{i=1}^{n} \left(\binom{\mu_i}{k_i} \prod_{p=0}^{k_i-1} (\nu_i' - p) a_i^{\nu_i+\nu_i'-k_i} d_i^{\mu_i+\mu_i'-k_i}\right)$$

$$= \sum_{\underline{k}\in\mathbb{I}} \left(\prod_{i=1}^{n} \binom{\mu_i}{k_i}\right)\left(\prod_{i=1}^{n}\prod_{p=0}^{k_i-1} (\nu_i' - p)\right) \prod_{i=1}^{n} a_i^{\nu_i+\nu_i'-k_i} d_i^{\mu_i+\mu_i'-k_i}$$

$$= \sum_{\underline{k}\in\mathbb{I}} \left(\prod_{i=1}^{n} \binom{\mu_i}{k_i}\right)\left(\prod_{i=1}^{n}\prod_{p=0}^{k_i-1} (\nu_i' - p)\right) \prod_{i=1}^{n} a_i^{\nu_i+\nu_i'-k_i} \prod_{i=1}^{n} d_i^{\mu_i+\mu_i'-k_i}$$

$$= \prod_{i=1}^{n} a_i^{\nu_i+\nu_i'} \prod_{i=1}^{n} d_i^{\mu_i+\mu_i'} + \sum_{\underline{k}\in\mathbb{I}\setminus\{\underline{0}\}} \lambda_{\underline{k}} \prod_{i=1}^{n} a_i^{\nu_i+\nu_i'-k_i} \prod_{i=1}^{n} d_i^{\mu_i+\mu_i'-k_i}$$

$$= \underline{a}^{\underline{\nu}+\underline{\nu}'} \underline{d}^{\underline{\mu}+\underline{\mu}'} + \sum_{\underline{k}\in\mathbb{I}\setminus\{\underline{0}\}} \lambda_{\underline{k}} \underline{a}^{\underline{\nu}+\underline{\nu}'-\underline{k}} \underline{d}^{\underline{\mu}+\underline{\mu}'-\underline{k}}.$$

As an application we now get the announced product formula for elementary differential operators.

Proposition 1.6.2 (The Product Formula for Elementary Differential Operators) *Let $n \in \mathbb{N}$, let K be a field and consider the standard Weyl algebra*

$$\mathbb{W}(K, n) = K[X_1, X_2, \ldots X_n, \partial_1, \partial_2 \ldots, \partial_n].$$

Moreover, let

$$\underline{\mu} := (\mu_1, \mu_1, \ldots, \mu_n), \quad \underline{\nu} := (\nu_1, \nu_2, \ldots, \nu_n) \text{ and}$$

$$\underline{\mu}' := (\mu_1', \mu_1', \ldots, \mu_n'), \quad \underline{\nu}' := (\nu_1', \nu_2', \ldots, \nu_n') \in \mathbb{N}_0^n.$$

Set

$$\mathbb{I} := \{\underline{k} := (k_1, k_2, \ldots, k_n) \in \mathbb{N}_0^n \mid k_i \leq \min\{\mu_i, \nu_i'\} \text{ for } i = 1, 2, \ldots, n\},$$

and let

$$\lambda_{\underline{k}} := \left(\prod_{i=1}^{n} \binom{\mu_i}{k_i}\right)\left(\prod_{i=1}^{n}\prod_{p=0}^{k_i-1} (\nu_i' - p)\right).$$

Then, we have the equality

$$(\underline{X}^{\underline{\nu}}\underline{\partial}^{\underline{\mu}})(\underline{X}^{\underline{\nu}'}\underline{\partial}^{\underline{\mu}'}) := (\prod_{i=1}^{n} X_i^{\nu_i} \prod_{j=1}^{n} \partial_j^{\mu_j})(\prod_{i=1}^{n} X_i^{\nu'_i} \prod_{j=1}^{n} \partial_j^{\mu'_j})$$

$$= \prod_{i=1}^{n} X_i^{\nu_i+\nu'_i} \prod_{i=1}^{n} \partial_i^{\mu_i+\mu'_i} + \sum_{\underline{k}\in\mathbb{I}\setminus\{\underline{0}\}} \lambda_{\underline{k}} \prod_{i=1}^{n} X_i^{\nu_i+\nu'_i-k_i} \prod_{i=1}^{n} \partial_i^{\mu_i+\mu'_i-k_i}$$

$$= \underline{X}^{\underline{\nu}+\underline{\nu}'}\underline{\partial}^{\underline{\mu}+\underline{\mu}'} + \sum_{\underline{k}\in\mathbb{I}\setminus\{\underline{0}\}} \lambda_{\underline{k}} \underline{X}^{\underline{\nu}+\underline{\nu}'-\underline{k}}\underline{\partial}^{\underline{\mu}+\underline{\mu}'-\underline{k}}.$$

Proof It suffices to apply Lemma 1.6.1 (b) with $a_i := X_i$ and $d_i := \partial_i$ for $i = 1, 2 \ldots, n$.

Now, we can prove the main result of the present section. To formulate it, we introduce another notation and suggest a further exercise.

Notation and Remark 1.6.3 (A) Let $n \in \mathbb{N}$ and let

$$\underline{\kappa} := (\kappa_1, \kappa_2, \ldots, \kappa_n) \text{ and } \underline{\lambda} := (\lambda_1, \lambda_2, \ldots, \lambda_n) \in \mathbb{N}_0^n.$$

We write

$$\underline{\kappa} \leq \underline{\lambda} \text{ if and only if } \kappa_i \leq \lambda_i \text{ for } i = 1, 2, \ldots, n$$

and

$$\underline{\kappa} < \underline{\lambda} \text{ if and only if } \underline{\kappa} \leq \underline{\lambda} \text{ and } \underline{\kappa} \neq \underline{\lambda}.$$

(B) Keep the notations of part (A). Observe that

$$\underline{\kappa} \leq \underline{\lambda} \text{ if and only if } \underline{\lambda} - \underline{\kappa} \in \mathbb{N}_0^n$$

and

$$\underline{\kappa} < \underline{\lambda} \text{ if and only if } \underline{\lambda} - \underline{\kappa} \in \mathbb{N}_0^n \setminus \{\underline{0}\}.$$

(C) We now introduce a few notations, which we will have to use later very frequently. Namely, for

$$\underline{\alpha} = (\alpha_1, \alpha_2, \ldots, \alpha_n), \underline{\beta} = (\beta_1, \beta_2, \ldots, \beta_n) \in \mathbb{N}_0^n$$

we set

$$\mathbb{M}(\underline{\alpha}, \underline{\beta}) := \{(\underline{\alpha} - \underline{k}, \underline{\beta} - \underline{k}) \mid \underline{k} \in \mathbb{N}_0^n \setminus \{\underline{0}\} \text{ with } \underline{k} \leq \underline{\alpha}, \underline{\beta}\}$$

and

$$\overline{\mathbb{M}}(\underline{\alpha},\underline{\beta}) := \{(\underline{\alpha}-\underline{k},\underline{\beta}-\underline{k}) \mid \underline{k}\in\mathbb{N}_0^n \text{ with } \underline{k}\leq\underline{\alpha},\underline{\beta}\} = \mathbb{M}(\underline{\alpha},\underline{\beta})\cup\{(\underline{\alpha},\underline{\beta})\}.$$

Moreover, we write

$$\mathbb{M}_{\leq}(\underline{\alpha},\underline{\beta}) := \{(\underline{\lambda},\underline{\kappa})\in\mathbb{N}_0^n\times\mathbb{N}_0^n \mid \underline{\lambda}\leq\underline{\nu} \text{ and } \underline{\kappa}\leq\underline{\mu} \text{ for some } (\underline{\nu},\underline{\mu})\in\mathbb{M}(\underline{\alpha},\underline{\beta})\}.$$

Observe that

$$\mathbb{M}(\underline{\alpha},\underline{\beta})\subseteq\mathbb{M}_{\leq}(\underline{\alpha},\underline{\beta}).$$

Exercise 1.6.4 (A) Let $n\in\mathbb{N}$, let K be a field and consider the standard Weyl algebra

$$\mathbb{W}(K,n) = K[X_1,X_2,\ldots,X_n,\partial_1,\partial_2,\ldots,\partial_n].$$

In addition, let

$$\underline{\mu} := (\mu_1,\mu_1,\ldots,\mu_n),\quad \underline{\nu} := (\nu_1,\nu_2,\ldots,\nu_n) \text{ and}$$

$$\underline{\mu}' := (\mu_1',\mu_1',\ldots,\mu_n'),\quad \underline{\nu}' := (\nu_1',\nu_2',\ldots,\nu_n')\in\mathbb{N}_0^n.$$

Moreover, let the sets

$$\mathbb{M}(\underline{\nu}+\underline{\nu}',\underline{\mu}+\underline{\mu}')\subset\overline{\mathbb{M}}(\underline{\nu}+\underline{\nu}',\underline{\mu}+\underline{\mu}')\subset\mathbb{N}_0^n\times\mathbb{N}_0^n$$

be defined according to Notation and Remark 1.6.3 (C). Prove that

$$(\underline{X}^{\underline{\nu}}\underline{\partial}^{\underline{\mu}})(\underline{X}^{\underline{\nu}'}\underline{\partial}^{\underline{\mu}'}) - \underline{X}^{\underline{\nu}+\underline{\nu}'}\underline{\partial}^{\underline{\mu}+\underline{\mu}'} \in \sum_{(\underline{\lambda},\underline{\kappa})\in\mathbb{M}(\underline{\nu}+\underline{\nu}',\underline{\mu}+\underline{\mu}')}\mathbb{Z}\underline{X}^{\underline{\lambda}}\underline{\partial}^{\underline{\kappa}}.$$

and

$$(\underline{X}^{\underline{\nu}}\underline{\partial}^{\underline{\mu}})(\underline{X}^{\underline{\nu}'}\underline{\partial}^{\underline{\mu}'}) \in \sum_{(\underline{\lambda},\underline{\kappa})\in\overline{\mathbb{M}}(\underline{\nu}+\underline{\nu}',\underline{\mu}+\underline{\mu}')}\mathbb{Z}\underline{X}^{\underline{\lambda}}\underline{\partial}^{\underline{\kappa}}.$$

(B) Let the notations be as in part (A) and let the set

$$\mathbb{M}(\underline{\nu}+\underline{\nu}',\underline{\mu}+\underline{\mu}')\subset\mathbb{N}_0^n\times\mathbb{N}_0^n$$

be defined according to Notation and Remark 1.6.3 (C). Prove that

$$\left[\underline{X}^{\underline{\nu}}\underline{\partial}^{\underline{\mu}},\underline{X}^{\underline{\nu}'}\underline{\partial}^{\underline{\mu}'}\right] \in \sum_{(\underline{\lambda},\underline{\kappa})\in\mathbb{M}(\underline{\nu}+\underline{\nu}',\underline{\mu}+\underline{\mu}')}\mathbb{Z}\underline{X}^{\underline{\lambda}}\underline{\partial}^{\underline{\kappa}}.$$

(C) To give a more precise statement than what was just said in part (B), keep the notations of Proposition 1.6.2 and set in addition

$$\mathbb{I}' := \{\underline{k}' := (k'_1, k'_2, \ldots, k'_n) \in \mathbb{N}_0^n \mid k'_i \leq \min\{\mu'_i, \nu_i\} \text{ for } i = 1, 2, \ldots, n\}.$$

Use the product formula of Proposition 1.6.2 to show that

$$\left[\underline{X}^{\underline{\nu}}\underline{\partial}^{\underline{\mu}}, \underline{X}^{\underline{\nu}'}\underline{\partial}^{\underline{\mu}'}\right] = \sum_{\underline{k}\in\mathbb{I}\setminus\{\underline{0}\}} \lambda_{\underline{k}}\underline{X}^{\underline{\nu}+\underline{\nu}'-\underline{k}}\underline{\partial}^{\underline{\mu}+\underline{\mu}'-\underline{k}} - \sum_{\underline{k}'\in\mathbb{I}'\setminus\{\underline{0}\}} \lambda_{\underline{k}'}\underline{X}^{\underline{\nu}+\underline{\nu}'-\underline{k}'}\underline{\partial}^{\underline{\mu}+\underline{\mu}'-\underline{k}'}.$$

(D) Let $i \in \{1, , 2, \ldots, n\}$ and consider the n-tuple $\underline{e}_i := (\delta_{i,j})_{j=1}^n = (0, \ldots, 0, 1, 0, \ldots, 0) \in \mathbb{N}_0^n$. Use what you have shown in part (C) to prove the following statements

(a) $\left[X_i, \underline{X}^{\underline{\nu}}\underline{\partial}^{\underline{\mu}}\right] = \begin{cases} -\mu_i\underline{X}^{\underline{\nu}}\underline{\partial}^{\underline{\mu}-\underline{e}_i} & , \text{ if } \mu_i > 0; \\ 0 & , \text{ if } \mu_i = 0. \end{cases}$

(b) $\left[\partial_i, \underline{X}^{\underline{\nu}}\underline{\partial}^{\underline{\mu}}\right] = \begin{cases} \nu_i\underline{X}^{\underline{\nu}-\underline{e}_i}\underline{\partial}^{\underline{\mu}} & , \text{ if } \nu_i > 0; \\ 0 & , \text{ if } \nu_i = 0. \end{cases}$

Theorem 1.6.5 (The Reduction Principle) *Let $n \in \mathbb{N}$, let K be a field and consider the standard Weyl algebra*

$$\mathbb{W}(K, n) = K[X_1, X_2, \ldots, X_n, \partial_1, \partial_2, \ldots, \partial_n].$$

Let $r \in \mathbb{N}$, let

$\underline{\nu}^{(i)} := (\nu_1^{(i)}, \nu_2^{(i)}, \ldots, \nu_n^{(i)})$ and $\underline{\mu}^{(i)} := (\mu_1^{(i)}, \mu_2^{(i)}, \ldots, \mu_n^{(i)}) \in \mathbb{N}_0^n$, for $i = 1, 2, \ldots, r$

and abbreviate

$$\underline{\nu} := \sum_{i=1}^r \underline{\nu}^{(i)}, \quad \underline{\mu} := \sum_{i=1}^r \underline{\mu}^{(i)}.$$

Moreover, let the set

$$\mathbb{M} := \mathbb{M}_{\leq}(\underline{\nu}, \underline{\mu}) \subset \mathbb{N}_0^n \times \mathbb{N}_0^n$$

be defined according to Notation and Remark 1.6.3 (C). Then, we have

$$\prod_{i=1}^r \underline{X}^{\underline{\nu}^{(i)}}\underline{\partial}^{\underline{\mu}^{(i)}} - \underline{X}^{\underline{\nu}}\underline{\partial}^{\underline{\mu}} \in \sum_{(\underline{\kappa},\underline{\lambda})\in\mathbb{M}} \mathbb{Z}\underline{X}^{\underline{\lambda}}\underline{\partial}^{\underline{\kappa}}.$$

Proof We proceed by induction on r. The case $r = 1$ is obvious. The case $r = 2$ follows from Proposition 1.6.2, more precisely from its consequence proved in Exercise 1.6.4 (A) (see also Notation and Remark 1.6.3 (C)) . So, let $r > 2$. We set

$$\underline{\nu}' := \sum_{i=1}^{r-1} \underline{\nu}^{(i)}, \quad \underline{\mu}' := \sum_{i=1}^{r-1} \underline{\mu}^{(i)} \text{ and } \mathbb{M}' := \mathbb{M}_{\leq}(\underline{\nu}', \underline{\mu}').$$

By induction we have

$$\varrho := \prod_{i=1}^{r-1} \underline{X}^{\underline{\nu}^{(i)}} \underline{\partial}^{\underline{\mu}^{(i)}} - \underline{X}^{\underline{\nu}'} \underline{\partial}^{\underline{\mu}'} \in \sum_{(\underline{\lambda}', \underline{\kappa}') \in \mathbb{M}'} \mathbb{Z} \underline{X}^{\underline{\lambda}'} \underline{\partial}^{\underline{\kappa}'} =: N.$$

By the case $r = 2$ we have (see once more Notation and Remark 1.6.3 (C) and Exercise 1.6.4 (A))

$$\sigma := \big(\underline{X}^{\underline{\nu}'} \underline{\partial}^{\underline{\mu}'}\big) \underline{X}^{\underline{\nu}^{(r)}} \underline{\partial}^{\underline{\mu}^{(r)}} - \underline{X}^{\underline{\nu}} \underline{\partial}^{\underline{\mu}} \in \sum_{(\underline{\lambda}, \underline{\kappa}) \in \mathbb{M}} \mathbb{Z} \underline{X}^{\underline{\lambda}} \underline{\partial}^{\underline{\kappa}} =: M.$$

As

$$\prod_{i=1}^{r} \underline{X}^{\underline{\nu}^{(i)}} \underline{\partial}^{\underline{\mu}^{(i)}} - \underline{X}^{\underline{\nu}} \underline{\partial}^{\underline{\mu}} = \sigma + \varrho \underline{X}^{\underline{\nu}^{(r)}} \underline{\partial}^{\underline{\mu}^{(r)}},$$

it remains to show that

$$\varrho \underline{X}^{\underline{\nu}^{(r)}} \underline{\partial}^{\underline{\mu}^{(r)}} \in M.$$

Observe that

$$\varrho \underline{X}^{\underline{\nu}^{(r)}} \underline{\partial}^{\underline{\mu}^{(r)}} \in N \underline{X}^{\underline{\nu}^{(r)}} \underline{\partial}^{\underline{\mu}^{(r)}} = \sum_{(\underline{\lambda}', \underline{\kappa}') \in \mathbb{M}'} \mathbb{Z} \underline{X}^{\underline{\lambda}'} \underline{\partial}^{\underline{\kappa}'} \underline{X}^{\underline{\nu}^{(r)}} \underline{\partial}^{\underline{\mu}^{(r)}}.$$

Observe also that

$$(\underline{\lambda}' + \underline{\nu}^{(r)}, \underline{\kappa}' + \underline{\mu}^{(r)}) \in \mathbb{M} \text{ for all } (\underline{\lambda}', \underline{\kappa}') \in \mathbb{M}',$$

so that in the notation introduced in Notation and Remark 1.6.3 (C) we have

$$\overline{\mathbb{M}}(\underline{\lambda}' + \underline{\nu}^{(r)}, \underline{\kappa}' + \underline{\mu}^{(r)}) \subseteq \mathbb{M} \text{ for all } (\underline{\lambda}', \underline{\kappa}') \in \mathbb{M}'.$$

Hence, on application of Exercise 1.6.4 (A) it follows that

$$\underline{X}^{\underline{\lambda}'}\underline{\partial}^{\underline{\kappa}'}\underline{X}^{\underline{\nu}^{(r)}}\underline{\partial}^{\underline{\mu}^{(r)}} \in \sum_{(\underline{\lambda},\underline{\kappa})\in\overline{\mathbb{M}}(\underline{\lambda}'+\underline{\nu}^{(r)},\underline{\kappa}'+\underline{\mu}^{(r)})} \mathbb{Z}\underline{X}^{\underline{\lambda}}\underline{\partial}^{\underline{\kappa}} \subseteq \sum_{(\underline{\lambda},\underline{\kappa})\in\mathbb{M}} \mathbb{Z}\underline{X}^{\underline{\lambda}}\underline{\partial}^{\underline{\kappa}} = M,$$

and this shows that indeed $\varrho\underline{X}^{\underline{\nu}^{(r)}}\underline{\partial}^{\underline{\mu}^{(r)}} \in M$.

Now, in the next section, we can show that the elementary differential operators form a K-basis of the standard Weyl algebra $\mathbb{W}(K, n)$, provided the field K has characteristic 0. To prepare this, we add an additional exercise.

Exercise 1.6.6 (A) Let $n \in \mathbb{N}$ and consider the polynomial ring $K[X_1, X_2, \ldots, X_n]$. Moreover, let

$$\underline{\mu} := (\mu_1, \mu_1, \ldots, \mu_n), \text{ and } \underline{\nu} := (\nu_1, \nu_2, \ldots, \nu_n) \in \mathbb{N}_0^n.$$

Fix $i \in \{1, 2, \ldots, n\}$ and prove by induction on μ_i, that

$$\partial_i^{\mu_i}(\underline{X}^{\underline{\nu}}) = \partial_i^{\mu_i}\Big(\prod_{j=1}^{n} X_j^{\nu_j}\Big) = \begin{cases} \prod_{k=0}^{\mu_i-1}(\nu_i - k)X_i^{\nu_i-\mu_i}\prod_{j\neq i} X_j^{\nu_j}, & \text{if } \nu_i \geq \mu_i; \\ 0, & \text{if } \nu_i < \mu_i. \end{cases}$$

(B) Let the notations and hypotheses be as in part (A) and use what you have shown there to prove that

$$\begin{aligned}
\underline{\partial}^{\underline{\mu}}(\underline{X}^{\nu}) &= \prod_{i=1}^{n}\partial_i^{\mu_i}\Big(\prod_{j=1}^{n} X^{\nu_j}\Big) \\
&= \begin{cases} \prod_{i=1}^{n}\prod_{k=0}^{\mu_i-1}(\nu_i - k)X_i^{\nu_i-\mu_i}, & \text{if } \nu_i \geq \mu_i \text{ for all } i \in \{1, 2, \ldots, n\}; \\ 0, & \text{if } \nu_i < \mu_i \text{ for some } i \in \{1, 2, \ldots, n\}. \end{cases} \\
&= \begin{cases} \prod_{i=1}^{n}\prod_{k=0}^{\mu_i-1}(\nu_i - k)\underline{X}^{\underline{\nu}-\underline{\mu}}, & \text{if } \underline{\nu} \geq \underline{\mu}; \\ 0, & \text{otherwise.} \end{cases}
\end{aligned}$$

1.7 The Standard Basis

Now, we are ready to prove the fact that over a base field of characteristic 0 the elementary differential operators form a vector space basis of the standard Weyl algebra.

Theorem 1.7.1 (The Standard Basis) *Let $n \in \mathbb{N}$ and let K be a field of characteristic 0. Then, the elementary differential operators*

$$\underline{X}^{\underline{\nu}}\underline{\partial}^{\underline{\mu}} = \prod_{i=1}^{n} X_i^{\nu_i}\prod_{i=1}^{n}\partial_i^{\mu_i} \text{ with } \underline{\mu} := (\mu_1, \mu_2, \ldots, \mu_n) \text{ and } \underline{\nu} := (\nu_1, \nu_2, \ldots, \nu_n) \in \mathbb{N}_0^n$$

form a K-vector space basis of the standard Weyl algebra

$$\mathbb{W}(K,n) = K[X_1, X_2, \ldots, X_n, \partial_1, \partial_2, \ldots, \partial_n].$$

So, in particular we can say

(a) $\mathbb{W}(K,n) = \bigoplus_{\underline{\nu},\underline{\mu}\in\mathbb{N}_0^n} K\underline{X}^{\underline{\nu}}\underline{\partial}^{\underline{\mu}} = \bigoplus_{\mu_1,\mu_2,\ldots,\mu_n,\nu_1,\nu_2,\ldots,\nu_n\in\mathbb{N}_0} K\prod_{i=1}^n X_i^{\nu_i}\prod_{i=1}^n \partial_i^{\mu_i}$.

(b) Each differential operator $d \in \mathbb{W}(K,n)$ can be written in the form

$$d = \sum_{\underline{\nu},\underline{\mu}\in\mathbb{N}_0^n} c^{(d)}_{\underline{\nu},\underline{\mu}}\underline{X}^{\underline{\nu}}\underline{\partial}^{\underline{\mu}}$$

with a unique family

$$\big(c^{(d)}_{\underline{\nu},\underline{\mu}}\big)_{\underline{\nu},\underline{\mu}\in\mathbb{N}_0^n} \in \prod_{\underline{\nu},\underline{\mu}\in\mathbb{N}_0^n} K = K^{\mathbb{N}_0^n\times\mathbb{N}_0^n},$$

whose support

$$\operatorname{supp}(d) = \operatorname{supp}\big((c^{(d)}_{\underline{\nu},\underline{\mu}})_{\underline{\nu},\underline{\mu}\in\mathbb{N}_0^n}\big) := \{(\underline{\nu},\underline{\mu}) \in \mathbb{N}_0^n \times \mathbb{N}_0^n \mid c^{(d)}_{\underline{\nu},\underline{\mu}} \neq 0\}$$

is a finite set. We thus can write

$$d = \sum_{(\underline{\nu},\underline{\mu})\in\operatorname{supp}(d)} c^{(d)}_{\underline{\nu},\underline{\mu}}\underline{X}^{\underline{\nu}}\underline{\partial}^{\underline{\mu}} \quad .$$

Proof We first show, that the elementary differential operators generate $\mathbb{W}(K,n)$ as a K-vector space, hence that

$$\mathbb{W}(K,n) = \sum_{\underline{\nu},\underline{\mu}\in\mathbb{N}_0^n} K\underline{X}^{\underline{\nu}}\underline{\partial}^{\underline{\mu}} =: M.$$

Observe, that by definition each element d of $\mathbb{W}(K,n)$ is a K-linear combination of products of elementary differential operators. But by the Reduction Principle of Theorem 1.6.5 each product of elementary differential operators is contained in the K-vector space M.

It remains to show, that the elementary differential operators are linearly independent among each other. Assume to the contrary, that there are linearly dependent elementary differential operators in $\mathbb{W}(K,n)$. Then, we find a positive integer $r \in \mathbb{N}$, families

$$\underline{\mu}^{(i)} := (\mu_1^{(i)}, \mu_2^{(i)}, \ldots, \mu_n^{(i)}), \quad \underline{\nu}^{(i)} := (\nu_1^{(i)}, \nu_2^{(i)}, \ldots, \nu_n^{(i)}) \in \mathbb{N}_0^n, \quad (i = 1, 2, \ldots, r)$$

with

$$(\underline{\mu}^{(i)}, \underline{\nu}^{(i)}) \neq (\underline{\mu}^{(j)}, \underline{\nu}^{(j)}) \text{ for all } i, j \in \{1, 2, \ldots, r\} \text{ with } i \neq j,$$

and elements

$$c^{(i)} \in K \setminus \{0\} \quad (i = 1, 2, \ldots, r),$$

such that

$$d := \sum_{i=1}^{r} c^{(i)} \underline{X}^{\underline{\nu}^{(i)}} \partial^{\underline{\mu}^{(i)}} = 0.$$

We may assume, that

$$|\underline{\mu}^{(r)}| = \max\{|\underline{\mu}^{(i)}| \mid i = 1, 2, \ldots, r\}$$

and that for some $s \in \{1, 2, \ldots, r\}$ we have

$$\underline{\mu}^{(i)} \neq \underline{\mu}^{(r)} \text{ for all } i < s \text{ and } \underline{\mu}^{(i)} = \underline{\mu}^{(r)} \text{ for all } i \geq s.$$

Then, it follows easily by what we have seen in Exercise 1.6.6 (B), that

$$\underline{X}^{\underline{\nu}^{(i)}} \partial^{\underline{\mu}^{(i)}} \big(\underline{X}^{\underline{\mu}^{(r)}}\big) = \begin{cases} \prod_{j=1}^{n} \mu_j^{(r)}! \underline{X}^{\underline{\nu}^{(r)}}, & \text{if } s \leq i \leq r \\ 0, & \text{if } i < s. \end{cases}$$

So, we get

$$0 = d\big(\underline{X}^{\underline{\mu}^{(r)}}\big) = \sum_{i=1}^{r} c^{(i)} \underline{X}^{\underline{\nu}^{(i)}} \partial^{\underline{\mu}^{(i)}} \big(\underline{X}^{\underline{\mu}^{(r)}}\big) = \sum_{i=s}^{r} c^{(i)} \prod_{j=1}^{n} \mu_j^{(r)}! \underline{X}^{\underline{\nu}^{(i)}}.$$

As $\mathrm{Char}(K) = 0$, and as the monomials $\underline{X}^{\underline{\nu}^{(i)}}$ are pairwise different for $i = s, s + 1, \ldots, r$, the last sum does not vanish, and we have a contradiction.

Definition and Remark 1.7.2 (A) Let the notations and hypotheses be as in Theorem 1.6.5. We call the basis of $\mathbb{W}(K, n)$ which consists of all elementary differential operators the *standard basis*. If we present a differential operator $d \in \mathbb{W}(K, n)$ with respect to the standard basis and write

$$d = \sum_{\underline{\nu}, \underline{\mu} \in \mathbb{N}_0^n} c_{\underline{\nu}, \underline{\mu}}^{(d)} \underline{X}^{\underline{\nu}} \partial^{\underline{\mu}}$$

as in statement (b) of Theorem 1.6.5, we say that d is written in *standard form*. The support of a differential operator d in $\mathbb{W}(K, n)$ is always defined with respect to the standard form as in statement (b) of Theorem 1.7.1. We therefore call the support of d also the *standard support* of d.

(B) Keep the above notations and hypotheses. It is a fundamental task, to write an arbitrarily given differential operator $d \in \mathbb{W}(K, n)$ in standard form. This task actually is reduced by the Reduction Principle of Theorem 1.6.5 to make explicit the coefficients of the differences

$$\Delta_{\underline{\nu}^{(\bullet)}\underline{\mu}^{(\bullet)}} := \prod_{i=1}^{r} \underline{X}^{\underline{\nu}^{(i)}} \partial^{\underline{\mu}^{(i)}} - \underline{X}^{\sum_{i=1}^{r} \underline{\nu}^{(i)}} \partial^{\sum_{i=1}^{r} \underline{\mu}^{(i)}} \in \sum_{(\underline{\lambda}, \underline{\kappa}) \in \mathbb{M}} \mathbb{Z}\underline{X}^{\underline{\lambda}} \partial^{\underline{\kappa}}.$$

This task can be solved by a repeated application of the Product Formula of Proposition 1.6.2 or—directly—by a repeated application of the Heisenberg relations. Clearly, this is a task which usually is performed by means of Computer Algebra systems.

We now prove the following application, a result on supports, which will turn out to be useful in the next section.

Proposition 1.7.3 (Behavior of Supports) *Let $n \in \mathbb{N}$, let K be a field of characteristic 0 and consider the differential operators*

$$d, e \in \mathbb{W}(K, n) = K[X_1, X_2, \ldots, X_n, \partial_1, \partial_2, \ldots, \partial_n].$$

For all $(\underline{\alpha}, \underline{\beta}) \in \mathbb{N}_0^n \times \mathbb{N}_0^n$, let the sets

$$\mathbb{M}(\underline{\alpha}, \underline{\beta}) \subset \overline{\mathbb{M}}(\underline{\alpha}, \underline{\beta}) \subset \mathbb{N}_0^n \times \mathbb{N}_0^n$$

be defined according to Notation and Remark 1.6.3 (C). Then, we have

(a) $\big(\text{supp}(d) \cup \text{supp}(e)\big) \setminus \big(\text{supp}(d) \cap \text{supp}(e)\big) \subseteq \text{supp}(d+e) \subseteq \text{supp}(d) \cup \text{supp}(e)$.
(b) $\text{supp}(cd) = \text{supp}(d)$ *for all* $c \in K \setminus \{0\}$.
(c) $\text{supp}(de) \subseteq \bigcup_{(\underline{\nu},\underline{\mu}) \in \text{supp}(d), (\underline{\nu}',\underline{\mu}') \in \text{supp}(e)} \overline{\mathbb{M}}(\underline{\nu} + \underline{\nu}', \underline{\mu} + \underline{\mu}')$.
(d) $\text{supp}\big([d, e]\big) \subseteq \bigcup_{(\underline{\nu},\underline{\mu}) \in \text{supp}(d), (\underline{\nu}',\underline{\mu}') \in \text{supp}(e)} \mathbb{M}(\underline{\nu} + \underline{\nu}', \underline{\mu} + \underline{\mu}')$.

Proof (a), (b) These statements follow in a straight forward way from our definition of support, and we leave it as an exercise to perform their proof.

(c) In the notations of Theorem 1.7.1 we write

$$d = \sum_{(\underline{\nu},\underline{\mu}) \in \text{supp}(d)} c^{(d)}_{\underline{\nu},\underline{\mu}} \underline{X}^{\underline{\nu}} \partial^{\underline{\mu}} \text{ and } e = \sum_{(\underline{\nu}',\underline{\mu}') \in \text{supp}(e)} c^{(e)}_{\underline{\nu}',\underline{\mu}'} \underline{X}^{\underline{\nu}'} \partial^{\underline{\mu}'}.$$

it follows that

$$de = \sum_{(\underline{\nu},\underline{\mu})\in\operatorname{supp}(d),(\underline{\nu}',\underline{\mu}')\in\operatorname{supp}(e)} c^{(d)}_{\underline{\nu},\underline{\mu}} c^{(e)}_{\underline{\nu}',\underline{\mu}'} \underline{X}^{\underline{\nu}} \underline{\partial}^{\underline{\mu}} \underline{X}^{\underline{\nu}'} \underline{\partial}^{\underline{\mu}'}.$$

But according to Exercise 1.6.4 (A) we have

$\operatorname{supp}(\underline{X}^{\underline{\nu}}\underline{\partial}^{\underline{\mu}}\underline{X}^{\underline{\nu}'}\underline{\partial}^{\underline{\mu}'}) \subseteq \overline{\mathbb{M}}(\underline{\nu}+\underline{\nu}', \underline{\mu}+\underline{\mu}')$ for all $(\underline{\nu}, \underline{\mu}) \in \operatorname{supp}(d)$ and all $(\underline{\nu}', \underline{\mu}') \in \operatorname{supp}(e)$.

Now, our claim follows easily on repeated application of statements (a) and (b).

(d) As in the proof of statement (c) we can write

$$de = \sum_{(\underline{\nu},\underline{\mu})\in\operatorname{supp}(d),(\underline{\nu}',\underline{\mu}')\in\operatorname{supp}(e)} c^{(d)}_{\underline{\nu},\underline{\mu}} c^{(e)}_{\underline{\nu}',\underline{\mu}'} \underline{X}^{\underline{\nu}} \underline{\partial}^{\underline{\mu}} \underline{X}^{\underline{\nu}'} \underline{\partial}^{\underline{\mu}'}$$

and, similarly

$$ed = \sum_{(\underline{\nu},\underline{\mu})\in\operatorname{supp}(d),(\underline{\nu}',\underline{\mu}')\in\operatorname{supp}(e)} c^{(d)}_{\underline{\nu},\underline{\mu}} c^{(e)}_{\underline{\nu}',\underline{\mu}'} \underline{X}^{\underline{\nu}'} \underline{\partial}^{\underline{\mu}'} \underline{X}^{\underline{\nu}} \underline{\partial}^{\underline{\mu}}.$$

It follows that

$$\begin{aligned}
\big[de, ed\big] &= de - ed \\
&= \sum_{(\underline{\nu},\underline{\mu})\in\operatorname{supp}(d),(\underline{\nu}',\underline{\mu}')\in\operatorname{supp}(e)} c^{(d)}_{\underline{\nu},\underline{\mu}} c^{(e)}_{\underline{\nu}',\underline{\mu}'} \underline{X}^{\underline{\nu}} \underline{\partial}^{\underline{\mu}} \underline{X}^{\underline{\nu}'} \underline{\partial}^{\underline{\mu}'} \\
&\quad - \sum_{(\underline{\nu},\underline{\mu})\in\operatorname{supp}(d),(\underline{\nu}',\underline{\mu}')\in\operatorname{supp}(e)} c^{(d)}_{\underline{\nu},\underline{\mu}} c^{(e)}_{\underline{\nu}',\underline{\mu}'} \underline{X}^{\underline{\nu}'} \underline{\partial}^{\underline{\mu}'} \underline{X}^{\underline{\nu}} \underline{\partial}^{\underline{\mu}} \\
&= \sum_{(\underline{\nu},\underline{\mu})\in\operatorname{supp}(d),(\underline{\nu}',\underline{\mu}')\in\operatorname{supp}(e)} c^{(d)}_{\underline{\nu},\underline{\mu}} c^{(e)}_{\underline{\nu}',\underline{\mu}'} \big(\underline{X}^{\underline{\nu}} \underline{\partial}^{\underline{\mu}} \underline{X}^{\underline{\nu}'} \underline{\partial}^{\underline{\mu}'} - \underline{X}^{\underline{\nu}'} \underline{\partial}^{\underline{\mu}'} \underline{X}^{\underline{\nu}} \underline{\partial}^{\underline{\mu}}\big) \\
&= \sum_{(\underline{\nu},\underline{\mu})\in\operatorname{supp}(d),(\underline{\nu}',\underline{\mu}')\in\operatorname{supp}(e)} c^{(d)}_{\underline{\nu},\underline{\mu}} c^{(e)}_{\underline{\nu}',\underline{\mu}'} \big[\underline{X}^{\underline{\nu}} \underline{\partial}^{\underline{\mu}}, \underline{X}^{\underline{\nu}'} \underline{\partial}^{\underline{\mu}'}\big].
\end{aligned}$$

By Exercise 1.6.4 (B) we have

$$\operatorname{supp}\big(\big[\underline{X}^{\underline{\nu}}\underline{\partial}^{\underline{\mu}}, \underline{X}^{\underline{\nu}'}\underline{\partial}^{\underline{\mu}'}\big]\big) \subseteq \mathbb{M}(\underline{\nu}+\underline{\nu}', \underline{\mu}+\underline{\mu}')$$

for all $(\underline{\nu}, \underline{\mu}) \in \operatorname{supp}(d)$ and all $(\underline{\nu}', \underline{\mu}') \in \operatorname{supp}(e)$.

Now, statement (d) follows easily on repeated application of statements (a) and (b).

Exercise 1.7.4

(A) Let $n \in \mathbb{N}$, let K be a field of characteristic 0 and consider the standard Weyl algebra

$$\mathbb{W} = \mathbb{W}(K, n) = K[X_1, X_2, \ldots, X_n, \partial_1, \partial_2, \ldots, \partial_n].$$

Prove in detail statements (a) and (b) of Proposition 1.7.3.

(B) Let the notations and hypotheses be as in part (A). Present in standard form the following differential operators:

$$\partial_1^2 X_1^2 - X_1\partial_1 X_1 - 1, \quad \partial_1^2 X_1^2 \partial_1^2 - \partial_1 X_1^2, \quad \partial_2 X_1 X_2 \partial_1 + \partial_1 X_1 X_2 \in \mathbb{W}(K, n).$$

(C) Keep the notations of part (A), but assume that $n = 1$ and $\mathrm{Char}(K) = 2$. Compute $\partial_1(X_1^\nu)$ for all $\nu \in \mathbb{N}_0$ and comment your findings in view of the Standard Basis Theorem.

(D) Keep the notations of part (A), let

$$d = \sum_{(\underline{\nu},\underline{\mu})\in \mathrm{supp}(d)} c^{(d)}_{\underline{\nu},\underline{\mu}} \underline{X}^{\underline{\nu}} \underline{\partial}^{\underline{\mu}} \in \mathbb{W}, \quad \big(c^{(d)}_{\underline{\nu},\underline{\mu}} \in K \setminus \{0\}, \forall (\underline{\nu}, \underline{\mu}) \in \mathrm{supp}(d)\big)$$

(see Theorem 1.7.1) and let $i \in \{1, 2, \ldots, n\}$. Use Exercise 1.6.4 (D) to prove the following equalities:

(a) $\big[X_i, d\big] = -\sum_{(\underline{\nu},\underline{\mu})\in \mathrm{supp}(d):\mu_i>0} \mu_i c^{(d)}_{\underline{\nu},\underline{\mu}} \underline{X}^{\underline{\nu}} \underline{\partial}^{\underline{\mu}-\underline{e}_i}$.

(b) $\big[\partial_i, d\big] = \sum_{(\underline{\nu},\underline{\mu})\in \mathrm{supp}(d):\nu_i>0} \nu_i c^{(d)}_{\underline{\nu},\underline{\mu}} \underline{X}^{\underline{\nu}-\underline{e}_i} \underline{\partial}^{\underline{\mu}}$.

Conclude that

(c) $d = 0 \Leftrightarrow \forall i \in \{1, 2, \ldots, n\} : \big[X_i, d\big] = \big[\partial_i, d\big] = 0.$

As another application of the Standard Basis Theorem we now can prove

Corollary 1.7.5 (The Universal Property of Weyl Algebras) *Let $n \geq 2$ and let the notations and hypotheses be as in Theorem 1.7.1. Let B be a K-algebra and let*

$$\phi : \{X_1, X_2, \ldots, X_n, \partial_1, \partial_2, \ldots, \partial_n\} \longrightarrow B$$

be a map "which respects the Heisenberg relations" and hence satisfies the requirements

(1) $[\phi(X_i), \phi(X_j)] = 0,$ *for all $i, j \in \{1, 2, \ldots, n\}$;*
(2) $[\phi(X_i), \phi(\partial_j)] = -\delta_{i,j},$ *for all $i, j \in \{1, 2, \ldots, n\}$;*
(3) $[\phi(\partial_i), \phi(\partial_j)] = 0,$ *for all $i, j \in \{1, 2, \ldots, n\}$.*

Then, there is a unique homomorphism of K-algebras

$$\widetilde{\phi} : \mathbb{W}(K, n) \longrightarrow B$$

such that

$$\widetilde{\phi}(X_i) = \phi(X_i) \textit{ and } \widetilde{\phi}(\partial_i) = \phi(\partial_i) \textit{ for all } i = 1, 2, \ldots, n.$$

Proof According to Theorem 1.7.1 there is a K-linear map

$$\widetilde{\phi} : \mathbb{W}(K, n) \longrightarrow B \text{ given by}$$

$$\widetilde{\phi}(\underline{X}^{\underline{\nu}}\underline{\partial}^{\underline{\mu}}) = \prod_{i=1}^{n} \phi(X_i)^{\nu_i} \prod_{i=1}^{n} \phi(\partial_i)^{\mu_i} \text{ for all}$$

$$\underline{\mu} = (\mu_1, \mu_2, \ldots, \mu_n) \text{ and } \underline{\nu} = (\nu_1, \nu_2, \ldots, \nu_n) \in \mathbb{N}_0^n.$$

Next, we show, that the previously defined K-linear map $\widetilde{\phi}$ is multiplicative, and hence satisfies the condition that

$$\widetilde{\phi}(de) = \widetilde{\phi}(d)\widetilde{\phi}(e) \text{ for all } d, e \in \mathbb{W}(K, n).$$

As the multiplication maps

$$\mathbb{W}(K, n) \times \mathbb{W}(K, n) \longrightarrow \mathbb{W}(K, n), (d, e) \mapsto de \quad \text{and} \quad B \times B \longrightarrow B, (a, b) \mapsto ab$$

are both K-bilinear, it suffices to verify the above multiplicativity condition in the special case where

$$d := \underline{X}^{\underline{\nu}}\underline{\partial}^{\underline{\mu}} \text{ and } e := \underline{X}^{\underline{\nu}'}\underline{\partial}^{\underline{\mu}'}$$

with

$$\underline{\mu} := (\mu_1, \mu_2, \ldots, \mu_n), \quad \underline{\nu} := (\nu_1, \nu_2, \ldots, \nu_n) \text{ and}$$

$$\underline{\mu}' := (\mu_1', \mu_2', \ldots, \mu_n'), \quad \underline{\nu}' := (\nu_1', \nu_2', \ldots, \nu_n') \in \mathbb{N}_0^n.$$

But this can be done by a straight forward computation, on use of the Product Formula of Proposition 1.6.2 and on application of Lemma 1.6.1 with

$$a_i : \phi(X_i) \text{ and } d_i := \phi(\partial_i) \text{ for all } i = 1, 2, \ldots, n.$$

It remains to show, that $\widetilde{\phi} : \mathbb{W}(K, n) \longrightarrow B$ is the only homomorphism of K-algebras which satisfies the requirement that

$$\widetilde{\phi}(X_i) = \phi(X_i) \quad \text{and} \quad \widetilde{\phi}(\partial_i) = \phi(\partial_i) \text{ for all } i = 1, 2, \ldots, n.$$

But indeed, if a map $\widetilde{\phi}$ satisfies this requirement and is multiplicative, it must be defined on the elementary differential operators as suggested above. This proves the requested uniqueness.

Exercise 1.7.6

(A) Let $n \in \mathbb{N}$, let K be a field of characteristic 0. Show, that there is a unique automorphism of K-algebras

$$\alpha : \mathbb{W}(K, n) \xrightarrow{\cong} \mathbb{W}(K, n) \text{ with } \alpha(X_i) = \partial_i \text{ and } \alpha(\partial_i) = -X_i \text{ for all } i = 1, 2, \ldots, n.$$

(B) Keep the notations and hypotheses of part (A). Present in standard form all elements $\alpha(X_i^\nu \partial_i^\mu) \in \mathbb{W}(K, n)$ with $\mu, \nu \in \mathbb{N}_0$.

1.8 Weighted Degrees and Filtrations

In this section we introduce and investigate a particularly nice class of filtrations of the standard Weyl algebras, the so-called weighted filtrations. To do so, we first will introduce the related notion of weighted degree of a differential operator.

Convention 1.8.1 Throughout this section we fix a positive integer n, a field K of characteristic 0 and we consider the standard Weyl algebra

$$\mathbb{W} := \mathbb{W}(K, n) = K[X_1, X_2, \ldots, X_n, \partial_1, \partial_2, \ldots, \partial_n]$$

Definition and Remark 1.8.2 (A) By a *weight* we mean a pair

$$(\underline{v}, \underline{w}) = \big((v_1, v_2, \ldots, v_n), (w_1, w_2, \ldots, w_n)\big) \in \mathbb{N}_0^n \times \mathbb{N}_0^n$$

such that

$$(v_i, w_i) \neq (0, 0) \text{ for all } i = 1, 2, \ldots, n.$$

For

$$\underline{a} := (a_1, a_2, \ldots, a_n), \quad \underline{b} := (b_1, b_2, \ldots, b_n) \in \mathbb{R}^n$$

we frequently shall use the *scalar product*

$$\underline{a} \cdot \underline{b} := \sum_{i=1}^{n} a_i b_i.$$

(B) Fix a weight $(\underline{v}, \underline{w}) \in \mathbb{N}_0^n \times \mathbb{N}_0^n$. We define the *degree associated to the weight* $(\underline{v}, \underline{w})$ (or just the *weighted degree*) of a differential form $d \in \mathbb{W}$ by

$$\deg^{\underline{vw}}(d) := \sup\{\underline{v} \cdot \underline{\nu} + \underline{w} \cdot \underline{\mu} \mid (\underline{\nu}, \underline{\mu}) \in \operatorname{supp}(d)\}.$$

with the usual convention that $\sup(\emptyset) = -\infty$.
Observe that by our definition of weight, for all $d \in \mathbb{W}$ and all $\underline{\mu}, \underline{\nu} \in \mathbb{N}_0$—and using the notations of Notation and Remark 1.6.3 (C)– we can say:

(a) $\deg^{\underline{vw}}(d) \in \mathbb{N}_0 \cup \{-\infty\}$ with $\deg^{\underline{vw}}(d) = -\infty$ if and only if $d = 0$.
(b) If $\underline{\lambda} \leq \underline{\nu}$ and $\underline{\kappa} \leq \underline{\mu}$ for all $(\underline{\lambda}, \underline{\kappa}) \in \operatorname{supp}(d)$, then

$$\deg^{\underline{vw}}(d) \leq \underline{v} \cdot \underline{\nu} + \underline{w} \cdot \underline{\mu}.$$

(c) If $\operatorname{supp}(d) \subseteq \mathbb{M}_{\leq}(\underline{\nu}, \underline{\mu})$, then

$$\deg^{\underline{vw}}(d) < \underline{v} \cdot \underline{\nu} + \underline{w} \cdot \underline{\mu}.$$

(C) Keep the notations and hypotheses of part (B). We fix some non-negative integer $i \in \mathbb{N}_0$ and set

$$\mathbb{W}_i^{\underline{vw}} := \{d \in \mathbb{W} \mid \deg^{\underline{vw}}(d) \leq i\}.$$

Observe, that we also may write

$$\mathbb{W}_i^{\underline{vw}} = \bigoplus_{\underline{\nu}, \underline{\mu} \in \mathbb{N}_0^n : \underline{v} \cdot \underline{\nu} + \underline{w} \cdot \underline{\mu} \leq i} K \underline{X}^{\underline{\nu}} \underline{\partial}^{\underline{\mu}}.$$

Lemma 1.8.3 *Let $(\underline{v}, \underline{w}) \in \mathbb{N}_0^n \times \mathbb{N}_0^n$ be a weight and let $d, e \in \mathbb{W}$. Then we have*

(a) $\deg^{\underline{vw}}(d + e) \leq \max\{\deg^{\underline{vw}}(d), \deg^{\underline{vw}}(e)\}$, *with equality if* $\deg^{\underline{vw}}(d) \neq \deg^{\underline{vw}}(e)$;
(b) $\deg^{\underline{vw}}(cd) = \deg^{\underline{vw}}(d)$ *for all* $c \in K \setminus \{0\}$.
(c) $\deg^{\underline{vw}}(de) \leq \deg^{\underline{vw}}(d) + \deg^{\underline{vw}}(e)$;
(d) $\deg^{\underline{vw}}([d, e]) < \deg^{\underline{vw}}(d) + \deg^{\underline{vw}}(e)$.

Notice: In statement (c) actually equality holds. We shall prove this later (see Corollary 1.9.5).

Proof (a) The stated inequality is clear by the second inclusion of the following relation (see Proposition 1.7.3 (a)):

$$\big(\operatorname{supp}(d) \cup \operatorname{supp}(e)\big) \setminus \big(\operatorname{supp}(d) \cap \operatorname{supp}(e)\big) \subseteq \operatorname{supp}(d + e) \subseteq \operatorname{supp}(d) \cup \operatorname{supp}(e).$$

It remains to establish the stated equality if $\deg^{\underline{vw}}(d) \neq \deg^{\underline{vw}}(e)$. It suffices to treat the case in which $\deg^{\underline{vw}}(d) < \deg^{\underline{vw}}(e)$. In this case, there is some

$$(\underline{\nu}, \underline{\mu}) \in \operatorname{supp}(e) \setminus \operatorname{supp}(d) \text{ with } \underline{v} \cdot \underline{\nu} + \underline{w} \cdot \underline{\mu} = \deg^{\underline{vw}}(e).$$

By the first of the previous inclusions we have $(\underline{\nu}, \underline{\mu}) \in \operatorname{supp}(d+e)$ and hence

$$\deg^{\underline{vw}}(d+e) \geq \underline{v} \cdot \underline{\nu} + \underline{w} \cdot \underline{\mu} = \deg^{\underline{vw}}(e).$$

By the already proved inequality $\deg^{\underline{vw}}(d+e) \leq \max\{\deg^{\underline{vw}}(d), \deg^{\underline{vw}}(e)\}$ it follows that $\deg^{\underline{vw}}(d+e) = \deg^{\underline{vw}}(e)$.

(b) This is obvious.

(c) This follows easily by Proposition 1.7.3 (c) and Definition and Remark 1.8.2 (B) (b).

(d) This follows in a straight forward manner by Proposition 1.7.3 (d) and Definition and Remark 1.8.2 (B) (c).

Theorem 1.8.4 (Weighted Filtrations) *Let*

$$\big((v_1, v_2, \ldots, v_n), (w_1, w_2, \ldots, w_n)\big) = (\underline{v}, \underline{w}) \in \mathbb{N}_0^n \times \mathbb{N}_0^n$$

be a weight. Then, the family

$$\mathbb{W}_\bullet^{\underline{vw}} := \big(\mathbb{W}_i^{\underline{vw}} = \{d \in \mathbb{W} \mid \deg^{\underline{vw}}(d) \leq i\}\big)_{i \in \mathbb{N}_0}$$

is a commutative filtration of the K-algebra $\mathbb{W} = \mathbb{W}(K, n)$.
Moreover, the following statements hold.

(a) $\mathbb{W}_0^{\underline{vw}} = K[X_i, \partial_j \mid v_i = 0, w_j = 0]$, *so that* $\mathbb{W}_0^{\underline{vw}}$ *is a commutative polynomial algebra in the variables X_i and ∂_j for which either $v_i = 0$ or else $w_j = 0$.*
(b) *Let $\delta = \delta(\underline{vw}) = \max\{v_1, v_2, \ldots, v_n, w_1, w_2, \ldots, w_n\}$. Then, for all $i > \delta$ it holds*

$$\mathbb{W}_i^{\underline{vw}} = \sum_{j=1}^{\delta} \mathbb{W}_j^{\underline{vw}} \mathbb{W}_{i-j}^{\underline{vw}}.$$

(c) *The filtration $\mathbb{W}_\bullet^{\underline{vw}} = \big(\mathbb{W}_i^{\underline{vw}}\big)_{i \in \mathbb{N}_0}$ is of finite type.*

Proof It is clear from our definitions, that

$$\mathbb{W}_i^{\underline{vw}} \subseteq \mathbb{W}_{i+1}^{\underline{vw}} \text{ for all } i \in \mathbb{N}_0, \quad 1 \in \mathbb{W}_0^{\underline{vw}} \quad \text{and } \mathbb{W} = \bigcup_{i \in \mathbb{N}_0} \mathbb{W}_i^{\underline{vw}}.$$

On use of Lemma 1.8.3 (c) it follows immediately that

$$\mathbb{W}_i^{\underline{vw}} \mathbb{W}_j^{\underline{vw}} \subseteq \mathbb{W}_{i+j}^{\underline{vw}} \text{ for all } i, j \in \mathbb{N}_0.$$

So the family $\big(\mathbb{W}_i^{\underline{vw}} := \{d \in \mathbb{W} \mid \deg^{\underline{vw}}(d) \leq i\}\big)_{i \in \mathbb{N}_0}$ constitutes indeed a filtration on the K-algebra $\mathbb{W}$.

Now, let $i, j \in \mathbb{N}_0$, let $d \in \mathbb{W}_i^{\underline{vw}}$ and let $e \in \mathbb{W}_j^{\underline{vw}}$. Then by Lemma 1.8.3 (d) we have

$$\deg^{\underline{vw}}\big(de - ed\big) = \deg^{\underline{vw}}\big([d, e]\big) \leq \deg^{\underline{vw}}(d) + \deg^{\underline{vw}}(e) - 1 \leq i + j - 1,$$

so that

$$de - ed \in \mathbb{W}_{i+j-1}^{\underline{vw}}.$$

This proves, that our filtration is commutative (see Definition 1.3.3).

(a) Set

$$\mathbb{S} := \{i = 1, 2, \dots, n \mid v_i \neq 0\} \text{ and } \mathbb{T} := \{j = 1, 2, \dots, n \mid w_j \neq 0\} \text{ and}$$

$$\overline{\mathbb{S}} := \{1, 2, \dots, n\} \setminus \mathbb{S} \text{ and } \overline{\mathbb{T}} := \{1, 2, \dots, n\} \setminus \mathbb{T}.$$

Let $\underline{\nu}, \underline{\mu} \in \mathbb{N}_0^n$. Then

$$\underline{v} \cdot \underline{\nu} + \underline{w} \cdot \underline{\mu} = 0 \text{ if and only if } \nu_i = 0 \text{ for all } i \in \mathbb{S} \text{ and } \mu_j = 0 \text{ for all } j \in \mathbb{T}.$$

But this means that

$$\begin{aligned}\mathbb{W}_0^{\underline{vw}} &= \sum_{(\nu_i)_{i\in\overline{\mathbb{S}}}, (\mu_j)_{j\in\overline{\mathbb{T}}}} K \prod_{i\in\overline{\mathbb{S}}, j\in\overline{\mathbb{T}}} X_i^{\nu_i} \partial_j^{\mu_j} \\ &= K[X_i, \partial_j \mid v_i = 0, w_j = 0].\end{aligned}$$

It remains to show, that this latter ring is a commutative polynomial algebra in all the variables X_i and ∂_j for which either $v_i = 0$ or else $w_j = 0$. In view of Theorem 1.7.1 it suffices to show that $X_i\partial_j = \partial_j X_i$ for all i, j with $v_i = v_j = 0$. But as $(v_k, w_k) \neq (0, 0)$ for all $k = 1, 2, \dots, n$ (see Definition and Remark 1.8.2 (A)), this is clear by the Heisenberg relations (see Proposition 1.5.4 (b)).

(b) Let $i > \delta$. Let

$$\underline{\nu} := (\nu_1, \nu_2, \dots, \nu_n), \quad \underline{\mu} := (\mu_1, \mu_2, \dots, \mu_n) \in \mathbb{N}_0^n \text{ with}$$

$$\sigma := \deg^{\underline{vw}}\big(\underline{X}^{\underline{\nu}}\underline{\partial}^{\underline{\mu}}\big) = \underline{v} \cdot \underline{\nu} + \underline{w} \cdot \underline{\mu} \leq i.$$

We aim to show that

$$\underline{X}^{\underline{\nu}}\underline{\partial}^{\underline{\mu}} \in \sum_{j=1}^{\delta} \mathbb{W}_j^{\underline{vw}} \mathbb{W}_{i-j}^{\underline{vw}} =: M.$$

If $\sigma \leq 0$ this is clear as $i > 0$ implies $i \geq 1$, so that

$$\mathbb{W}_0^{\underline{v}\underline{w}} = \mathbb{W}_0^{\underline{v}\underline{w}}\mathbb{W}_0^{\underline{v}\underline{w}} \subseteq \mathbb{W}_1^{\underline{v}\underline{w}}\mathbb{W}_{i-1}^{\underline{v}\underline{w}} \subseteq M.$$

So, let $\sigma > 0$. Then either

(1) there is some $p \in \{1, 2, \ldots, n\}$ with $v_p > 0$ and $\nu_p > 0$, or else,
(2) there is some $q \in \{1, 2, \ldots, n\}$ with $w_q > 0$ and $\mu_q > 0$.

In the above case (1) we can write

$$\underline{X}^{\underline{\nu}}\underline{\partial}^{\underline{\mu}} = X_p d, \text{ with } d := \big(\prod_{k=1}^{n} X_k^{\nu_k - \delta_{k,p}}\big)\underline{\partial}^{\underline{\mu}}.$$

As $\deg^{\underline{v}\underline{w}}(X_p) = v_p \leq \delta$ and $\deg^{\underline{v}\underline{w}}(d) = \sigma - v_p$ it follows that

$$\underline{X}^{\underline{\nu}}\underline{\partial}^{\underline{\mu}} = X_p d \in \mathbb{W}_{v_p}^{\underline{v}\underline{w}}\mathbb{W}_{\sigma - v_p}^{\underline{v}\underline{w}} \subseteq \mathbb{W}_{v_p}^{\underline{v}\underline{w}}\mathbb{W}_{i - v_p}^{\underline{v}\underline{w}} \subseteq M.$$

In the above case (2) we may first assume, that we are not in the case (1). This means in particular that either $v_q = 0$ or $\nu_q = 0$, hence $v_q \nu_q = 0$, so that

$$\deg^{\underline{v}\underline{w}}(X_q^{\nu_q}\partial_q) = w_q \leq \delta.$$

Now, in view of the Heisenberg relations, we may write

$$\underline{X}^{\underline{\nu}}\underline{\partial}^{\underline{\mu}} = X_q^{\nu_q}\partial_q e \text{ with } e := \prod_{s \neq q} X_s^{\nu_s} \prod_{k=1}^{n} \partial_k^{\mu_k - \delta_{k,q}}.$$

As $v_q \nu_q = 0$, we have $\deg^{\underline{v}\underline{w}}(e) = \sigma - w_q$, and it follows that

$$\underline{X}^{\underline{\nu}}\underline{\partial}^{\underline{\mu}} = X_q^{\nu_q}\partial_q e \in \mathbb{W}_{w_q}^{\underline{v}\underline{w}}\mathbb{W}_{\sigma - w_q}^{\underline{v}\underline{w}} \subseteq \mathbb{W}_{w_q}^{\underline{v}\underline{w}}\mathbb{W}_{i - w_q}^{\underline{v}\underline{w}} \subseteq M.$$

But this shows, what we were aiming for, hence that

$$\underline{X}^{\underline{\nu}}\underline{\partial}^{\underline{\mu}} \in M \text{ whenever } \underline{v} \cdot \underline{\nu} + \underline{w} \cdot \underline{\mu} \leq i.$$

But this means that

$$\mathbb{W}_i^{\underline{v}\underline{w}} \subseteq M = \sum_{j=1}^{\delta} \mathbb{W}_j^{\underline{v}\underline{w}}\mathbb{W}_{i-j}^{\underline{v}\underline{w}}$$

and hence proves statement (b).

(c) This is an immediate consequence of statements (a) and (b) (see Definition and Remark 1.3.4 (C)).

Definition 1.8.5 Let the notations and hypotheses be as in Theorem 1.8.4. In particular, let

$$\big((v_1, v_2, \ldots, v_n), (w_1, w_2, \ldots, w_n)\big) = (\underline{v}, \underline{w}) \in \mathbb{N}_0^n \times \mathbb{N}_0^n$$

be a weight. Then, the filtration

$$\mathbb{W}_\bullet^{\underline{vw}} = \big(\mathbb{W}_i^{\underline{vw}}\big)_{i \in \mathbb{N}_0} = \big(\{d \in \mathbb{W} \mid \deg^{\underline{vw}}(d) \leq i\}\big)_{i \in \mathbb{N}_0}$$

is called the *filtration induced by the weight* $(\underline{v}, \underline{w})$. Generally, we call *weighted filtrations* all filtrations which are induced in this way by a weight.

Definition and Remark 1.8.6 (A) We consider the strings

$$\underline{0} := (0, 0, \ldots, 0), \quad \underline{1} := (1, 1, \ldots, 1) \in \mathbb{N}_0^n$$

and a differential form $d \in \mathbb{W}$. We define the *standard degree* or just the *degree* $\deg(d)$ of d as the weighted degree with respect to the weight $(\underline{1}, \underline{1}) \in \mathbb{N}_0^n \times \mathbb{N}_0^n$, hence

$$\deg(d) := \deg^{\underline{11}}(d).$$

Observe that

$$\deg(d) := \sup\{|\underline{\nu}| + |\underline{\mu}| \mid (\underline{\nu}, \underline{\mu}) \in \operatorname{supp}(d)\}.$$

The corresponding induced weighted filtration

$$\mathbb{W}^{\deg_\bullet} := \mathbb{W}_\bullet^{\underline{11}} = \big(\mathbb{W}_i^{\underline{11}}\big)_{i \in \mathbb{N}_0} = \big(\{d \in \mathbb{W} \mid \deg(d) \leq i\}\big)_{i \in \mathbb{N}_0}$$

is called the *standard degree filtration* or just the *degree filtration* of $\mathbb{W}$.

(B) Keep the notations and hypotheses of part (A). The *order* of the differential operator d is defined by

$$\operatorname{ord}(d) := \deg^{\underline{01}}(d).$$

Observe that

$$\operatorname{ord}(d) = \sup\{|\underline{\mu}| \mid (\underline{\nu}, \underline{\mu}) \in \operatorname{supp}(d)\}.$$

The corresponding induced weighted filtration

$$\mathbb{W}_\bullet^{\mathrm{ord}} := \mathbb{W}_\bullet^{\underline{01}} = \big(\mathbb{W}_i^{\underline{01}}\big)_{i \in \mathbb{N}_0} = \big(\{d \in \mathbb{W} \mid \operatorname{ord}(d) \leq i\}\big)_{i \in \mathbb{N}_0}$$

is called the *order filtration* of $\mathbb{W}$.

Now, as an immediate application of Theorem 1.8.4 we obtain:

Corollary 1.8.7 *Let the notations be as in Convention 1.8.1. Then it holds*

(a) *The degree filtration* $\mathbb{W}_\bullet^{\deg}$ *is very good.*
(b) *The order filtration* $\mathbb{W}_\bullet^{\mathrm{ord}}$ *is good and* $\mathbb{W}_0^{\mathrm{ord}} = K[X_1, X_2, \ldots, X_n]$.

Proof In the notations of Theorem 1.8.4 (b) we have

$$\delta(\underline{1}, \underline{1}) = 1 \text{ and } \delta(\underline{0}, \underline{1}) = 1.$$

Moreover, by Theorem 1.8.4 (a) we have

$$\mathbb{W}_0^{\underline{11}} = K \text{ and } \mathbb{W}_0^{\underline{01}} = K[X_1, X_2, \ldots, X_n]$$

This proves our claim (see Definition and Remark 1.3.4 (C)).

Exercise 1.8.8

(A) Show that the degree filtration is the only very good filtration on $\mathbb{W}$.
(B) Write down all weights $(\underline{v}, \underline{w}) \in \mathbb{N}_0^n \times \mathbb{N}_0^n$ for which the induced filtration $\mathbb{W}_\bullet^{\underline{vw}}$ is good.

1.9 Weighted Associated Graded Rings

This section is devoted to the study of the associated graded rings of weighted filtrations of standard Weyl algebras. We shall see, that these are all naturally isomorphic to polynomial rings.

Convention 1.9.1 Again, throughout this section we fix a positive integer n, a field K of characteristic 0 and consider the standard Weyl algebra

$$\mathbb{W} := \mathbb{W}(K, n) = K[X_1, X_2, \ldots, X_n, \partial_1, \partial_2, \ldots, \partial_n].$$

In addition, we introduce the polynomial ring

$$\mathbb{P} := K[Y_1, Y_2, \ldots, Y_n, Z_1, Z_2, \ldots, Z_n]$$

in the indeterminates $Y_1, Y_2, \ldots, Y_n, Z_1, Z_2, \ldots, Z_n$ with coefficients in the field K.

Definition and Remark 1.9.2 (A) Fix a weight $(\underline{v}, \underline{w}) \in \mathbb{N}_0^n \times \mathbb{N}_0^n$ and consider the induced weighted filtration $\mathbb{W}_\bullet^{\underline{vw}}$. To write down the corresponding associated graded ring, we introduce the following notation:

$$\mathbb{G}^{\underline{vw}} = \bigoplus_{i \in \mathbb{N}_0} \mathbb{G}_i^{\underline{vw}} := \mathrm{Gr}_{\mathbb{W}_\bullet^{\underline{vw}}}(\mathbb{W}) = \bigoplus_{i \in \mathbb{N}_0} \mathrm{Gr}_{\mathbb{W}_\bullet^{\underline{vw}}}(\mathbb{W})_i.$$

(B) Keep the above notations and hypotheses. For each $j \in \mathbb{Z}$ we introduce the notations:

$$\mathbb{I}^{\underline{v}\underline{w}}_{\leq j} := \{(\underline{\nu}, \underline{\mu}) \in \mathbb{N}_0^n \times \mathbb{N}_0^n \mid \underline{v} \cdot \underline{\nu} + \underline{w} \cdot \underline{\mu} \leq j\};$$

$$\mathbb{I}^{\underline{v}\underline{w}}_{=j} := \{(\underline{\nu}, \underline{\mu}) \in \mathbb{N}_0^n \times \mathbb{N}_0^n \mid \underline{v} \cdot \underline{\nu} + \underline{w} \cdot \underline{\mu} = j\}.$$

Fix some $i \in \mathbb{N}_0$. Observe that

$$\begin{aligned}
\mathbb{G}^{\underline{v}\underline{w}}_i &= \mathbb{W}^{\underline{v}\underline{w}}_i / \mathbb{W}^{\underline{v}\underline{w}}_{i-1} \\
&= \Big(\bigoplus_{(\underline{\nu},\underline{\mu}) \in \mathbb{I}^{\underline{v}\underline{w}}_{\leq i}} K \underline{X}^{\underline{\nu}} \underline{\partial}^{\underline{\mu}} \Big) / \Big(\bigoplus_{(\underline{\nu},\underline{\mu}) \in \mathbb{I}^{\underline{v}\underline{w}}_{\leq i-1}} K \underline{X}^{\underline{\nu}} \underline{\partial}^{\underline{\mu}} \Big) \\
&= \Big[\Big(\bigoplus_{(\underline{\nu},\underline{\mu}) \in \mathbb{I}^{\underline{v}\underline{w}}_{\leq i-1}} K \underline{X}^{\underline{\nu}} \underline{\partial}^{\underline{\mu}} \Big) \oplus \Big(\bigoplus_{(\underline{\nu},\underline{\mu}) \in \mathbb{I}^{\underline{v}\underline{w}}_{=i}} K \underline{X}^{\underline{\nu}} \underline{\partial}^{\underline{\mu}} \Big) \Big] / \Big(\bigoplus_{(\underline{\nu},\underline{\mu}) \in \mathbb{I}^{\underline{v}\underline{w}}_{\leq i-1}} K \underline{X}^{\underline{\nu}} \underline{\partial}^{\underline{\mu}} \Big).
\end{aligned}$$

As a consequence, we get an isomorphism of K-vector spaces

$$\varepsilon^{\underline{v}\underline{w}}_i : \bigoplus_{(\underline{\nu},\underline{\mu}) \in \mathbb{I}^{\underline{v}\underline{w}}_{=i}} K \underline{X}^{\underline{\nu}} \underline{\partial}^{\underline{\mu}} \xrightarrow{\cong} \mathbb{G}^{\underline{v}\underline{w}}_i$$

such that

$$\varepsilon^{\underline{v}\underline{w}}_i \big(\underline{X}^{\underline{\nu}} \underline{\partial}^{\underline{\mu}} \big) = \big(\underline{X}^{\underline{\nu}} \underline{\partial}^{\underline{\mu}} + \mathbb{W}^{\underline{v}\underline{w}}_{i-1} \big) \in \mathbb{W}^{\underline{v}\underline{w}}_i / \mathbb{W}^{\underline{v}\underline{w}}_{i-1} = \mathbb{G}^{\underline{v}\underline{w}}_i \text{ for all } (\underline{\nu}, \underline{\mu}) \in \mathbb{I}^{\underline{v}\underline{w}}_{=i}.$$

In particular we can say:

The family $\big((\underline{X}^{\underline{\nu}} \underline{\partial}^{\underline{\mu}})^* := \varepsilon^{\underline{v}\underline{w}}_i (\underline{X}^{\underline{\nu}} \underline{\partial}^{\underline{\mu}}) \big)_{(\underline{\nu},\underline{\mu}) \in \mathbb{I}^{\underline{v}\underline{w}}_{=i}}$ is a K-basis of $\mathbb{G}^{\underline{v}\underline{w}}_i$.

We call this basis the *standard basis* of $\mathbb{G}^{\underline{v}\underline{w}}_i$. Its elements are called *standard basis elements* of the associated graded ring $\mathbb{G}^{\underline{v}\underline{w}}$.

(C) Keep the previously introduced notation. We add a few more useful observations on standard basis elements. First, observe that we may write

(a) $(\underline{X}^{\underline{\nu}} \underline{\partial}^{\underline{\mu}})^* \in \mathbb{G}^{\underline{v}\underline{w}}_{\underline{v} \cdot \underline{\nu} + \underline{w} \cdot \underline{\mu}}$ for all $(\underline{\nu}, \underline{\mu}) \in \mathbb{N}_0^n \times \mathbb{N}_0^n$.
(b) $X_i^* \in \mathbb{G}^{\underline{v}\underline{w}}_{v_i}$ and $\partial_j^* \in \mathbb{G}^{\underline{v}\underline{w}}_{w_j}$ for all $i, j \in \{1, 2, \ldots, n\}$.

Moreover, by the observations made in part (B) we also can say that all standard basis elements form a K-basis of the whole associated graded ring, thus:

(c) The family $\big((\underline{X}^{\underline{\nu}} \underline{\partial}^{\underline{\mu}})^* \big)_{(\underline{\nu},\underline{\mu}) \in \mathbb{N}_0^n \times \mathbb{N}_0^n}$ is a K-basis of $\mathbb{G}^{\underline{v}\underline{w}}$.

Finally, as the associated graded ring is commutative, and keeping in mind how the multiplication in this ring is defined (see Remark and Definition 1.3.1 (B)) we get the following product formula

(d) $(\underline{X}^{\underline{\nu}}\partial^{\underline{\mu}})^* = \left(\prod_{i=1}^n X_i^{\nu_i} \prod_{j=1}^n \partial^{\mu_j}\right)^* = \prod_{i=1}^n \left(X_i^*\right)^{\nu_i} \prod_{j=1}^n \left(\partial_j^*\right)^{\mu_j} =: \left(\underline{X}^*\right)^{\underline{\nu}}\left(\underline{\partial}^*\right)^{\underline{\mu}}$.

Exercise and Definition 1.9.3 (A) We fix a weight $(\underline{v}, \underline{w}) \in \mathbb{N}_0^n \times \mathbb{N}_0^n$. As in Definition and Remark 1.9.2 (A) we use again the notation

$$\mathbb{I}_{=i}^{\underline{vw}} := \{(\underline{\nu}, \underline{\mu}) \in \mathbb{N}_0^n \times \mathbb{N}_0^n \mid \underline{v} \cdot \underline{\nu} + \underline{w} \cdot \underline{\mu} = i\}$$

and consider the K-subspace

$$\mathbb{P}_i^{\underline{vw}} := \bigoplus_{(\underline{\nu}, \underline{\mu}) \in \mathbb{I}_{=i}^{\underline{vw}}} K\underline{Y}^{\underline{\nu}}\underline{Z}^{\underline{\mu}} \subseteq \mathbb{P} \text{ for all } i \in \mathbb{N}_0.$$

of our polynomial ring $\mathbb{P} = K[Y_1, Y_2, \ldots, Y_n, Z_1, Z_2, \ldots, Z_n]$. Prove the following statements:

(a) $K \subseteq \mathbb{P}_0^{\underline{vw}}$;
(b) $\mathbb{P}_i^{\underline{vw}}\mathbb{P}_j^{\underline{vw}} \subseteq \mathbb{P}_{i+j}^{\underline{vw}}$ for all $i, j \in \mathbb{N}_0$.
(c) $\mathbb{P} = \bigoplus_{i \in \mathbb{N}_0} \mathbb{P}_i^{\underline{vw}}$.

(B) Let the hypotheses and notations be as in part (A). Conclude that

$$\text{the family } \left(\mathbb{P}_i^{\underline{vw}}\right)_{i \in \mathbb{N}_0} \text{ defines a grading of the ring } \mathbb{P}.$$

We call this grading the *grading induced by the weight* $(\underline{v}, \underline{w}) \in \mathbb{N}_0^n \times \mathbb{N}_0^n$. If we endow our polynomial ring with this grading we write it as $\mathbb{P}^{\underline{vw}}$, thus

$$\mathbb{P} = \mathbb{P}^{\underline{vw}} = \bigoplus_{i \in \mathbb{N}_0} \mathbb{P}_i^{\underline{vw}}.$$

Theorem 1.9.4 (Structure of Weighted Associated Graded Rings) *Let* $(\underline{v}, \underline{w}) \in \mathbb{N}_0^n \times \mathbb{N}_0^n$ *be a weight. Then there exists an isomorphism of K-algebras, which preserves gradings (see Convention, Reminders and Notations 1.1.1 (I)).*

$$\eta^{\underline{vw}} : \mathbb{P} = \mathbb{P}^{\underline{vw}} \xrightarrow{\cong} \mathbb{G}^{\underline{vw}}$$

given by

$$Y_i \mapsto \eta^{\underline{vw}}(Y_i) := X_i^*, \text{ for all } i = 1, 2, \ldots, n;$$

$$Z_j \mapsto \eta^{\underline{vw}}(Z_j) := \partial_j^*, \text{ for all } j = 1, 2, \ldots, n.$$

Proof According to the universal property of the polynomial ring $\mathbb{P}$ there is a unique homomorphism of K-algebras

$$\eta^{\underline{v}\underline{w}} : \mathbb{P} \longrightarrow \mathbb{G}^{\underline{v}\underline{w}}$$

such that

$$Y_i \mapsto \eta^{\underline{v}\underline{w}}(Y_i) := X_i^*, \text{ for all } i = 1, 2, \ldots, n;$$
$$Z_j \mapsto \eta^{\underline{v}\underline{w}}(Z_j) := \partial_j^*, \text{ for all } j = 1, 2, \ldots, n.$$

In view of the product formula of Definition and Remark 1.9.2 (C) we obtain

$$\eta^{\underline{v}\underline{w}}\big(\underline{Y}^{\underline{\nu}}\underline{Z}^{\underline{\mu}}\big) = \big(\underline{X}^{\underline{\nu}}\underline{\partial}^{\underline{\mu}}\big)^* \text{ for all } \underline{\nu}, \underline{\mu} \in \mathbb{N}_0^n.$$

In particular $\eta^{\underline{v}\underline{w}}$ yields a bijection between the monomial basis of the polynomial ring $\mathbb{P}$ and the standard basis of the associated graded ring $\mathbb{G}^{\underline{v}\underline{w}}$. So, $\eta^{\underline{v}\underline{w}}$ is indeed an isomorphism. But moreover, for each $i \in \mathbb{N}_0$ it also follows that $\eta^{\underline{v}\underline{w}}$ yields an bijection between the monomial basis of the subspace $\mathbb{P}_i^{\underline{v}\underline{w}} \subseteq \mathbb{P}$ and the standard basis of $\mathbb{G}_i^{\underline{v}\underline{w}}$. But this means, that $\eta^{\underline{v}\underline{w}}$ preserves the gradings.

In Lemma 1.8.3 (c) we have seen that weighted degrees are *sub-additive*, which means that $\deg^{\underline{v}\underline{w}}(de) \leq \deg^{\underline{v}\underline{w}}(d) + \deg^{\underline{v}\underline{w}}(e)$ for all $d, e \in \mathbb{W}$. As an application of Theorem 1.9.4 we now shall improve on this and show, that weighted degrees are indeed *additive*, which means that the above inequality is in fact always an equality.

Corollary 1.9.5 (Additivity of Weighted Degrees) *Let $(\underline{v}, \underline{w}) \in \mathbb{N}_0^n \times \mathbb{N}_0^n$ be a weight and let $d, e \in \mathbb{W}$. Then*

$$\deg^{\underline{v}\underline{w}}(de) = \deg^{\underline{v}\underline{w}}(d) + \deg^{\underline{v}\underline{w}}(e).$$

Proof If $d = 0$ or $e = 0$ our claim is clear. So let $d, e \neq 0$. We have

$$i := \deg^{\underline{v}\underline{w}}(d) \in \mathbb{N}_0 \text{ and } j := \deg^{\underline{v}\underline{w}}(e).$$

We use again the notation

$$\mathbb{I}_{=k}^{\underline{v}\underline{w}} := \{(\underline{\nu}, \underline{\mu}) \in \mathbb{N}_0^n \times \mathbb{N}_0^n \mid \underline{v} \cdot \underline{\nu} + \underline{w} \cdot \underline{\mu} = k\} \text{ for all } k \in \mathbb{N}_0$$

and set

$$M := \bigoplus_{(\underline{\nu},\underline{\mu}) \in \mathbb{I}_{=i}^{\underline{v}\underline{w}}} K\underline{X}^{\underline{\nu}}\underline{\partial}^{\underline{\mu}} \quad \text{and } N := \bigoplus_{(\underline{\nu},\underline{\mu}) \in \mathbb{I}_{=j}^{\underline{v}\underline{w}}} K\underline{X}^{\underline{\nu}}\underline{\partial}^{\underline{\mu}}.$$

We then may write

$$d = a + r \text{ with } a \in M \setminus \{0\} \text{ and } \deg^{\underline{v}\underline{w}}(r) < i;$$
$$e = b + s \text{ with } a \in N \setminus \{0\} \text{ and } \deg^{\underline{v}\underline{w}}(s) < j.$$

We thus have

$$de = ab + (as + rb + rs)$$

By what we know already about degrees we have $\deg^{\underline{v}\underline{w}}(as + rb + rs) < i + j$ (see Lemma 1.8.3 (a), (c)). So, in view of Lemma 1.8.3 (a) it suffices to show that

$$\deg^{\underline{v}\underline{w}}(ab) = i + j.$$

To do so, we write

$$a = \sum_{(\underline{\nu},\underline{\mu})\in\text{supp}(a)} c^{(a)}_{\underline{\nu},\underline{\mu}} \underline{X}^{\underline{\nu}} \underline{\partial}^{\underline{\mu}}, \text{ with } c^{(a)}_{\underline{\nu},\underline{\mu}} \in K \setminus \{0\} \text{ for all } (\underline{\nu}, \underline{\mu}) \in \text{supp}(a) \text{ and}$$

$$b = \sum_{(\underline{\nu}',\underline{\mu}')\in\text{supp}(b)} c^{(b)}_{\underline{\nu}',\underline{\mu}'} \underline{X}^{\underline{\nu}'} \underline{\partial}^{\underline{\mu}'}, \text{ with } c^{(b)}_{\underline{\nu}',\underline{\mu}'} \in K \setminus \{0\} \text{ for all } (\underline{\nu}', \underline{\mu}') \in \text{supp}(b).$$

It follows that

$$ab = \sum_{(\underline{\nu},\underline{\mu})\in\text{supp}(a) \text{ and } (\underline{\nu}',\underline{\mu}')\in\text{supp}(b)} c^{(a)}_{\underline{\nu},\underline{\mu}} c^{(b)}_{\underline{\nu}',\underline{\mu}'} \underline{X}^{\underline{\nu}} \underline{\partial}^{\underline{\mu}} \underline{X}^{\underline{\nu}'} \underline{\partial}^{\underline{\mu}'}.$$

By Exercise 1.6.4 (A) and in the notation of Notation and Remark 1.6.3 (C), it follows that

$$\underline{X}^{\underline{\nu}} \underline{\partial}^{\underline{\mu}} \underline{X}^{\underline{\nu}'} \underline{\partial}^{\underline{\mu}'} - \underline{X}^{\underline{\nu}+\underline{\nu}'} \underline{\partial}^{\underline{\mu}+\underline{\mu}'} \in \sum_{(\underline{\lambda},\underline{\kappa})\in\mathbb{M}(\underline{\nu}+\underline{\nu}',\underline{\mu}+\underline{\mu}')} K \underline{X}^{\underline{\lambda}} \underline{\partial}^{\underline{\kappa}}$$

for all $(\underline{\nu}, \underline{\mu}) \in \text{supp}(a)$ and all $(\underline{\nu}', \underline{\mu}') \in \text{supp}(b)$. Observe that

$$(\underline{\nu} + \underline{\nu}', \underline{\mu} + \underline{\mu}') \in \mathbb{I}^{\underline{v}\underline{w}}_{=i+j} \text{ for all } (\underline{\nu}, \underline{\mu}) \in \text{supp}(a) \text{ and all } (\underline{\nu}', \underline{\mu}') \in \text{supp}(b).$$

So, by Definition and Remark 1.8.2 (B)(c) it follows that

$$\deg^{\underline{v}\underline{w}}\left(\underline{X}^{\underline{\nu}} \underline{\partial}^{\underline{\mu}} \underline{X}^{\underline{\nu}'} \underline{\partial}^{\underline{\mu}'} - \underline{X}^{\underline{\nu}+\underline{\nu}'} \underline{\partial}^{\underline{\mu}+\underline{\mu}'}\right) < i + j$$

for all $(\underline{\nu}, \underline{\mu}) \in \mathrm{supp}(a)$ and all $(\underline{\nu}', \underline{\mu}') \in \mathrm{supp}(b)$. If we set

$$h := \sum_{(\underline{\nu},\underline{\mu})\in\mathrm{supp}(a),(\underline{\nu}',\underline{\mu}')\in\mathrm{supp}(b)} c^{(a)}_{\underline{\nu},\underline{\mu}} c^{(b)}_{\underline{\nu}',\underline{\mu}'} \underline{X}^{\underline{\nu}+\underline{\nu}'} \underline{\partial}^{\underline{\mu}+\underline{\mu}'}.$$

and on repeated use of Lemma 1.8.3 (a) and (b) we thus get

$$\deg^{\underline{vw}}(ab - h) =$$

$$\deg^{\underline{vw}}[\sum_{(\underline{\nu},\underline{\mu})\in\mathrm{supp}(a),(\underline{\nu}',\underline{\mu}')\in\mathrm{supp}(b)} c^{(a)}_{\underline{\nu},\underline{\mu}} c^{(b)}_{\underline{\nu}',\underline{\mu}'} (\underline{X}^{\underline{\nu}}\underline{\partial}^{\underline{\mu}}\underline{X}^{\underline{\nu}'}\underline{\partial}^{\underline{\mu}'} - \underline{X}^{\underline{\nu}+\underline{\nu}'}\underline{\partial}^{\underline{\mu}+\underline{\mu}'})] < i + j.$$

So, we may write

$$ab = h + u \text{ with } \deg^{\underline{vw}}(u) < i + j.$$

By Lemma 1.8.3 (a) it thus suffices to show that $\deg^{\underline{vw}}(h) = i + j$. As

$$h = \sum_{(\underline{\nu},\underline{\mu})\in\mathrm{supp}(a),(\underline{\nu}',\underline{\mu}')\in\mathrm{supp}(b)} c^{(a)}_{\underline{\nu},\underline{\mu}} c^{(b)}_{\underline{\nu}',\underline{\mu}'} \underline{X}^{\underline{\nu}+\underline{\nu}'} \underline{\partial}^{\underline{\mu}+\underline{\mu}'} \in \bigoplus_{(\underline{\nu},\underline{\mu})\in\mathbb{I}^{\underline{vw}}_{=i+j}} K\underline{X}^{\underline{\nu}}\underline{\partial}^{\underline{\mu}}$$

It suffices to show that $h \neq 0$. To do so, we consider the two polynomials

$$f := \sum_{(\underline{\nu},\underline{\mu})\in\mathrm{supp}(a)} c^{(a)}_{\underline{\nu},\underline{\mu}} \underline{Y}^{\underline{\nu}} \underline{Z}^{\underline{\mu}} \in \mathbb{P}^{\underline{vw}}_i \text{ and}$$

$$g := \sum_{(\underline{\nu}',\underline{\mu}')\in\mathrm{supp}(b)} c^{(b)}_{\underline{\nu}',\underline{\mu}'} \underline{Y}^{\underline{\nu}'} \underline{Z}^{\underline{\mu}'} \in \mathbb{P}^{\underline{vw}}_j.$$

As $\mathrm{supp}(a)$ and $\mathrm{supp}(b)$ are non-empty, and all coefficients of f and g are non-zero, we have $f \neq 0$ and $g \neq 0$. As $\mathbb{P}$ is an integral domain. it follows that $fg \neq 0$. We set

$$h^* := (h + \mathbb{W}^{\underline{vw}}_{i+j-1}) \in \mathbb{W}^{\underline{vw}}_{i+j}/\mathbb{W}^{\underline{vw}}_{i+j-1} = \mathbb{G}^{\underline{vw}}_{i+j},$$

so that

$$h^* = \sum_{(\underline{\nu},\underline{\mu})\in\mathrm{supp}(a),(\underline{\nu}',\underline{\mu}')\in\mathrm{supp}(b)} c^{(a)}_{\underline{\nu},\underline{\mu}} c^{(b)}_{\underline{\nu}',\underline{\mu}'} \left(\underline{X}^{\underline{\nu}+\underline{\nu}'} \underline{\partial}^{\underline{\mu}+\underline{\mu}'}\right)^*.$$

Applying the isomorphism

$$\eta^{\underline{vw}} : \mathbb{P} = \mathbb{P}^{\underline{vw}} \xrightarrow{\cong} \mathbb{G}^{\underline{vw}}$$

of Theorem 1.9.4, we now get

$$0 \neq \eta^{\underline{v}\underline{w}}(fg) = \eta^{\underline{v}\underline{w}}\Big([\sum_{(\underline{\nu},\underline{\mu})\in\operatorname{supp}(a)} c^{(a)}_{\underline{\nu},\underline{\mu}} \underline{Y}^{\underline{\nu}} \underline{Z}^{\underline{\mu}}][\sum_{(\underline{\nu}',\underline{\mu}')\in\operatorname{supp}(b)} c^{(b)}_{\underline{\nu}',\underline{\mu}'} \underline{Y}^{\underline{\nu}'} \underline{Z}^{\underline{\mu}'}]\Big)$$

$$= \eta^{\underline{v}\underline{w}}\Big(\sum_{(\underline{\nu},\underline{\mu})\in\operatorname{supp}(a),(\underline{\nu}',\underline{\mu}')\in\operatorname{supp}(b)} c^{(a)}_{\underline{\nu},\underline{\mu}} c^{(b)}_{\underline{\nu}',\underline{\mu}'} \underline{Y}^{\underline{\nu}+\underline{\nu}'} \underline{Z}^{\underline{\mu}+\underline{\mu}'}\Big)$$

$$= \sum_{(\underline{\nu},\underline{\mu})\in\operatorname{supp}(a),(\underline{\nu}',\underline{\mu}')\in\operatorname{supp}(b)} c^{(a)}_{\underline{\nu},\underline{\mu}} c^{(b)}_{\underline{\nu}',\underline{\mu}'} \eta^{\underline{v}\underline{w}}\big(\underline{Y}^{\underline{\nu}+\underline{\nu}'} \underline{Z}^{\underline{\mu}+\underline{\mu}'}\big)$$

$$= \sum_{(\underline{\nu},\underline{\mu})\in\operatorname{supp}(a),(\underline{\nu}',\underline{\mu}')\in\operatorname{supp}(b)} c^{(a)}_{\underline{\nu},\underline{\mu}} c^{(b)}_{\underline{\nu}',\underline{\mu}'} \big(\underline{X}^{\underline{\nu}+\underline{\nu}'} \underline{\partial}^{\underline{\mu}+\underline{\mu}'}\big)^* = h^*.$$

But this clearly implies that $h \neq 0$.

Corollary 1.9.6 (Integrity of Standard Weyl Algebras) *The standard Weyl algebra $\mathbb{W}$ is an integral domain:*

$$\text{If } d, e \in \mathbb{W} \setminus \{0\}, \text{ then } de \neq 0.$$

Proof Apply Theorem 1.9.4 and keep in mind that an element of $\mathbb{W}$ vanishes if and only if its degree (with respect to any weight) equals $-\infty$.

Exercise 1.9.7

(A) We fix a weight $(\underline{v}, \underline{w}) \in \mathbb{N}_0^n \times \mathbb{N}_0^n$ and set

$$\Gamma^{\underline{v},\underline{w}} := \{\underline{v} \cdot \underline{\nu} + \underline{w} \cdot \underline{\mu} \mid \underline{\nu}, \underline{\mu} \in \mathbb{N}_0^n\}.$$

Prove the following statements

(a) $0 \in \Gamma^{\underline{v}\underline{w}} \subseteq \mathbb{N}_0$.
(b) If $i, j \in \Gamma^{\underline{v}\underline{w}}$, then $i + j \in \Gamma^{\underline{v}\underline{w}}$.
(c) $\mathbb{G}_i^{\underline{v}\underline{w}} \neq 0 \Leftrightarrow \mathbb{P}_i^{\underline{v}\underline{w}} \neq 0 \Leftrightarrow i \in \Gamma^{\underline{v}\underline{w}}$.

$\Gamma^{\underline{v},\underline{w}}$ is called the *degree semigroup* associated to the weight $(\underline{v}, \underline{w})$.

(B) Let $n = 1$, $\underline{v} = (p)$ and $\underline{w} = (q)$, where $p, q \in \mathbb{N}$ are two distinct prime numbers. Determine $\Gamma^{\underline{v},\underline{w}}$ and the standard bases of all K-vector spaces

$$\mathbb{P}_i^{\underline{v}\underline{w}} \text{ and } \mathbb{G}_i^{\underline{v}\underline{w}} \text{ for } i \in \Gamma^{\underline{v}\underline{w}},$$

at least for some specified pairs like $(p, q) = (2, 3), (2, 5), (5, 7), \ldots$

(C) Show, that the ring $\operatorname{End}_K\big(K[X_1, X_2, \ldots, X_n]\big)$ is not an integral domain.

1.10 Filtered Modules

Now, we aim to consider finitely generated left-modules over standard Weyl algebras: the so-called D-modules. Our basic aim is to endow such modules with appropriate filtrations, which are compatible with a given weighted filtration of the underlying Weyl algebra. This will allow us to define associated graded modules over the corresponding associated graded ring of the Weyl algebra—hence over a weight graded polynomial ring. We approach the subject in a more general setting.

Definition and Remark 1.10.1 (A) Let K be a field and let $A = (A, A_\bullet)$ be a filtered K-algebra. Let U be a left-module over A. By a *filtration of U compatible with $A_\bullet$* or just an *$A_\bullet$-filtration* of U we mean a family

$$U_\bullet = (U_i)_{i \in \mathbb{Z}}$$

such that the following conditions hold:

(a) Each U_i is a K-vector subspace of U;
(b) $U_i \subseteq U_{i+1}$ for all $i \in \mathbb{Z}$;
(c) $U = \bigcup_{i \in \mathbb{Z}} U_i$;
(d) $A_i U_j \subseteq U_{i+j}$ for all $i \in \mathbb{N}_0$ and all $j \in \mathbb{Z}$.

In requirement (d) we have used the standard notation

$$A_i U_j := \sum_{(f,u) \in A_i \times U_j} K f u \text{ for all } i \in \mathbb{N}_0 \text{ and all } j \in \mathbb{Z},$$

which we shall use from now on without further mention. If an $A_\bullet$-filtration $U_\bullet$ of U is given, we say that $(U, U_\bullet)$ or—by abuse of language—that U is a $A_\bullet$ *filtered A-module* or just that U is a *filtered A-module*.

(B) Keep the notations and hypotheses of part (A) and let $U_\bullet = (U_i)_{i \in \mathbb{Z}}$ be a filtered A-module. Observe that

For all $i \in \mathbb{Z}$ the K-vector space U_i is a left A_0-submodule of U.

(C) We say that two $A_\bullet$-filtrations $U_\bullet^{(1)}$, $U_\bullet^{(2)}$ are *equivalent* if there is some $r \in \mathbb{N}_0$ such that

(a) $U_{i-r}^{(1)} \subseteq U_i^{(2)} \subseteq U_{i+r}^{(1)}$ for all $i \in \mathbb{Z}$.

Later, we shall use the following observation.

Assume that the above condition (a) holds, let $i \in \mathbb{N}$ and let $a \in A_i$. Then we have

(b) $aU_j^{(1)} \subseteq U_{j+i-1}^{(1)}$ for all $j \in \mathbb{Z} \quad \Rightarrow \quad a^k U_j^{(1)} \subseteq U_{j+k(i-1)}^{(1)}$ for all $j \in \mathbb{Z}$ and all $k \in \mathbb{N}_0$.
(c) $aU_j^{(1)} \subseteq U_{j+i-1}^{(1)}$ for all $j \in \mathbb{Z} \quad \Rightarrow \quad a^{2r+1} U_j^{(2)} \subseteq U_{j+(2r+1)i-1}^{(2)}$ for all $j \in \mathbb{Z}$.

To prove statement (b), we assume that $aU_j^{(1)} \subseteq U_{j+i-1}^{(1)}$ for all $j \in \mathbb{Z}$ and proceed by induction on k. If $k = 0$ our claim is obvious. If $k > 0$, we may assume by induction that $a^{k-1}U_j^{(1)} \subseteq U_{j+(k-1)(i-1)}^{(1)}$ for all $j \in \mathbb{Z}$, so that indeed

$$a^k U_j^{(1)} = aa^{k-1}U_j^{(1)} \subseteq aU_{j+(k-1)(i-1)}^{(1)} \subseteq U_{j+(k-1)(i-1)+(i-1)}^{(1)} = U_{j+k(i-1)}^{(1)}$$
$$\text{for all } j \in \mathbb{Z},$$

and this proves statement (b). If we apply statement (b) with $k = 2r+1$ and observe condition (a), we get

$$a^{2r+1}U_j^{(2)} \subseteq a^{2r+1}U_{j+r}^{(1)} \subseteq U_{j+r+(2r+1)(i-1)}^{(1)} \subseteq U_{j+2r+(2r+1)(i-1)}^{(2)}$$
$$= U_{j+2r+2ri-2r+i-1}^{(1)} = U_{j+2ri+i-1}^{(2)} = U_{j+(2r+1)i-1}^{(2)} \text{ for all } j \in \mathbb{Z},$$

and this proves statement (c).

Remark and Definition 1.10.2 (A) Let K be a field and let $A = (A, A_\bullet)$ be a filtered K-algebra and let $U = (U, U_\bullet)$ be an $A_\bullet$-filtered A-module. We consider the corresponding associated graded ring

$$\mathrm{Gr}(A) = \mathrm{Gr}_{A_\bullet}(A) = \bigoplus_{i \in \mathbb{N}_0} A_i / A_{i-1}.$$

and the K-vector space

$$\mathrm{Gr}(U) = \mathrm{Gr}_{U_\bullet}(U) = \bigoplus_{i \in \mathbb{Z}} U_i / U_{i-1}.$$

For all $i \in \mathbb{Z}$ we also use the notation

$$\mathrm{Gr}(U)_i = \mathrm{Gr}_{U_\bullet}(U)_i := U_i / U_{i-1},$$

so that we may write

$$\mathrm{Gr}(U) = \mathrm{Gr}_{U_\bullet}(U) = \bigoplus_{i \in \mathbb{Z}} \mathrm{Gr}_{U_\bullet}(U)_i.$$

(B) Let $i \in \mathbb{N}_0$, let $j \in \mathbb{Z}$ let $f, f' \in A_i$ and let $g, g' \in U_j$ such that

$$h := f - f' \in A_{i-1} \text{ and } k := g - g' \in U_{j-1}.$$

It follows that

$$\begin{aligned} fg - f'g' = fg - (f-h)(g-k) &= fk + hg - hk \\ &\in A_iU_{j-1} + A_{i-1}U_j + A_{i-1}U_{j-1} \\ &\subseteq U_{i+(j-1)} + U_{j+(i-1)} + U_{(i-1)+(j-1)} \subseteq U_{i+j-1}. \end{aligned}$$

So in $U_{i+j}/U_{i+j-1} = \mathrm{Gr}_{U_\bullet}(U)_{i+j} \subset \mathrm{Gr}_{U_\bullet}(U)$ we get the relation

$$fg + U_{i+j-1} = f'g' + U_{i+j-1}.$$

This allows to define a $\mathrm{Gr}_{A_\bullet}(A)$-*scalar multiplication* on the K-space $\mathrm{Gr}_{U_\bullet}(U)$ which is induced by

$$(f + A_{i-1})(g + U_{j-1}) := fg + U_{i+j-1}$$

for all $i \in \mathbb{N}_0$, all $j \in \mathbb{Z}$, all $f \in A_i$ $g \in U_j$. More generally, if $r, s \in \mathbb{N}_0$, $t \in \mathbb{Z}$,

$$\overline{f} = \sum_{i=0}^{r} \overline{f_i}, \text{ with } f_i \in A_i \text{ and } \overline{f_i} = (f_i + A_{i-1}) \in \mathrm{Gr}_{A_\bullet}(A)_i \text{ for all } i = 0, 1, \ldots, r,$$

and

$$\overline{g} = \sum_{j=t}^{t+s} \overline{g_j}, \text{ with } g_j \in U_j \text{ and } \overline{g_j} = (g_j + U_{j-1}) \in \mathrm{Gr}_{U_\bullet}(U)_j$$

for all $j = t, t+1, \ldots, t+s$,

then

$$\overline{f}\overline{g} = \sum_{k=t}^{r+t+s} \sum_{i+j=k} \overline{f_i}\overline{g_j} = \sum_{k=t}^{r+t+s} \sum_{i+j=k} (f_i g_j + U_{i+j-1}).$$

(C) Keep the above notations and hypotheses. With respect to our scalar multiplication on $\mathrm{Gr}_{U_\bullet}(U)$ we have the relations

$$\mathrm{Gr}_{A_\bullet}(A)_i \mathrm{Gr}_{U_\bullet}(U)_j \subseteq \mathrm{Gr}_{U_\bullet}(U)_{i+j} \text{ for all } i, j \in \mathbb{Z}.$$

So, the K-vector space $\mathrm{Gr}_{U_\bullet}(U)$ is turned into a graded $\mathrm{Gr}_{A_\bullet}(A)$-module

$$\mathrm{Gr}_{U_\bullet}(U) = \big(\mathrm{Gr}_{U_\bullet}(U), (\mathrm{Gr}_{U_\bullet}(U)_i)_{i\in\mathbb{Z}}\big) = \bigoplus_{i\in\mathbb{Z}} \mathrm{Gr}_{U_\bullet}(U)_i$$

by means of the above multiplication. We call this $\mathrm{Gr}_{A_\bullet}(A)$-module $\mathrm{Gr}_{U_\bullet}(U)$ the *associated graded module* of U with respect to the filtration $U_\bullet$. From now on, we always furnish $\mathrm{Gr}_{U_\bullet}(U)$ with this structure of graded $\mathrm{Gr}_{A_\bullet}(A)$-module.

Definition 1.10.3 Let K be a field and let $A = (A, A_\bullet)$ be a filtered K-algebra. Assume that the filtration $A_\bullet$ is commutative, so that the corresponding associated graded ring

$$\mathrm{Gr}(A) = \mathrm{Gr}_{A_\bullet}(A) = \bigoplus_{i \in \mathbb{N}_0} A_i/A_{i-1}$$

is commutative.
Moreover, let $U = (U, U_\bullet)$ be an $A_\bullet$-filtered A-module and consider the corresponding associated graded module

$$\mathrm{Gr}(U) = \mathrm{Gr}_{U_\bullet}(U) = \bigoplus_{i \in \mathbb{Z}} U_i/U_{i-1}.$$

in addition, consider the *annihilator* ideal

$$\mathrm{Ann}_{\mathrm{Gr}_{A_\bullet}(A)}\big(\mathrm{Gr}_{U_\bullet}(U)\big) := \{f \in \mathrm{Gr}_{A_\bullet}(A) \mid f\mathrm{Gr}_{U_\bullet}(U) = 0\}$$

of the $\mathrm{Gr}_{A_\bullet}(A)$-module $\mathrm{Gr}_{U_\bullet}(U)$. We define the *characteristic variety* $\mathbb{V}_{U_\bullet}(U)$ of the $A_\bullet$-filtered A-module $U = (U, U_\bullet)$ as the *prime variety* of the annihilator ideal of $\mathrm{Gr}_{U_\bullet}(U)$, hence

$$\mathbb{V}_{U_\bullet}(U) := \mathrm{Var}\big(\mathrm{Ann}_{\mathrm{Gr}_{A_\bullet}(A)}(\mathrm{Gr}_{U_\bullet}(U))\big) \subseteq \mathrm{Spec}(\mathrm{Gr}_{A_\bullet}(A)).$$

We also call this variety the *characteristic variety of the left A-module U with respect to the $A_\bullet$-filtration $U_\bullet$* or just the *characteristic variety of U with respect to $U_\bullet$*.

Proposition 1.10.4 (Equality of Characteristic Varieties for Equivalent Filtrations) *Let K be a field and let $A = (A, A_\bullet)$ be a filtered K-algebra. Assume that the filtration $A_\bullet$ is commutative (see Definition 1.3.3). Let U be an A-module which is endowed with two equivalent $A_\bullet$-filtrations $U_\bullet^{(1)}$ and $U_\bullet^{(2)}$. Then*

$$\mathbb{V}_{U_\bullet^{(1)}}(U) = \mathbb{V}_{U_\bullet^{(2)}}(U).$$

Proof We have to show that

$$\sqrt{\mathrm{Ann}_{\mathrm{Gr}_{A_\bullet}(A)}\big(\mathrm{Gr}_{U_\bullet^{(1)}}(U)\big)} = \sqrt{\mathrm{Ann}_{\mathrm{Gr}_{A_\bullet}(A)}\big(\mathrm{Gr}_{U_\bullet^{(2)}}(U)\big)}.$$

By symmetry, it suffices to show that

$$\sqrt{\mathrm{Ann}_{\mathrm{Gr}_{A_\bullet}(A)}\big(\mathrm{Gr}_{U^{(1)}_\bullet}(U)\big)} \subseteq \sqrt{\mathrm{Ann}_{\mathrm{Gr}_{A_\bullet}(A)}\big(\mathrm{Gr}_{U^{(2)}_\bullet}(U)\big)}.$$

In view of the fact that the formation of radicals of ideals is idempotent, it suffices even to show that

$$\mathrm{Ann}_{\mathrm{Gr}_{A_\bullet}(A)}\big(\mathrm{Gr}_{U^{(1)}_\bullet}(U)\big) \subseteq \sqrt{\mathrm{Ann}_{\mathrm{Gr}_{A_\bullet}(A)}\big(\mathrm{Gr}_{U^{(2)}_\bullet}(U)\big)}.$$

As $\mathrm{Gr}_{U^{(1)}_\bullet}(U)$ is a graded $\mathrm{Gr}_{A_\bullet}(A)$-module, its annihilator is a graded ideal of $\mathrm{Gr}_{A_\bullet}(A)$. So, it finally is enough to show, that

$$\overline{a} \in \sqrt{\mathrm{Ann}_{\mathrm{Gr}_{A_\bullet}(A)}\big(\mathrm{Gr}_{U^{(2)}_\bullet}(U)\big)} \text{ for all } i \in \mathbb{N}_0 \text{ and all } \overline{a} \in \mathrm{Ann}_{\mathrm{Gr}_{A_\bullet}(A)}\big(\mathrm{Gr}_{U^{(1)}_\bullet}(U)\big)_i.$$

So, fix some $i \in \mathbb{N}_0$ and some

$$\overline{a} \in \mathrm{Ann}_{\mathrm{Gr}_{A_\bullet}(A)}\big(\mathrm{Gr}_{U^{(1)}_\bullet}(U)\big)_i \subseteq \mathrm{Gr}_{A_\bullet}(A)_i = A_i/A_{i-1}.$$

We chose some $a \in A_i$ with $\overline{a} = a + A_{i-1} \in A_i/A_{i-1}$.. For all $j \in \mathbb{Z}$ we have in $\mathrm{Gr}_{U^{(1)}_\bullet}(U)$ the relation

$$aU^{(1)}_j + U^{(1)}_{j+i-1} = (a + A_{i-1})(U^{(1)}_j/U^{(1)}_{j-1}) = \overline{a}(U^{(1)}_j/U^{(1)}_{j-1}) = \overline{a}\mathrm{Gr}_{U^{(1)}_\bullet}(U)_j = 0,$$

and hence

$$aU^{(1)}_j \subseteq U^{(1)}_{j+i-1} \text{ for all } j \in \mathbb{Z}.$$

According to our hypotheses we find some $r \in \mathbb{N}_0$ such that $U^{(1)}_{k-r} \subseteq U^{(2)}_k \subseteq U^{(1)}_{k+r}$ for all $k \in \mathbb{Z}$. By Definition and Remark 1.10.1 (C)(c) we therefore have

$$a^{2r+1}U^{(2)}_j \subseteq U^{(2)}_{j+(2r+1)i-1} \text{ for all } j \in \mathbb{Z}.$$

So, for all $j \in \mathbb{Z}$ we get in $U^{(2)}_{j+(2r+1)i}/U^{(2)}_{j+(2r+1)i-1} = \mathrm{Gr}_{U^{(2)}_\bullet}(U)_{j+(2r+1)i}$ the relation:

$$\overline{a}^{2r+1}\mathrm{Gr}_{U_\bullet}(U)_j = (a^{2r+1} + A_{(2r+1)i-1})(U^{(2)}_j/U^{(2)}_{j-1}) \subseteq a^{2r+1}U^{(2)}_j/U^{(2)}_{j+(2r+1)i-1} = 0.$$

This shows that $\overline{a}^{2r+1} \in \mathrm{Ann}_{\mathrm{Gr}_{A_\bullet}(A)}\big(\mathrm{Gr}_{U^{(2)}_\bullet}(U)\big)$ and hence that indeed

$$\overline{a} \in \sqrt{\mathrm{Ann}_{\mathrm{Gr}_{A_\bullet}(A)}\big(\mathrm{Gr}_{U^{(2)}_\bullet}(U)\big)}.$$

So, provided $(A, A_\bullet)$ is a commutatively filtered K-algebra (see Definition 1.3.3), the characteristic variety of an $A_\bullet$-graded A-module $(U, U_\bullet)$ depends only on the equivalence class of the filtration $U_\bullet$. This allows us to define in an intrinsic way the notion of characteristic variety of a finitely generated (left-) module over the filtered ring A. We work this out in the following combined exercise and definition.

Exercise and Definition 1.10.5 (A) Let $(A, A_\bullet)$ be a filtered K-algebra and let U be a (left) module over A.

$$\text{Let } V \subseteq U \text{ be a } K\text{-subspace such that } U = AV.$$

Prove the following claims:

(a) $A_i V = 0$ for all $i < 0$.
(b) The family $A_\bullet V := \big(A_i V\big)_{i \in \mathbb{Z}}$ is an $A_\bullet$-filtration of U.

The above filtration $A_\bullet V$ is called the $A_\bullet$-filtration of U *induced* by the subspace V.

(B) Let the notations and hypotheses be as in part (A). Assume in addition that

$$s := \dim_K(V) < \infty.$$

Prove that

(a) U is finitely generated as an A-module;
(b) $A_i V$ is a finitely generated (left-) module over A_0.
(c) The graded $\mathrm{Gr}_{A_\bullet}(A)$-module $\mathrm{Gr}_{A_\bullet V}(U)$ is generated by finitely many elements $\overline{g}_1, \overline{g}_2, \ldots, \overline{g}_s \in \mathrm{Gr}_{A_\bullet V}(U)_0$.

Keep in mind that we can always find a vector space $V \subseteq U$ of finite dimension with $AV = U$ if the A-module U is finitely generated.

(C) Let the notations and hypotheses be as above. Let $V^{(1)}, V^{(2)} \subseteq U$ be two K-subspaces such that

$$AV^{(1)} = AV^{(2)} = U \text{ and } \dim_K(V^{(1)}), \dim_K(V^{(2)}) < \infty.$$

Prove that

(a) The two induced $A_\bullet$-filtrations $A_\bullet V^{(1)}$ and $A_\bullet V^{(2)}$ are equivalent.
(b) If the filtration $A_\bullet$ is commutative, it holds

$$\mathbb{V}_{A_\bullet V^{(1)}}(U) = \mathbb{V}_{A_\bullet V^{(2)}}(U).$$

(D) Keep the above notations and hypotheses. Assume that the filtration $A_\bullet$ is commutative and that the (left) A-module U is finitely generated. By what we have learned by the previous considerations, we find a K-subspace $V \subseteq U$ of finite dimension such that $AV = U$, and the characteristic variety $\mathbb{V}_{A_\bullet V}(U)$ of U with respect to the induced filtration $A_\bullet V$ is independent of the choice of V. So, we may just write

$$\mathbb{V}_{A_\bullet}(U) := \mathbb{V}_{A_\bullet V}(U),$$

and we call $\mathbb{V}_{A_\bullet}(U)$ the *characteristic variety of U with respect to the (commutative !) filtration $A_\bullet$* of A. This is the announced notion of *intrinsic characteristic variety*.

(E) Keep the above notations. Assume that the filtration $A_\bullet$ is of finite type (see Definition and Remark 1.3.4 (C)) and that the (left) A-module U is finitely generated. The $A_\bullet$ filtration $U_\bullet$ of U is said to be *of finite type* if

(a) There is some $j_0 \in \mathbb{Z}$ such that $U_j = 0$ for all $j \leq j_0$;
(b) There is an integer σ such that:

  (1) U_j is finitely generated as a (left) A_0-module for all $j \leq \sigma$ and
  (2) $U_i = \sum_{j \leq \sigma} A_j U_{i-j}$ for all $i > \sigma$.

In this situation σ is again called a *generating degree* of the $A_\bullet$-filtration $U_\bullet$ (compare Definition and Remark 1.3.4 (C)). Prove that in this situation, we have

$$A_{i-\sigma} U_\sigma \subseteq U_i = \sum_{j=j_0}^{\sigma} A_{i-j} U_j \subseteq A_{i-j_0} U_\sigma \text{ for all } i > \sigma.$$

As U_σ is a finitely generated A_0-module, we may chose a K-subspace $V \subseteq U$ such that

$$\dim_K(V) < \infty \text{ and } A_0 V = U_\sigma.$$

Prove that for this choice of V we have:

$$U = AV \text{ and the filtrations } U_\bullet \text{ and } A_\bullet V \text{ are equivalent.}$$

As a consequence it follows by Proposition 1.10.4 and the observations made in part (D), that

$$\mathbb{V}_{U_\bullet}(U) = \mathbb{V}_{A_\bullet}(U) \text{ for each } A_\bullet\text{-filtration } U_\bullet \text{ which is of finite type.}$$

1.11 *D*-Modules

Convention 1.11.1 (A) As in Sect. 1.9, we fix a positive integer n, a field K of characteristic 0 and consider the standard Weyl algebra

$$\mathbb{W} := \mathbb{W}(K, n) = K[X_1, X_2, \ldots, X_n, \partial_1, \partial_2, \ldots, \partial_n].$$

In addition, we consider the polynomial ring

$$\mathbb{P} := K[Y_1, Y_2, \ldots, Y_n, Z_1, Z_2, \ldots, Z_n]$$

in the indeterminates $Y_1, Y_2, \ldots, Y_n, Z_1, Z_2, \ldots, Z_n$ with coefficients in the field K.

(B) Let $(\underline{v}, \underline{w}) \in \mathbb{N}_0^n \times \mathbb{N}_0^n$ be a weight. We consider the induced weighted filtration $\mathbb{W}_\bullet^{\underline{v}\underline{w}}$ and also the corresponding associated graded ring.

$$\mathbb{G}^{\underline{v}\underline{w}} = \bigoplus_{i \in \mathbb{N}_0} \mathbb{G}_i^{\underline{v}\underline{w}} := \mathrm{Gr}_{\mathbb{W}_\bullet^{\underline{v}\underline{w}}}(\mathbb{W}^{\underline{v}\underline{w}}) = \bigoplus_{i \in \mathbb{N}_0} \mathrm{Gr}_{\mathbb{W}_\bullet^{\underline{v}\underline{w}}}(\mathbb{W}^{\underline{v}\underline{w}})_i.$$

(see Definition and Remark 1.9.2 (A)).

(C) Moreover, we shall consider the polynomial ring

$$\mathbb{P} = \mathbb{P}^{\underline{v}\underline{w}} = \bigoplus_{i \in \mathbb{N}_0} \mathbb{P}_i^{\underline{v}\underline{w}}.$$

furnished with the grading induced by our given weight $(\underline{v}, \underline{w})$ (see Exercise and Definition 1.9.3 (B)), as well as the canonical isomorphism of graded rings (see Theorem 1.9.4):

$$\eta^{\underline{v}\underline{w}} : \mathbb{P} = \mathbb{P}^{\underline{v}\underline{w}} \xrightarrow{\cong} \mathbb{G}^{\underline{v}\underline{w}}.$$

Definition and Remark 1.11.2 (A) By a *D-module* we mean a finitely generated left module over the standard Weyl algebra $\mathbb{W}$.

(B) Let U be a D-module. If $U_\bullet$ is a $\mathbb{W}_\bullet^{\underline{v}\underline{w}}$-filtration of U, we may again introduce the corresponding *associated graded module* of U with respect to the filtration $U_\bullet$ (see Definition 1.10.3):

$$\mathrm{Gr}_{U_\bullet}(U) = \bigoplus_{i \in \mathbb{Z}} U_i/U_{i-1},$$

which is indeed a graded module over the associated graded ring $\mathbb{G}^{\underline{v}\underline{w}}$. But, in fact, we prefer to consider $\mathrm{Gr}_{U_\bullet}(U)$ as a graded $\mathbb{P}^{\underline{v}\underline{w}}$-module by means of the canonical isomorphism $\eta^{\underline{v}\underline{w}} : \mathbb{P} = \mathbb{P}^{\underline{v}\underline{w}} \xrightarrow{\cong} \mathbb{G}^{\underline{v}\underline{w}}$.

(C) Keep the notations and hypotheses of part (B). Then, we may again consider the *characteristic variety* of U with respect to the filtration $U_\bullet$, but under the previous view, that $\mathrm{Gr}_{U_\bullet}(U)$ is a graded module over the graded polynomial ring $\mathbb{P} = \mathbb{P}^{\underline{v}\underline{w}}$. So, we define this characteristic variety by

$$\mathbb{V}_{U_\bullet}(U) := \mathrm{Var}\big(\mathrm{Ann}_{\mathbb{P}^{\underline{v}\underline{w}}}(\mathrm{Gr}_{U_\bullet}(U))\big) = \mathrm{Var}\big((\eta^{\underline{v}\underline{w}})^{-1}\big[\mathrm{Ann}_{\mathbb{G}^{\underline{v}\underline{w}}}(\mathrm{Gr}_{U_\bullet}(U))\big]\big) \subseteq \mathrm{Spec}(\mathbb{P}).$$

Observe in particular, that the ideal

$$\mathrm{Ann}_{\mathbb{P}^{\underline{v}\underline{w}}}\big(\mathrm{Gr}_{U_\bullet}(U)\big) = (\eta^{\underline{v}\underline{w}})^{-1}\big[\mathrm{Ann}_{\mathbb{G}^{\underline{v}\underline{w}}}\big(\mathrm{Gr}_{U_\bullet}(U)\big)\big] \subseteq \mathbb{P}^{\underline{v}\underline{w}}$$

is graded.

(D) Finally, as U is finitely generated, we may again chose a finite dimensional K-subspace $V \subseteq U$ such that $\mathbb{W}V = U$, and then consider the induced filtration $\mathbb{W}_\bullet^{\underline{v}\underline{w}}V$ of U and the corresponding *intrinsic characteristic variety* (see Exercise and Definition 1.10.5 (D)) of U with respect to the weight $(\underline{v}, \underline{w})$, hence:

$$\mathbb{V}^{\underline{v}\underline{w}}(U) := \mathbb{V}_{\mathbb{W}_\bullet^{\underline{v}\underline{w}}}(U) = \mathbb{V}_{\mathbb{W}_\bullet^{\underline{v}\underline{w}}V}(U).$$

Example 1.11.3 (A) Keep the above notations and let

$$d := \sum_{(\underline{\nu},\underline{\mu})\in\mathrm{supp}(d)} c_{\underline{\nu}\underline{\mu}}^{(d)} \underline{X}^{\underline{\nu}} \underline{\partial}^{\underline{\mu}} \in \mathbb{W}\setminus\{0\} \text{ and } \delta := \deg^{\underline{v}\underline{w}}(d),$$

with $c_{\underline{\nu}\underline{\mu}}^{(d)} \in K\setminus\{0\}$ for all $(\underline{\nu}, \underline{\mu}) \in \mathrm{supp}(d)$. We also consider the so-called *leading differential* form of d with respect to the weight $(\underline{v}, \underline{w})$, which is given by

$$h^{\underline{v}\underline{w}} := \sum_{(\underline{\nu},\underline{\mu})\in\mathrm{supp}(d):\underline{v}\cdot\underline{\nu}+\underline{w}\cdot\underline{\mu}=\delta} c_{\underline{\nu}\underline{\mu}}^{(d)} \underline{X}^{\underline{\nu}} \underline{\partial}^{\underline{\mu}} \in \mathbb{W}\setminus\{0\}.$$

Moreover, we introduce the polynomial

$$f^{\underline{v}\underline{w}} := \sum_{(\underline{\nu},\underline{\mu})\in\mathrm{supp}(d):\underline{v}\cdot\underline{\nu}+\underline{w}\cdot\underline{\mu}=\delta} c_{\underline{\nu}\underline{\mu}}^{(d)} \underline{Y}^{\underline{\nu}} \underline{Z}^{\underline{\mu}} \in \mathbb{P}\setminus\{0\}.$$

Now, consider the cyclic left $\mathbb{W}$-module

$U := \mathbb{W}/\mathbb{W}d$, the element $\overline{1} := (1 + \mathbb{W}_d)/\mathbb{W}_d \in U$ and the K-subspace $K\overline{1} \subseteq U$.

Endow U with the $\mathbb{W}_\bullet^{\underline{v}\underline{w}}$-filtration (see Exercise and Definition 1.10.5 (A)):

$$U_\bullet := \mathbb{W}_\bullet^{\underline{v}\underline{w}}K\overline{1} = \big(U_i := (\mathbb{W}_i^{\underline{v}\underline{w}} + \mathbb{W}d)/\mathbb{W}d\big)_{i\in\mathbb{Z}}.$$

(B) Keep the above notations and hypotheses. Observe first, that for all $i \in \mathbb{Z}$ we may write

$$U_i/U_{i-1} = \mathbb{W}_i^{\underline{v}\underline{w}}/(\mathbb{W}_{i-1}^{\underline{v}\underline{w}} + (\mathbb{W}d \cap \mathbb{W}_i^{\underline{v}\underline{w}})).$$

By the additivity of weighted degrees (see Corollary 1.9.5) we have

$$\mathbb{W}d \cap \mathbb{W}_i^{\underline{v}\underline{w}} = \mathbb{W}_{i-\delta}^{\underline{v}\underline{w}}d \text{ for all } i \in \mathbb{Z}.$$

So, we obtain

$$\mathrm{Gr}_{U_\bullet}(U)_i = U_i/U_{i-1} = \mathbb{W}_i^{\underline{v}\underline{w}}/\big(\mathbb{W}_{i-1}^{\underline{v}\underline{w}} + \mathbb{W}_{i-\delta}^{\underline{v}\underline{w}} d\big) \text{ for all } i \in \mathbb{N}_0.$$

Consequently, there is a surjective homomorphism of graded $\mathbb{G}^{\underline{v}\underline{w}}$-modules

$$\pi : \mathbb{G}^{\underline{v}\underline{w}} = \bigoplus_{i\in\mathbb{Z}} \mathbb{W}_i^{\underline{v}\underline{w}}/\mathbb{W}_{i-1}^{\underline{v}\underline{w}} \twoheadrightarrow \mathrm{Gr}_{U_\bullet}(U) = \bigoplus_{i\in\mathbb{Z}} \mathbb{W}_i^{\underline{v}\underline{w}}/\big(\mathbb{W}_{i-1}^{\underline{v}\underline{w}} + \mathbb{W}_{i-\delta}^{\underline{v}\underline{w}} d\big).$$

If we set

$$\overline{h}^{\underline{v}\underline{w}} := h^{\underline{v}\underline{w}} + \mathbb{W}_{\delta-1}^{\underline{v}\underline{w}} \in \mathbb{W}_\delta^{\underline{v}\underline{w}}/\mathbb{W}_{\delta-1}^{\underline{v}\underline{w}} = \mathbb{G}_\delta^{\underline{v}\underline{w}}$$

it follows that

$$\begin{aligned}\mathrm{Ann}_{\mathbb{G}^{\underline{v}\underline{w}}}\big(\mathrm{Gr}_{U_\bullet}(U)\big) &= \mathrm{Ker}(\pi) = \bigoplus_{i\in\mathbb{Z}} \big(\mathbb{W}_{i-1}^{\underline{v}\underline{w}} + \mathbb{W}_{i-\delta}^{\underline{v}\underline{w}} d\big)/\mathbb{W}_{i-1}^{\underline{v}\underline{w}} \\ &= \bigoplus_{i\in\mathbb{Z}} \big(\mathbb{W}_{i-1}^{\underline{v}\underline{w}} + \mathbb{W}_{i-\delta}^{\underline{v}\underline{w}} h^{\underline{v}\underline{w}}\big)/\mathbb{W}_{i-1}^{\underline{v}\underline{w}} = \mathbb{G}^{\underline{v}\underline{w}}\overline{h}^{\underline{v}\underline{w}}.\end{aligned}$$

Consequently we get

$$\mathrm{Gr}_{U_\bullet}(U) \cong \mathbb{G}^{\underline{v}\underline{w}}/\mathbb{G}^{\underline{v}\underline{w}}\overline{h}^{\underline{v}\underline{w}}.$$

As $\eta^{\underline{v}\underline{w}}(f^{\underline{v}\underline{w}}) = \overline{h}^{\underline{v}\underline{w}}$ and if we consider $\mathrm{Gr}_{U_\bullet}(U)$ as a graded $\mathbb{P}^{\underline{v}\underline{w}}$-module by means of $\eta^{\underline{v}\underline{w}}$, we thus may write

$$\mathrm{Gr}_{U_\bullet}(U) \cong \mathbb{P}^{\underline{v}\underline{w}}/\mathbb{P}^{\underline{v}\underline{w}} f^{\underline{v}\underline{w}} \text{ and } \mathrm{Ann}_{\mathbb{P}}\big(\mathrm{Gr}_{U_\bullet}(U)\big) = \mathbb{P} f^{\underline{v}\underline{w}}.$$

In particular we obtain:

$$\mathbb{V}_{U_\bullet}(U) = \mathbb{V}^{\underline{v}\underline{w}}(U) = \mathbb{V}^{\underline{v}\underline{w}}\big(\mathbb{W}/\mathbb{W}d\big) = \mathrm{Var}(\mathbb{P} f^{\underline{v}\underline{w}}) \subseteq \mathrm{Spec}(\mathbb{P}).$$

Exercise 1.11.4

(A) Let $n = 1$, $K = \mathbb{R}$ and let $d := X_1^4 + \partial_1^2 - X_1^2\partial_1^2$. Determine the two characteristic varieties

$$\mathbb{V}^{\underline{v}\underline{w}}(\mathbb{W}/\mathbb{W}d) \text{ for } (\underline{v}, \underline{w}) = (1, 1) \text{ and } (\underline{v}, \underline{w}) = (0, 1).$$

(B) To make more apparent what you have done in part (A), determine and sketch the *real traces*

$$\mathbb{V}_{\mathbb{R}}^{\underline{v}\underline{w}}(\mathbb{W}/\mathbb{W}d) := \{(y, z) \in \mathbb{R}^2 \mid (Y_1 - y, Z_1 - z)K[Y_1, Z_1] \in \mathbb{V}^{\underline{v}\underline{w}}(\mathbb{W}/\mathbb{W}d)\}$$

for $(\underline{v}, \underline{w}) = (1, 1)$ and $(\underline{v}, \underline{w}) = (0, 1)$. Comment your findings.

Now, we shall establish the fact that D-modules are finitely presentable. To do so we first will show that standard Weyl algebras are left Noetherian (see Conventions, Reminders and Notations 1.1.1 (G) and (H)). We begin with the following preparation.

Definition and Remark 1.11.5 (A) Let $I \subseteq \mathbb{W}$ be a left ideal. We consider the following K-subspace of $\mathbb{G}^{\underline{v}\underline{w}}$:

$$\mathbb{G}^{\underline{v}\underline{w}}(I) := \bigoplus_{i \in \mathbb{N}_0} \big(I \cap \mathbb{W}_i^{\underline{v}\underline{w}} + \mathbb{W}_{i-1}^{\underline{v}\underline{w}}\big)/\mathbb{W}_{i-1}^{\underline{v}\underline{w}} \subseteq \bigoplus_{i \in \mathbb{N}_0} \mathbb{W}_i^{\underline{v}\underline{w}}/\mathbb{W}_{i-1}^{\underline{v}\underline{w}} = \mathbb{G}^{\underline{v}\underline{w}}.$$

It is immediate to see, that $\mathbb{G}^{\underline{v}\underline{w}}(I) \subseteq \mathbb{G}^{\underline{v}\underline{w}}$ is graded ideal. We call this ideal the *graded ideal induced by I* in $\mathbb{G}^{\underline{v}\underline{w}}$.

(B) Let the notations and hypotheses as in part (A). It is straight forward to see, that the family

$$I_\bullet^{\underline{v}\underline{w}} := \big(I \cap \mathbb{W}_i^{\underline{v}\underline{w}}\big)_{i \in \mathbb{Z}}$$

is a filtration of the (left) $\mathbb{W}$-module I, which we call the *filtration induced by* $\mathbb{W}_\bullet^{\underline{v}\underline{w}}$. Observe, that for all $i \in \mathbb{Z}$ we have a canonical isomorphism of K-vector spaces

$$\mathbb{G}^{\underline{v}\underline{w}}(I)_i := \big(I \cap \mathbb{W}_i^{\underline{v}\underline{w}} + \mathbb{W}_{i-1}^{\underline{v}\underline{w}}\big)/\mathbb{W}_{i-1}^{\underline{v}\underline{w}} \cong I \cap \mathbb{W}_i^{\underline{v}\underline{w}}/I \cap \mathbb{W}_{i-1}^{\underline{v}\underline{w}} = I_i^{\underline{v}\underline{w}}/I_{i-1}^{\underline{v}\underline{w}} = \mathrm{Gr}_{I_\bullet^{\underline{v}\underline{w}}}(I)_i.$$

It is easy to see, that these isomorphisms of K-vector spaces actually give rise to a canonical isomorphism of graded $\mathbb{G}^{\underline{v}\underline{w}}$-modules

$$\mathbb{G}^{\underline{v}\underline{w}}(I) := \bigoplus_{i \in \mathbb{Z}} \big((I \cap \mathbb{W}_i^{\underline{v}\underline{w}}) + \mathbb{W}_{i-1}^{\underline{v}\underline{w}}\big)/\mathbb{W}_{i-1}^{\underline{v}\underline{w}} \cong \bigoplus_{i \in \mathbb{Z}} I_i^{\underline{v}\underline{w}}/I_{i-1}^{\underline{v}\underline{w}} = \mathrm{Gr}_{I_\bullet^{\underline{v}\underline{w}}}(I).$$

So, by means of this canonical isomorphism we may identify

$$\mathbb{G}^{\underline{v}\underline{w}}(I) = \mathrm{Gr}_{I_\bullet^{\underline{v}\underline{w}}}(I).$$

Lemma 1.11.6 *Let $I, J \subseteq \mathbb{W}$ be two left ideals with $I \subseteq J$. Then we can say:*

(a) *There is an inclusion of graded ideals $\mathbb{G}^{\underline{v}\underline{w}}(I) \subseteq \mathbb{G}^{\underline{v}\underline{w}}(J)$ in the graded ring $\mathbb{G}^{\underline{v}\underline{w}}$.*
(b) *If $\mathbb{G}^{\underline{v}\underline{w}}(I) = \mathbb{G}^{\underline{v}\underline{w}}(J)$, then $I = J$.*

Proof (a): This is immediate by Definition and Remark 1.11.5 (A).
(b): Assume that $I \subsetneq J$. Then, there is a least integer $i \in \mathbb{N}_0$ such that

$$I_i^{\underline{v}\underline{w}} = I \cap \mathbb{W}_i^{\underline{v}\underline{w}} \subsetneq J_i^{\underline{v}\underline{w}} = J \cap \mathbb{W}_i^{\underline{v}\underline{w}}.$$

As $I_{i-1}^{\underline{vw}} = J_{i-1}^{\underline{vw}}$ it follows that

$$\mathbb{G}^{\underline{vw}}(I)_i \cong I_i^{\underline{vw}}/I_{i-1}^{\underline{vw}} \text{ is not isomorphic to } I_i^{\underline{vw}}/I_{i-1}^{\underline{vw}} \cong \mathbb{G}^{\underline{vw}}(J)_i,$$

so that indeed

$$\mathbb{G}^{\underline{vw}}(I) \neq \mathbb{G}^{\underline{vw}}(J).$$

Theorem 1.11.7 (Noetherianness of Weyl Algebras) *The Weyl algebra* $\mathbb{W}$ *is left Noetherian.*

Proof Otherwise $\mathbb{W}$ would contain an infinite strictly ascending chain of left ideals $I^{(1)} \subsetneq I^{(2)} \subsetneq I^{(3)} \subsetneq \cdots$. But then, by Lemma 1.11.6 we would have an infinite strictly ascending chain $\mathbb{G}^{\underline{vw}}(I^{(1)}) \subsetneq \mathbb{G}^{\underline{vw}}(I^{(2)}) \subsetneq \mathbb{G}^{\underline{vw}}(I^{(3)}) \subsetneq \cdots$ of ideals in the Noetherian ring $\mathbb{G}^{\underline{vw}} \cong \mathbb{P}^{\underline{vw}} = \mathbb{P}$, a contradiction.

Corollary 1.11.8 (Finite Presentability of D**-Modules)** *Each* D*-module* U *admits a finite presentation*

$$\mathbb{W}^s \longrightarrow \mathbb{W}^r \longrightarrow U \longrightarrow 0.$$

Proof This follows immediately by Theorem 1.11.7 and the observations made in Conventions, Reminders and Notations 1.1.1 (H).

Example 1.11.9 (A) Consider the polynomial ring $U := K[X_1, X_2, \ldots, X_n]$. As

$$\mathbb{W} \subseteq \mathrm{End}_K\big(K[X_1, X_2, \ldots, X_n]\big) = \mathrm{End}_K(U),$$

this polynomial ring can be viewed in a canonical way as a left module over $\mathbb{W}$, the scalar being multiplication given by

$$d \cdot f := d(f) \text{ for all } d \in \mathbb{W} \text{ and all } f \in U.$$

As $f \cdot 1 = f$ for all $f \in U$ it follows that

$$U = \mathbb{W}1_U.$$

So, the $\mathbb{W}$-module $U := K[X_1, X_2, \ldots, X_n]$ is generated by a single element, and hence in particular a D-module.

(B) Keep the previous notations and hypotheses. Observe that

$$\sum_{i=1}^{n} \mathbb{W}\partial_i = \bigoplus_{\underline{\nu},\underline{\mu}\in\mathbb{N}_0^n:\underline{\mu}\neq\underline{0}} K\underline{X}^{\underline{\nu}}\underline{\partial}^{\underline{\mu}}$$

and hence

$$\mathbb{W} = K[X_1, X_2, \ldots, X_n] \oplus \sum_{i=1}^{n} \mathbb{W}\partial_i = U \oplus \sum_{i=1}^{n} \mathbb{W}\partial_i.$$

We thus have an exact sequence of K-vector spaces

$$0 \longrightarrow \sum_{i=1}^{n} \mathbb{W}\partial_i \longrightarrow \mathbb{W} \xrightarrow{\pi} U \longrightarrow 0,$$

in which $\mathbb{W} \xrightarrow{\pi} U$ is the *canonical projection* map given by

$$\pi\left(\underline{X}^{\underline{\nu}}\underline{\partial}^{\underline{\mu}}\right) = \begin{cases} \underline{X}^{\underline{\nu}}, & \text{if } \underline{\mu} = \underline{0}, \\ 0, & \text{if } \underline{\mu} \neq \underline{0} \end{cases}.$$

Our aim is to show:

$$\mathbb{W} \xrightarrow{\pi} U \text{ is a homomorphism of left } \mathbb{W}\text{-modules.}$$

To do so, it suffices to show that for all $\underline{\nu}, \underline{\mu}, \underline{\nu}', \underline{\mu}' \in \mathbb{N}_0^n$ it holds

$$\pi(dd') = d\pi(d'), \text{ where } d := \underline{X}^{\underline{\nu}}\underline{\partial}^{\underline{\mu}} \text{ and } d' := \underline{X}^{\underline{\nu}'}\underline{\partial}^{\underline{\mu}'}.$$

If $\underline{\mu} = \underline{\mu}' = \underline{0}$, we have

$$\pi(dd') = \pi\left(\underline{X}^{\underline{\nu}}\underline{X}^{\underline{\nu}'}\right) = \pi\left(\underline{X}^{\underline{\nu}+\underline{\nu}'}\right) = \underline{X}^{\underline{\nu}+\underline{\nu}'} = \underline{X}^{\underline{\nu}}\underline{X}^{\underline{\nu}'} = \underline{X}^{\underline{\nu}}\pi\left(\underline{X}^{\underline{\nu}'}\right) = d\pi(d').$$

If $\underline{\mu} = 0$ and $\underline{\mu}' \neq 0$ we have

$$\pi(dd') = \pi\left(\underline{X}^{\underline{\nu}}\underline{X}^{\underline{\nu}'}\underline{\partial}^{\underline{\mu}'}\right) = \pi\left(\underline{X}^{\underline{\nu}+\underline{\nu}'}\underline{\partial}^{\underline{\mu}'}\right) = 0 = \underline{X}^{\underline{\nu}}\pi\left(\underline{X}^{\underline{\nu}'}\underline{\partial}^{\underline{\mu}'}\right) = d\pi(d').$$

So, let $\underline{\mu} \neq \underline{0}$. By the Product Formula of Proposition 1.6.2 we have

$$dd' = \underline{X}^{\underline{\nu}}\underline{\partial}^{\underline{\mu}}\underline{X}^{\underline{\nu}'}\underline{\partial}^{\underline{\mu}'} = \underline{X}^{\underline{\nu}+\underline{\nu}'}\underline{\partial}^{\underline{\mu}+\underline{\mu}'} + s,$$

with

$$s := \sum_{\underline{k}\in\mathbb{N}_0^n : \underline{0}<\underline{k}\leq\underline{\mu},\underline{\nu}'} \lambda_{\underline{k}}\underline{X}^{\underline{\nu}+\underline{\nu}'-\underline{k}}\underline{\partial}^{\underline{\mu}+\underline{\mu}'-\underline{k}}$$

and

$$\lambda_{\underline{k}} = \Big(\prod_{i=1}^{n}\binom{\mu_i}{k_i}\Big)\Big(\prod_{i=1}^{n}\prod_{p=0}^{k_i-1}(\nu'_i - p)\Big).$$

Assume first, that $\underline{\mu}' \neq 0$. Then we have

$$\pi\big(\underline{X}^{\underline{\nu}+\underline{\nu}'}\underline{\partial}^{\underline{\mu}+\underline{\mu}'}\big) = 0 \text{ and } \pi\big(\underline{X}^{\underline{\nu}+\underline{\nu}'-\underline{k}}\underline{\partial}^{\underline{\nu}+\underline{\nu}'-\underline{k}}\big) = 0 \text{ for all } \underline{k} \in \mathbb{N}_0^n \text{ with } \underline{0} < \underline{k} \leq \underline{\mu}, \underline{\nu}'.$$

It thus follows, that

$$\pi(dd') = 0 = d0 = d\pi\big(\underline{X}^{\underline{\nu}'}\underline{\partial}^{\underline{\mu}'}\big) = d\pi(d').$$

So, finally let $\underline{\mu}' = 0$. Then $dd' = \underline{X}^{\underline{\nu}+\underline{\nu}'}\underline{\partial}^{\underline{\mu}} + s$, and

$$s = \begin{cases} \prod_{i=1}^{n}\prod_{p=0}^{\mu_i-1}(\nu'_i - p)\underline{X}^{\underline{\nu}+\underline{\nu}'-\underline{\mu}}, & \text{if } \underline{\mu} \leq \underline{\nu}'; \\ 0, & \text{otherwise.} \end{cases}$$

So, by what we have learned in Exercise 1.6.6 (B), we have

$$s = \underline{X}^{\underline{\nu}}\underline{\partial}^{\underline{\mu}}\big(\underline{X}^{\underline{\nu}'}\big).$$

As s is a K-multiple of a monomial in the X_i's we have $\pi(s) = s$. It thus follows

$$\pi(dd') = \pi\big(\underline{X}^{\underline{\nu}+\underline{\nu}'}\underline{\partial}^{\underline{\mu}'}\big) + \pi(s) = s = \underline{X}^{\underline{\nu}}\underline{\partial}^{\underline{\mu}}\big(\underline{X}^{\underline{\nu}'}\big) = \underline{X}^{\underline{\nu}}\underline{\partial}^{\underline{\mu}}\underline{X}^{\underline{\nu}'} = d\pi(d').$$

This proves, that π is indeed a homomorphism of left $\mathbb{W}$-modules.

(C) Keep the previous notations and hypotheses. Then, according the above observations, we have an exact sequence of left $\mathbb{W}$-modules

$$0 \longrightarrow \mathbb{W}^n \xrightarrow{h} \mathbb{W} \xrightarrow{\pi} U \longrightarrow 0,$$

in which h is given by

$$(d_1, d_2, \ldots, d_n) \mapsto h(d_1, d_2, \ldots, d_n) = \sum_{i=1}^{n} d_i\partial_i.$$

This sequence clearly constitutes a presentation of the left $\mathbb{W}$-module U (see Conventions, Reminders and Notations 1.1.1 (H)) and the corresponding presentation matrix for U is the row

$$\partial := \begin{pmatrix} \partial_1 \\ \partial_2 \\ \vdots \\ \partial_n \end{pmatrix} \in \mathbb{W}^{n\times 1}.$$

Exercise 1.11.10 (A) We consider the polynomial ring $U = K[X_1, X_2, \ldots, X_n]$ canonically as a D-module, as done in Example 1.11.9. Fix a weight $(\underline{v}, \underline{w}) \in \mathbb{N}_0^n \times \mathbb{N}_0^n$. Consider the K-subspace $K \subset U$, observe that $\mathbb{W}K = U$ and endow U with the induced filtration

$$U_\bullet := \mathbb{W}_\bullet^{\underline{v}\underline{w}} K.$$

Show, that there is an isomorphism of graded $\mathbb{P}$-modules

$$\mathrm{Gr}_{U_\bullet}(U) = \mathrm{Gr}_{\mathbb{W}_\bullet^{\underline{v}\underline{w}} K}(U) \cong U^{\underline{v}},$$

where

$$U^{\underline{v}} := \bigoplus_{i \in \mathbb{N}_0} U_i^{\underline{v}} \text{ with } U_i^{\underline{v}} := \sum_{\underline{v} \cdot \underline{\nu} = i} K\underline{X}^{\underline{\nu}} \text{ for all } i \in \mathbb{N}_0$$

is the polynomial ring U endowed with the grading associated to the weight $\underline{v} \in \mathbb{N}_0^n$. Determine the characteristic variety

$$\mathbb{V}^{\underline{v}\underline{w}}(U) \subseteq \mathrm{Spec}(\mathbb{P}).$$

(B) Keep the notations and hypotheses of part (A). Show, the left $\mathbb{W}$-module U is simple: If $V \subsetneq U$ is a proper left $\mathbb{W}$-submodule, then $V = 0$. (Hint: Let $f \in U \setminus \{0\}$ be of degree r and assume that $\underline{\nu} = (\nu_1, \nu_2, \ldots, \nu_n) \in \mathrm{supp}(f)$ with $\sum_{i=1}^n \nu_i = r$ and show that $\partial^{\underline{\nu}} \in K \setminus \{0\}$. Conclude that $\mathbb{W}f = U$.)

Remark and Definition 1.11.11 (A) We furnish the polynomial ring $K[X_1, X_2, \ldots, X_n]$ with its *canonical structure of D-module* (see Example 1.11.9). We now consider a ring $\mathscr{A}$ with the following properties

(1) $\mathscr{A}$ is commutative;
(2) $\mathscr{A}$ is a left $\mathbb{W}$-module;
(3) $K[X_1, X_2, \ldots, X_n] \subseteq \mathscr{A}$ is a left submodule.

In this situation, we call $\mathscr{A}$ a ring of *good functions* in $X_1, X_2, \ldots, X_n$ over K. The idea covered by this concept is that for all $d \in \mathbb{W}$ and all $f \in$ the product $df \in \mathscr{A}$ should be viewed as the result of the application of the differential operator d to the function f. Therefore, one often writes

$$d(f) := df \text{ for all } d \in \mathbb{W} \text{ and all } f \in \mathscr{A}.$$

(B) Let the notations and hypotheses be as in part (A). By a *system of polynomial differential equations* in $\mathscr{A}$ we mean a system of equations

$$d_{11}(f_1) + d_{12}(f_2) + \ldots + d_{1r}(f_r) = 0$$
$$d_{21}(f_1) + d_{22}(f_2) + \ldots + d_{2r}(f_r) = 0$$
$$\vdots$$
$$d_{s1}(f_1) + d_{s2}(f_2) + \ldots + d_{sr}(f_r) = 0$$

with $r, s \in \mathbb{N}$ such that

$$d_{ij} \in \mathbb{W} \text{ and } f_j \in \mathscr{A} \text{ for all } i, j \in \mathbb{N} \text{ with } i \leq s \text{ and } j \leq r.$$

The above system of differential equations can be understood as a linear system of equations over the ring $\mathscr{A}$. We namely may consider the matrix

$$\mathscr{D} := \begin{pmatrix} d_{11} & d_{12} & \dots & d_{1r} \\ d_{21} & d_{22} & \dots & d_{2r} \\ \vdots & \vdots & & \vdots \\ d_{s1} & d_{s2} & \dots & d_{sr} \end{pmatrix} \in \mathbb{W}^{s \times r}.$$

Then, the above system may be written in matrix form as

$$\mathscr{D} \begin{pmatrix} f_1 \\ f_2 \\ \vdots \\ f_r \end{pmatrix} = \begin{pmatrix} 0 \\ 0 \\ \vdots \\ 0 \end{pmatrix}.$$

We call $\mathscr{D}$ the *matrix of differential operators* associated to our system of linear differential equations. So, systems of differential equations correspond to matrices with entries in a standard Weyl algebra.

(C) Keep the previous notations and hypotheses, then the matrix of differential operators $\mathscr{D} \in \mathbb{W}^{s \times r}$ gives rise to an exact sequence of left $\mathbb{W}$-modules

$$0 \longrightarrow \mathbb{W}^s \xrightarrow{h_{\mathscr{D}}} \mathbb{W}^r \xrightarrow{\pi_{\mathscr{D}}} U_{\mathscr{D}} \longrightarrow 0.$$

In particular $U_{\mathscr{D}}$ is a D-module and the previous sequence is a finite presentation of $U_{\mathscr{D}}$. We call this presentation the *presentation induced by the matrix* $\mathscr{D}$ and we call $U_{\mathscr{D}}$ the D-module *defined* by the matrix $\mathscr{D}$—or the D-module associated with our system of differential equations. So, each system of differential equations defines a D-module. Obviously, one is particularly interested in the *solution space* of our system of differential equations, hence in the K-vector space

$$\mathbb{S}_{\mathscr{D}}(\mathscr{A}) := \{(f_1, f_2, \dots, f_r) \in \mathscr{A}^r \mid \mathscr{D} \begin{pmatrix} f_1 \\ f_2 \\ \vdots \\ f_r \end{pmatrix} = \begin{pmatrix} 0 \\ 0 \\ \vdots \\ 0 \end{pmatrix}\}.$$

Observe, that $\mathbb{S}_{\mathscr{D}}(\mathscr{A})$ is a K-subspace of $\mathscr{A}^r$.

Proposition 1.11.12 *Let $r, s \in \mathbb{N}$, let*

$$\mathscr{D} = \begin{pmatrix} d_{11} & d_{12} & \dots & d_{1r} \\ d_{21} & d_{22} & \dots & d_{2r} \\ \vdots & \vdots & & \vdots \\ d_{s1} & d_{s2} & \dots & d_{sr} \end{pmatrix} \in \mathbb{W}^{s \times r}$$

be a matrix of differential operators, consider the induced presentation

$$0 \longrightarrow \mathbb{W}^s \xrightarrow{h=h_{\mathscr{D}}} \mathbb{W}^r \xrightarrow{\pi=\pi_{\mathscr{D}}} U_{\mathscr{D}} \longrightarrow 0$$

and the corresponding solution space $\mathbb{S}_{\mathscr{D}}(\mathscr{A})$.
For all $i = 1, 2, \dots, r$ let $\mathrm{e}_i := (\delta_{i,j})_{j=1}^r \in \mathbb{W}^r$ be the i-th canonical basis element. Then, there is an isomorphism of K-vector spaces

$$\varepsilon_{\mathscr{D}} : \mathrm{Hom}_{\mathbb{W}}\big(U_{\mathscr{D}}, \mathscr{A}\big) \xrightarrow{\cong} \mathbb{S}_{\mathscr{D}}(\mathscr{A}),$$

given by

$$m \mapsto \varepsilon_{\mathscr{D}}(m) := \big(m(\pi(\mathrm{e}_1)), m(\pi(\mathrm{e}_2)), \dots, m(\pi(\mathrm{e}_r))\big) \textit{ for all } m \in \mathrm{Hom}_{\mathbb{W}}\big(U_{\mathscr{D}}, \mathscr{A}\big).$$

Proof Observe, that there is indeed a K-linear map

$$\varepsilon := \varepsilon_{\mathscr{D}} : \mathrm{Hom}_{\mathbb{W}}\big(U_{\mathscr{D}}, \mathscr{A}\big) \longrightarrow \mathscr{A}^r$$

given by

$$m \mapsto \varepsilon_{\mathscr{D}}(m) := \big(m(\pi(\mathrm{e}_1)), m(\pi(\mathrm{e}_2)), \dots, m(\pi(\mathrm{e}_r))\big) \text{ for all } m \in \mathrm{Hom}_{\mathbb{W}}\big(U_{\mathscr{D}}, \mathscr{A}\big).$$

If $\varepsilon(m) = 0$, then $m(\pi(\mathrm{e}_i)) = 0$ for all $i = 1, 2, \dots, r$. As π is surjective, the elements $\pi(\mathrm{e}_i)$ $(i = 1, 2, \dots, r)$ generate the left $\mathbb{W}$-module $U = U_{\mathscr{D}}$. So, it follows that $m = 0$ and this proves, that the map ε is injective.
It remains to show that

$$\varepsilon\big(\mathrm{Hom}_{\mathbb{W}}\big(U_{\mathscr{D}}, \mathscr{A}\big)\big) = \mathbb{S}_{\mathscr{D}}(\mathscr{A}).$$

To do so, let

$$\mathrm{b}_j := (\delta_{j,k})_{k=1}^s \in \mathbb{W}^s \quad (j = 1, 2, \dots, s)$$

be the canonical basis elements of $\mathbb{W}^s$.

First, let $m \in \mathrm{Hom}_{\mathbb{W}}\big(U_{\mathscr{D}}, \mathscr{A}\big)$. We aim to show, that $\varepsilon(m) \in \mathbb{S}_{\mathscr{D}}(\mathscr{A})$. We have to show, that the column

$$\begin{pmatrix} g_1 \\ g_2 \\ \vdots \\ g_s \end{pmatrix} := \mathscr{D} \begin{pmatrix} m(\mathrm{e}_1) \\ m(\mathrm{e}_2) \\ \vdots \\ m(\mathrm{e}_r) \end{pmatrix}$$

vanishes. For each $i = 1, 2, \ldots, s$ we can write $\sum_{j=1}^{r} d_{ij}\mathrm{e}_j = \mathrm{b}_i\mathscr{D} = h(\mathrm{b}_i)$, and hence get indeed

$$g_i = \sum_{j=1}^{r} d_{ij} m(\pi(\mathrm{e}_j)) = m\Big(\sum_{j=1}^{r} d_{ij}\pi(\mathrm{e}_j)\Big) = m\big(\pi(\sum_{j=1}^{r} d_{ij}\mathrm{e}_j)\big) = m\big(\pi(h(\mathrm{b}_i))\big)$$
$$= m(0) = 0.$$

Conversely, let $(f_1, f_2, \ldots, f_r) \in \mathbb{S}_{\mathscr{D}}(\mathscr{A})$, so that $\sum_{j=1}^{r} d_{ij} f_j = 0$. We aim to show that $(f_1, f_2, \ldots, f_r) \in \varepsilon\big(\mathrm{Hom}_{\mathbb{W}}(U, \mathscr{A})\big)$.
To this end, we consider the homomorphism of left $\mathbb{W}$-modules

$$k : \mathbb{W}^r \longrightarrow \mathscr{A}, \text{ given by } (u_1, u_2, \ldots, u_r) \mapsto \sum_{j+1}^{r} u_j f_j.$$

Observe that

$$k(h(\mathrm{b}_i)) = k(\mathrm{b}_i\mathscr{D}) = k(d_{i1}, d_{i2}, \ldots, d_{ir}) = \sum_{j=1}^{r} d_{ij} f_j = 0 \text{ for all } i = 1, 2, \ldots, s.$$

It follows that $k \circ h = 0$. Therefore k induces a homomorphism of left $\mathbb{W}$-modules

$$m : U \longrightarrow \mathscr{A}, \text{ such that } m \circ \pi = k.$$

It follows that $m(\pi(\mathrm{e}_j)) = k(\mathrm{e}_j) = f_j$ for all $j = 1, 2, \ldots, r$. But this means that $(f_1, f_2, \ldots, f_r) = \varepsilon(m) \in \varepsilon\big(\mathrm{Hom}_{\mathbb{W}}(U, \mathscr{A})\big)$.

Exercise 1.11.13 (A) Let $n = 1$, $K = \mathbb{R}$ and let $\mathscr{A} := \mathscr{C}^\infty(\mathbb{R})$ be set of smooth functions on $\mathbb{R}$. Fix $d \in \mathbb{W} = \mathbb{W}(\mathbb{R}, 1) = \mathbb{R}[X, \partial]$ and consider the matrix $\mathscr{D} = (d) \in \mathbb{W}^{1\times 1}$. Determine

$$U_{\mathscr{D}}, \quad \mathbb{S}_{\mathscr{D}}(\mathscr{A}) \text{ and } \mathbb{V}^{\underline{v},\underline{w}}(U_{\mathscr{D}})$$

for all weights $(\underline{v}, \underline{w}) = (v, w) \in \mathbb{N}_0 \times \mathbb{N}_0 \setminus \{(0, 0)\}$ and for

$$d = \partial, \quad d = \partial^2 - 1, \quad d = \partial - x^2 \text{ and } d = \partial^2 + c\partial - b \text{ with } c, b \in \mathbb{R} \setminus \{0\}.$$

(B) Let $n, m \in \mathbb{N}$, $\mathscr{A} := K[X_1, X_2, \ldots, X_n]$ and consider the matrix

$$\mathscr{D} := \begin{pmatrix} \partial_1^m \\ \partial_2^m \\ \vdots \\ \partial_n^m \end{pmatrix} \in \mathbb{W}^{n\times 1}.$$

Determine

$$U_{\mathscr{D}}, \quad \mathbb{S}_{\mathscr{D}}(\mathscr{A}) \text{ and } \mathbb{V}^{\underline{1}\underline{1}}(U_{\mathscr{D}}).$$

1.12 Gröbner Bases

In this section, we introduce and treat Gröbner bases of left ideals in standard Weyl algebras with respect to so-called admissible orderings of the set of elementary differential operators. What we get is a theory very similar to the theory of Gröbner bases of ideals in polynomial rings. A theory many readers may be familiar with already. Indeed a great deal of what we shall present in the sequel could also be deduced from the theory of Gröbner in polynomial rings. Nevertheless, we prefer to introduce the subject in a self contained way so that readers who are not familiar with Gröbner in polynomial rings can follow our approach without further prerequisites. As for Gröbner bases in (commutative) polynomial rings and their applications, there are indeed many introductory and advanced textbooks and monograph. So, we mention only a sample of possible references for this subject, namely [1, 6, 19, 25, 26, 30, 36] and [42].

In general, Gröbner bases are intimately related to Division Theorems, which generalize Euclid's Division Theorem for univariate polynomial rings over a field. Gröbner bases and Division Theorems for rings of linear differential operators were introduced by Briançon and Maisonobe [14] in the univariate case and by Castro-Jiménez [21] in the multivariate case. Two more recent basic references in the field of are the textbook of Bueso,Gómez-Torricellas and Verschoren [20] and the PhD thesis [31] of Levandovskyy.

The main goal of the present section is to prove that left ideals in Weyl algebras admit so-called universal Gröbner bases. This existence result can actually be proved in the more general setting of admissible algebras. Readers, who are interested in this, should consult for example Boldini's thesis [10] or else [38], [41] or [43].

Convention 1.12.1 (A) As previously, we fix a positive integer n, a field K of characteristic 0 and consider the standard Weyl algebra

$$\mathbb{W} := \mathbb{W}(K, n) = K[X_1, X_2, \ldots, X_n, \partial_1, \partial_2, \ldots, \partial_n].$$

Moreover, we consider the polynomial ring

$$\mathbb{P} := K[Y_1, Y_2, \ldots, Y_n, Z_1, Z_2, \ldots, Z_n]$$

in the indeterminates $Y_1, Y_2, \ldots, Y_n, Z_1, Z_2, \ldots, Z_n$ with coefficients in the field K.

(B) In addition, we fix the isomorphism of K-vector spaces

$$\Phi : \mathbb{W} \xrightarrow{\cong} \mathbb{P} \text{ given by } \underline{X}^{\underline{\nu}}\underline{\partial}^{\underline{\mu}} \mapsto \underline{Y}^{\underline{\nu}}\underline{Z}^{\underline{\mu}} \text{ for all } \underline{\nu}, \underline{\mu} \in \mathbb{N}_0^n.$$

Moreover we respectively consider the set $\mathbb{E}$ of all elementary differential operators in $\mathbb{W}$ and the set $\mathbb{M}$ of all monomials in $\mathbb{P}$, thus:

$$\mathbb{E} := \{\underline{X}^{\underline{\nu}}\underline{\partial}^{\underline{\mu}} \mid \underline{\nu}, \underline{\mu} \in \mathbb{N}_0^n\} \text{ and } \mathbb{M} := \{\underline{Y}^{\underline{\nu}}\underline{Z}^{\underline{\mu}} \mid \underline{\nu}, \underline{\mu} \in \mathbb{N}_0^n\} = \Phi(\mathbb{E}).$$

In a first step we now introduce some basic notions of our subject, namely: admissible orderings (of the set $\mathbb{E}$ of elementary differential operators, leading (elementary) differential operators and (in the polynomial ring $\mathbb{P}$) leading monomials and leading terms. Mainly for those readers who have not met these concepts in the framework of polynomial rings, we shall add below a number of examples and exercises on these new notions.

Definition, Reminder and Exercise 1.12.2 (A) *(Total Orderings)* Let S be any set. A *total ordering* of S is a binary relation $\leq \subseteq S \times S$ such that for all $a, b, c \in S$ the following requirements are satisfied:

(a) *(Reflexivity)* $a \leq a$.
(b) *(Antisymmetry)* If $a \leq b$ and $b \leq a$, then $a = b$.
(c) *(Transitivity)* If $a \leq b$ and $b \leq c$, then $a \leq c$.
(b) *(Totality)* Either $a \leq b$ or $b \leq a$.

We write $\mathrm{TO}(S)$ for the set of total orderings on S.
If $\leq \in \mathrm{TO}(S)$ and $a, b \in S$, we write

$$a < b \text{ if } a \leq b \text{ and } a \neq b, \quad b \geq a \text{ if } a \leq b, \quad b > a \text{ if } a < b.$$

(B) *(Well Orderings)* Keep the above notations and hypotheses. A total ordering $\leq \in \mathrm{TO}(S)$ is said to be a *well ordering* of S, if it satisfies the following additional requirement:

(e) *(Existence of Least Elements)* For each non-empty subset $T \subseteq S$ there is an element $t \in T$ such that $t \leq t'$ for all $t' \in T$.

In the situation mentioned in statement (e), the element $t \in T$—if it exists at all—is uniquely determined by T and called the *least element* or the *minimum* of T with respect to $\leq$ and denoted by $\min_{\leq}(T)$, thus

$$t = \min_{\leq}(T) \text{ if } t \in T \text{ and } t \leq t' \text{ for all } t' \in T.$$

We write $\mathrm{WO}(S)$ for the set of all well orderings of S.

(C) *(Admissible Orderings)* A total ordering $\leq \in \mathrm{TO}(\mathbb{E})$ of the set of all elementary differential operators is called an *admissible ordering* of $\mathbb{E}$ if it satisfies the following requirements:

(a) *(Foundedness)* $1 \leq \underline{X}^{\underline{\nu}}\underline{\partial}^{\underline{\mu}}$ for all $\underline{\nu}, \underline{\mu} \in \mathbb{N}_0^n$
(b) *(Compatibility)* For all $\underline{\lambda}, \underline{\lambda}', \underline{\kappa}, \underline{\kappa}', \underline{\nu}, \underline{\mu} \in \mathbb{N}_0^n$ we have the implication:

$$\text{If } \underline{X}^{\underline{\lambda}}\underline{\partial}^{\underline{\kappa}} \leq \underline{X}^{\underline{\lambda}'}\underline{\partial}^{\underline{\kappa}'}, \text{ then } \underline{X}^{\underline{\lambda}+\underline{\nu}}\underline{\partial}^{\underline{\kappa}+\underline{\mu}} \leq \underline{X}^{\underline{\lambda}'+\underline{\nu}}\underline{\partial}^{\underline{\kappa}'+\underline{\mu}}.$$

We write $\mathrm{AO}(\mathbb{E})$ for the set of all admissible orderings of $\mathbb{E}$.
Prove the following facts:

(c) If $\underline{\nu}, \underline{\nu}', \underline{\mu}, \underline{\mu}', \underline{\lambda}, \underline{\lambda}', \underline{\kappa}, \underline{\kappa}', \in \mathbb{N}_0^n$ with $\underline{X}^{\underline{\nu}}\underline{\partial}^{\underline{\mu}} \leq \underline{X}^{\underline{\nu}'}\underline{\partial}^{\underline{\mu}'}$ and $\underline{X}^{\underline{\lambda}}\underline{\partial}^{\underline{\kappa}} < \underline{X}^{\underline{\lambda}'}\underline{\partial}^{\underline{\kappa}'}$, then

$$\underline{X}^{\underline{\lambda}+\underline{\nu}}\underline{\partial}^{\underline{\kappa}+\underline{\mu}} < \underline{X}^{\underline{\lambda}'+\underline{\nu}'}\underline{\partial}^{\underline{\kappa}'+\underline{\mu}'}.$$

(d) $\mathrm{AO}(\mathbb{E}) \subseteq \mathrm{WO}(\mathbb{E})$.

(D) *(Leading Elementary Differential Operators and Related Concepts)* From now on, for all $d \in \mathbb{W}$, we use the notation

$$\mathrm{Supp}(d) := \{\underline{X}^{\underline{\nu}}\underline{\partial}^{\underline{\mu}} \mid (\underline{\nu}, \underline{\mu}) \in \mathrm{supp}(d)\}.$$

Keep the above notations and hypotheses. If $\leq \in \mathrm{AO}(\mathbb{E})$ and $d \in \mathbb{W} \setminus \{0\}$, we define the *leading elementary differential operator* of d with respect to $\leq$ by:

$$\mathrm{LE}_{\leq}(d) := \max_{\leq}\mathrm{Supp}(d),$$

so that

$$\mathrm{LE}_{\leq}(d) \in \mathrm{Supp}(d) \text{ and } e \leq \mathrm{LE}_{\leq}(d) \text{ for all } e \in \mathrm{Supp}(d).$$

Moreover, we define the *leading coefficient* $\mathrm{LC}_{\leq}(d)$ of d with respect to $\leq$ as the coefficient of d with respect to $\mathrm{LE}_{\leq}(d)$, and the *leading differential operator* $\mathrm{LD}_{\leq}(d)$ of d with respect to $\leq$ as the product of the leading elementary differential operator with the leading coefficient, so that:

(a) $\mathrm{LC}_{\leq}(d) \in K \setminus \{0\}$ with $\mathrm{LE}_{\leq}\big(d - \mathrm{LC}_{\leq}(d)\mathrm{LE}_{\leq}(d)\big) < \mathrm{LE}_{\leq}(d)$.
(b) $\mathrm{LD}_{\leq}(d) = \mathrm{LC}_{\leq}(d)\mathrm{LE}_{\leq}(d)$.
(c) $\mathrm{LE}_{\leq}\big(d - \mathrm{LD}_{\leq}(d)\big) < \mathrm{LE}_{\leq}(d)$.

Finally, we define the *leading monomial* and the *leading term* of d with respect to $\leq$ respectively by

$$\mathrm{LM}_{\leq}(d) := \Phi\big(\mathrm{LE}_{\leq}(d)\big) \text{ and } \mathrm{LT}_{\leq}(d) := \Phi\big(\mathrm{LD}_{\leq}(d)\big) = \mathrm{LC}_{\leq}(d)\mathrm{LM}_{\leq}(d).$$

Prove the following statements:

(d) If $d, e \in \mathbb{W} \setminus \{0\}$, with $d \neq -e$, then $\mathrm{LE}_{\leq}(d+e) \leq \max_{\leq}\{\mathrm{LE}_{\leq}(d), \mathrm{LE}_{\leq}(e)\}$, with equality if and only if $\mathrm{LD}_{\leq}(d) \neq -\mathrm{LD}_{\leq}(e)$.

The previously introduced notions are of basic significance for this and the next section. So, we hope to illuminate their meaning in the following series of examples and exercises, which were already announced prior to the definition of these concepts.

Examples and Exercises 1.12.3 (A) *(Well Orderings)* Keep the above notations and hypotheses. Prove the following statements:

(a) Let $\varphi : \mathbb{N}_0 \longrightarrow \mathbb{N}_0^n \times \mathbb{N}_0^n$ be a bijective map. Show that the binary relation $\leq_{\varphi} \subseteq \mathbb{E} \times \mathbb{E}$ defined by

$$\underline{X}^{\underline{\nu}} \underline{\partial}^{\underline{\mu}} \leq_{\varphi} \underline{X}^{\underline{\nu}'} \underline{\partial}^{\underline{\mu}'} \Leftrightarrow \varphi^{-1}(\underline{\nu}, \underline{\mu}) \leq \varphi^{-1}(\underline{\nu}, \underline{\mu})$$

for all $\underline{\nu}, \underline{\mu}, \underline{\nu}', \underline{\mu}' \in \mathbb{N}_0^n$ is a well ordering of $\mathbb{E}$.

(b) Show that in the notations of exercise (a) the well ordering $\leq_{\varphi}$ is *discrete*, which means that the set $\{e \in \mathbb{E} \mid e \leq_{\varphi} d\}$ is finite for all $d \in \mathbb{E}$.

(c) Show, that there uncountably many discrete well orderings of $\mathbb{E}$.

(d) Let $n = 1$, set $X_1 =: X$, $\partial_1 =: \partial$ and define the binary relation $\leq$ on the set of elementary differential operators $\mathbb{E} = \{X^{\nu}\partial^{\mu} \mid \nu, \mu \in \mathbb{N}_0\}$ by

$$X^{\nu}\partial^{\mu} \leq X^{\nu'}\partial^{\mu'} \text{ if either } \begin{cases} \nu < \nu' \text{ or else} \\ \nu = \nu' \text{ and } \mu < \mu' \end{cases}$$

for all $\nu, \mu \in \mathbb{N}_0$. Show, that $\leq$ is a non-discrete well ordering of $\mathbb{E}$.

(B) *(Admissible Orderings)* Keep the above notations and hypotheses.

(a) We define the binary relation $\leq_{\mathrm{lex}} \subseteq \mathbb{E} \times \mathbb{E}$ by setting (again for all $\underline{\nu}, \underline{\mu}, \underline{\nu}', \underline{\mu}' \in \mathbb{N}_0^n$):

$$\underline{X}^{\underline{\nu}} \underline{\partial}^{\underline{\mu}} \leq_{\mathrm{lex}} \underline{X}^{\underline{\nu}'} \underline{\partial}^{\underline{\mu}'} \text{ if either}$$

(1) $\underline{\nu} = \underline{\nu}'$ and $\underline{\mu} = \underline{\mu}'$, or
(2) $\underline{\nu} = \underline{\nu}'$ and $\exists j \in \{1, 2, \ldots, n\} : \big[\mu_j < \mu'_j \text{ and } \mu_k = \mu'_k, \forall k < j\big]$, or else
(3) $\exists i \in \{1, 2, \ldots, n\} : \big[\nu_i < \nu'_i \text{ and } \nu_k = \nu'_k, \forall k < i\big]$.

Prove that $\leq_{\text{lex}} \in \text{AO(E)}$. The admissible ordering $\leq_{\text{lex}}$ is called the *lexicographic ordering* of the set of elementary differential operators.

(b) Set $n = 1$, $X_1 =: X$, $\partial_1 =: \partial$ and write down the first 20 elementary differential operators $d \in \mathbb{E} = \{X^{\nu}\partial^{\mu} \mid \nu, \mu \in \mathbb{N}_0\}$ with respect to the ordering $\leq_{\text{lex}}$.

(c) Solve the similar task as in exercise (b), but with $n = 2$ instead of $n = 1$ and with 30 instead of 20.

(d) We define another binary relation $\leq_{\text{deglex}} \subseteq \mathbb{E} \times \mathbb{E}$ by setting

$$d \leq_{\text{deglex}} e \text{ if either } \begin{cases} \deg(d) < \deg(e) \text{ or else} \\ \deg(d) = \deg(e) \text{ and } d \leq_{\text{lex}} e. \end{cases}$$

Show, that $\leq_{\text{deglex}} \in \text{AO}(\mathbb{E})$. This admissible ordering is called the *degree-lexicographic ordering* of the set of elementary differential operators.

(e) Solve the previous exercises (b) and (c) but this time with the ordering $\leq_{\text{deglex}}$.

(f) We introduce a further binary relation $\leq_{\text{degrevlex}} \subseteq \mathbb{E} \times \mathbb{E}$ by setting (again for all $\underline{\nu}, \underline{\mu}, \underline{\nu}', \underline{\mu}' \in \mathbb{N}_0^n$):

$$\underline{X}^{\underline{\nu}}\underline{\partial}^{\underline{\mu}} \leq_{\text{degrevlex}} \underline{X}^{\underline{\nu}'}\underline{\partial}^{\underline{\mu}'} \text{ if either}$$

(1) $\deg(\underline{X}^{\underline{\nu}}\underline{\partial}^{\underline{\mu}}) < \deg(\underline{X}^{\underline{\nu}'}\underline{\partial}^{\underline{\mu}'})$, or else

(2) $\deg(\underline{X}^{\underline{\nu}}\underline{\partial}^{\underline{\mu}}) = \deg(\underline{X}^{\underline{\nu}'}\underline{\partial}^{\underline{\mu}'})$ and either

   (i) $\underline{\nu} = \underline{\nu}'$ and $\underline{\mu} = \underline{\mu}'$, or

   (ii) $\underline{\mu} = \underline{\mu}'$ and $\exists i \in \{1, 2, \ldots, n\} : [\nu_i > \nu_i' \text{ and } \nu_k = \nu_k', \forall k > i]$, or else

   (iii) $\exists j \in \{1, 2, \ldots, n\} : [\mu_j > \mu_j' \text{ and } \mu_k = \mu_k', \forall k > j]$.

Prove, that $\leq_{\text{degrevlex}} \in \text{AO}(\mathbb{E})$. This admissible ordering is called the *degree-reverse-lexicographic ordering* of the set of elementary differential operators.

(g) Solve the previous exercise (e) but with $\leq_{\text{degrevlex}}$ instead of $\leq_{\text{deglex}}$.

(h) An *admissible ordering* of the set $\mathbb{M} = \{\underline{Y}^{\underline{\nu}}\underline{Z}^{\underline{\mu}} \mid \underline{\nu}, \underline{\mu} \in \mathbb{N}_0^n\}$ of all monomials in $\mathbb{P}$ is a total ordering of $\mathbb{M}$ which satisfies the requirements

   (1) *(Foundedness)* $1 \leq m$ for all $m \in \mathbb{M}$.

   (2) *(Compatibility)* For all m, m' and $t \in \mathbb{M}$ we have the implication:

$$\text{If } m \leq m', \text{ then } mt \leq m't.$$

For any $\leq \in \text{AO}(\mathbb{E})$ we define the binary relation $\leq_{\Phi} \subseteq \mathbb{M} \times \mathbb{M}$ by setting

$$m \leq_{\Phi} m' \Leftrightarrow \Phi^{-1}(m) \leq \Phi^{-1}(m') \text{ for all } m, m' \in \mathbb{M}.$$

Prove, that $\leq_\Phi \in \mathrm{AO}(\mathbb{M})$ and that there is indeed a bijection

$$\bullet_\Phi : \mathrm{AO}(\mathbb{E}) \xrightarrow{\cong} \mathrm{AO}(\mathbb{M}), \text{ given by } \leq \mapsto \leq_\Phi .$$

The names given in the previous exercises (a), (d) and (f) to the three admissible orderings of $\mathbb{E}$ introduced in these exercises are "inherited" from the "classical" designations used in polynomial rings, via the above bijection.

(i) Prove, that $\leq_{\mathrm{deglex}}$ and $\leq_{\mathrm{degrevlex}}$ are both discrete in the sense of exercise (A) (b), where as $\leq_{\mathrm{lex}}$ is not.

(C) *(Leading Elementary Differential Operators and Related Concepts)* Keep the previous notations and hypotheses.

(a) Let $n = 1$, set $X_1 =: X$, $\partial_1 =: \partial$, $Y_1 =: Y$ and $Z_1 =: Z$. Write down the leading elementary differential operator, the leading differential operator, the leading coefficient, the leading monomial and the leading term of each of the following differential operators, with respect to each of the admissible orderings $\leq_{\mathrm{lex}}$, $\leq_{\mathrm{deglex}}$ and $\leq_{\mathrm{degrevlex}}$:

(1) $5X^6 + 4X^4\partial - 2X^2\partial^3 + X\partial^4 - 3\partial^6$.
(2) $\partial^4 - 4X\partial^3 + 6X^2\partial^2 - 4X\partial + X^4$.
(3) $\partial^{12} - X^5\partial^7 + X^7\partial^5 - X^9\partial^3 + X^{12}$.

(b) Let $n = 2$ solve the task corresponding to exercise (a) above for the differential operators

(1) $X_1^3X_2^2 + 2\partial_1^3\partial_2^2$.
(2) $X_1^2X_2^3\partial_1^2\partial_2^3 - \partial_1^4\partial_2^6$.
(3) $X_1^k + X_2^k + \partial_1^k + \partial_2^k$ with $k \in \mathbb{N}$.

The next proposition will play a crucial role for our further considerations. it tells us essentially, that "leading differential operators behave as leading terms of polynomials". It is precisely this property, which will allow us to introduce a fertile notion of Gröbner bases for left ideals in Weyl algebras.

Proposition 1.12.4 (Multiplicativity of Leading Terms) *Let* $\leq \in \mathrm{AO}(\mathbb{E})$ *and let* $d, e \in \mathbb{W} \setminus \{0\}$. *Then it holds*

(a) $\mathrm{LT}_\leq(de) = \mathrm{LT}_\leq(d)\mathrm{LT}_\leq(e)$.
(b) $\mathrm{LM}_\leq(de) = \mathrm{LM}_\leq(d)\mathrm{LM}_\leq(e)$.

Proof The product formula for elementary differential operators of Proposition 1.6.2 yields that

$$\mathrm{LE}_\leq\left(\underline{X}^{\underline{\nu}}\underline{\partial}^{\underline{\mu}}\underline{X}^{\underline{\nu}'}\underline{\partial}^{\underline{\mu}'}\right) = \underline{X}^{\underline{\nu}+\underline{\nu}'}\underline{\partial}^{\underline{\mu}'+\underline{\mu}'} \text{ for all } \underline{\nu}, \underline{\nu}', \underline{\mu}, \underline{\mu}' \in \mathbb{N}_0^n.$$

We may write

$$d = \sum_{(\underline{\nu},\underline{\mu})\in\text{supp}(d)} c^{(d)}_{\underline{\nu}\underline{\mu}} \underline{X}^{\underline{\nu}} \underline{\partial}^{\underline{\mu}} \text{ and } e = \sum_{(\underline{\nu}',\underline{\mu}')\in\text{supp}(e)} c^{(e)}_{\underline{\nu}'\underline{\mu}'} \underline{X}^{\underline{\nu}'} \underline{\partial}^{\underline{\mu}'}$$

with $c^{(d)}_{\underline{\nu}\underline{\mu}}, c^{(e)}_{\underline{\nu}'\underline{\mu}'} \in K \setminus \{0\}$ for all $(\underline{\nu}, \underline{\mu}) \in \text{supp}(d)$ and all $(\underline{\nu}', \underline{\mu}') \in \text{supp}(e)$. With appropriate pairs $(\underline{\nu}^{(0)}, \underline{\mu}^{(0)}) \in \text{supp}(d)$ and $(\underline{\nu}'^{(0)}, \underline{\mu}'^{(0)}) \in \text{supp}(e)$ we also may write

$$\text{LE}_{\leq}(d) = \underline{X}^{\underline{\nu}^{(0)}} \underline{\partial}^{\underline{\mu}^{(0)}} \text{ and } \text{LE}_{\leq}(e) = \underline{X}^{\underline{\nu}'^{(0)}} \underline{\partial}^{\underline{\mu}'^{(0)}}, \text{ hence also}$$

$$\text{LC}_{\leq}(d) = c^{(d)}_{\underline{\nu}^{(0)}\underline{\mu}^{(0)}} \text{ and } \text{LC}_{\leq}(e) = c^{(e)}_{\underline{\nu}'^{(0)}\underline{\mu}'^{(0)}}.$$

Now, bearing in mind the previous observation on leading elementary differential operators we may write

$$\begin{aligned} de &= \sum_{(\underline{\nu},\underline{\mu})\in\text{supp}(d),(\underline{\nu}',\underline{\mu}')\in\text{supp}(e)} c^{(d)}_{\underline{\nu}\underline{\mu}} \underline{X}^{\underline{\nu}} \underline{\partial}^{\underline{\mu}} c^{(e)}_{\underline{\nu}'\underline{\mu}'} \underline{X}^{\underline{\nu}'} \underline{\partial}^{\underline{\mu}'} \\ &= \sum_{(\underline{\nu},\underline{\mu})\in\text{supp}(d),(\underline{\nu}',\underline{\mu}')\in\text{supp}(e)} c^{(d)}_{\underline{\nu}\underline{\mu}} c^{(e)}_{\underline{\nu}'\underline{\mu}'} \underline{X}^{\underline{\nu}} \underline{\partial}^{\underline{\mu}} \underline{X}^{\underline{\nu}'} \underline{\partial}^{\underline{\mu}'} \\ &= \sum_{(\underline{\nu},\underline{\mu})\in\text{supp}(d),(\underline{\nu}',\underline{\mu}')\in\text{supp}(e)} \left[c^{(d)}_{\underline{\nu}\underline{\mu}} c^{(e)}_{\underline{\nu}'\underline{\mu}'} \underline{X}^{\underline{\nu}+\underline{\nu}'} \underline{\partial}^{\underline{\mu}+\underline{\mu}'} + r_{\underline{\nu}\underline{\nu}'\underline{\mu}\underline{\mu}'}\right], \end{aligned}$$

with $r_{\underline{\nu}\underline{\nu}'\underline{\mu}\underline{\mu}'} \in \mathbb{W}$, such that for all $(\underline{\nu}, \underline{\mu}) \in \text{supp}(d)$ and all $(\underline{\nu}', \underline{\mu}') \in \text{supp}(e)$ it holds

$$\text{LE}_{\leq}(r_{\underline{\nu}\underline{\nu}'\underline{\mu}\underline{\mu}'}) < \underline{X}^{\underline{\nu}+\underline{\nu}'} \underline{\partial}^{\underline{\mu}+\underline{\mu}'}, \text{ whenever } r_{\underline{\nu}\underline{\nu}'\underline{\mu}\underline{\mu}'} \neq 0.$$

By Definition, Reminder and Exercise 1.12.2 (C)(c) we have

$$\underline{X}^{\underline{\nu}+\underline{\nu}'} \underline{\partial}^{\underline{\mu}+\underline{\mu}'} < \underline{X}^{\underline{\nu}^{(0)}+\underline{\nu}'^{(0)}} \underline{\partial}^{\underline{\mu}^{(0)}+\underline{\mu}'^{(0)}}, \text{ for all}$$

$$\big((\underline{\nu}, \underline{\mu}), (\underline{\nu}', \underline{\mu}')\big) \in \text{supp}(d) \times \text{supp}(e) \setminus \{((\underline{\nu}^{(0)}, \underline{\mu}^{(0)}), (\underline{\nu}'^{(0)}, \underline{\mu}'^{(0)}))\}.$$

By Definition, Reminder and Exercise 1.12.2 (D)(d) it now follows easily that

$$\text{LE}_{\leq}(de) = \underline{X}^{\underline{\nu}^{(0)}+\underline{\nu}'^{(0)}} \underline{\partial}^{\underline{\mu}^{(0)}+\underline{\mu}'^{(0)}} \text{ and}$$

$$\text{LC}_{\leq}(de) = c^{(d)}_{\underline{\nu}^{(0)}\underline{\mu}^{(0)}} c^{(e)}_{\underline{\nu}'^{(0)}\underline{\mu}'^{(0)}} = \text{LC}_{\leq}(d)\text{LC}_{\leq}(e).$$

We thus obtain

$$\mathrm{LM}_{\leq}(de) = \Phi\big(\underline{X}^{\underline{\nu}^{(0)}+\underline{\nu}'^{(0)}}\underline{\partial}^{\underline{\mu}^{(0)}+\underline{\mu}'^{(0)}}\big) = \underline{Y}^{\underline{\nu}^{(0)}+\underline{\nu}'^{(0)}}\underline{Z}^{\underline{\mu}^{(0)}+\underline{\mu}'^{(0)}} = \underline{Y}^{\underline{\nu}^{(0)}}\underline{Z}^{\underline{\mu}^{(0)}}\underline{Y}^{\underline{\nu}'^{(0)}}\underline{Z}^{\underline{\mu}'^{(0)}}$$
$$= \Phi\big(\underline{X}^{\underline{\nu}^{(0)}}\underline{\partial}^{\underline{\mu}^{(0)}}\big)\Phi\big(\underline{X}^{\underline{\nu}'^{(0)}}\underline{\partial}^{\underline{\mu}'^{(0)}}\big) = \Phi\big(\mathrm{LE}_{\leq}(d)\big)\Phi\big(\mathrm{LE}_{\leq}(e)\big) = \mathrm{LM}_{\leq}(d)\mathrm{LM}_{\leq}(e).$$

But now it follows

$$\mathrm{LT}_{\leq}(de) = \mathrm{LC}_{\leq}(de)\mathrm{LM}_{\leq}(de) = \mathrm{LC}_{\leq}(d)\mathrm{LC}_{\leq}(e)\mathrm{LM}_{\leq}(d)\mathrm{LM}_{\leq}(e)$$
$$= \mathrm{LC}_{\leq}(d)\mathrm{LM}_{\leq}(d)\mathrm{LC}_{\leq}(e)\mathrm{LM}_{\leq}(e) = \mathrm{LT}_{\leq}(d)\mathrm{LT}_{\leq}(e).$$

The next result may be understood as an extension of the classical division algorithms of Euclid for univariate polynomials to the case of differential operators. It was first proved in 1984 by Briançon-Maisonobe in the univariate case and by Castro-Jiménez in the multivariate case.
Those readers, who are familiar with the Buchberger algorithm in multivariate polynomial rings will realize that our result corresponds to the division algorithm in multi-variate polynomial rings. Observe in particular that—as in the case of multi-variate polynomials—we will divide "by a family of denominators" and that the presented division procedure depends on an admissible ordering.

Proposition 1.12.5 (The Division Property, Briançon-Maisonobe [14] and Castro-Jiménez [21]) *Let $\leq \in \mathrm{AO}(\mathbb{E})$, let $d \in \mathbb{W}$ and let $F \subset \mathbb{W}$ be a finite set. Then, there is an element $r \in \mathbb{W}$ and a family $(q_f)_{f\in F} \in \mathbb{W}^F$ such that (in the notations of Convention 1.12.1 (B) and Definition, Reminder and Exercise 1.12.2 (D))*

(a) $d = \sum_{f\in F} q_f f + r$;
(b) $\Phi(s) \notin \mathbb{P}\mathrm{LM}_{\leq}(f)$ *for all* $f \in F \setminus \{0\}$ *and all* $s \in \mathrm{Supp}(r)$.
(c) $\mathrm{LE}_{\leq}(q_f f) \leq \mathrm{LE}_{\leq}(d)$ *for all* $f \in F$ *with* $q_f f \neq 0$.

Proof We clearly may assume that $F \subset \mathbb{W} \setminus \{0\}$. If $d = 0$, we choose $r = 0$ and $q_f = 0$ for all $f \in F$. Assume, that our claim is wrong, and let $U \subsetneq \mathbb{W}$ be the non-empty set of all differential operators $d \in \mathbb{W}$ which do not admit a presentation of the requested form. As $\leq \in \mathrm{WO}(\mathbb{E})$ and $U \subset \mathbb{W} \setminus \{0\}$, we find some $d \in U$ such that

$$\mathrm{LE}_{\leq}(d) = \min_{\leq}\{\mathrm{LE}_{\leq}(u) \mid u \in U\}.$$

We distinguish the following two cases:

(1) There is some $f \in F$ such that $\mathrm{LM}_{\leq}(d) \in \mathbb{P}\mathrm{LM}_{\leq}(f)$.
(2) $\mathrm{LM}_{\leq}(d) \notin \bigcup_{f\in F} \mathbb{P}\mathrm{LM}_{\leq}(f)$.

In the case (1) we find some $e \in \mathbb{E}$ such that $\mathrm{LM}_{\leq}(d) = \Phi(e)\mathrm{LM}_{\leq}(f)$ and so we can introduce the element

$$d' := d - \frac{\mathrm{LC}_{\leq}(d)}{\mathrm{LC}_{\leq}(f)} ef \in \mathbb{W}.$$

If $d' = 0$, we set

$$r = 0, \quad q_f := \frac{\mathrm{LC}_{\leq}(d)}{\mathrm{LC}_{\leq}(f)} e, \text{ and } q_{f'} = 0 \text{ for all } f' \in F \setminus \{f\}.$$

But then

$$d = \frac{\mathrm{LC}_{\leq}(d)}{\mathrm{LC}_{\leq}(f)} ef = q_f f + r$$

is a presentation of d with the requested properties.
So, let $d' \neq 0$. Observe, that by Proposition 1.12.4 (a) we can write

$$\mathrm{LT}_{\leq}\big(\frac{\mathrm{LC}_{\leq}(d)}{\mathrm{LC}_{\leq}(f)} ef\big) = \frac{\mathrm{LC}_{\leq}(d)}{\mathrm{LC}_{\leq}(f)} \mathrm{LT}_{\leq}(ef) = \frac{\mathrm{LC}_{\leq}(d)}{\mathrm{LC}_{\leq}(f)} \mathrm{LT}_{\leq}(e)\mathrm{LT}_{\leq}(f) =$$

$$\mathrm{LC}_{\leq}(d)\mathrm{LM}_{\leq}(e)\mathrm{LM}_{\leq}(f) = \mathrm{LC}_{\leq}(d)\Phi(e)\mathrm{LM}_{\leq}(f) = \mathrm{LC}_{\leq}(d)\mathrm{LM}_{\leq}(d) = \mathrm{LT}_{\leq}(d).$$

If follows that $\mathrm{LD}_{\leq}\big(\frac{\mathrm{LC}_{\leq}(d)}{\mathrm{LC}_{\leq}(f)} ef\big) = \mathrm{LD}_{\leq}(d)$, and hence by Definition, Reminder and Exercise 1.12.2 (D)(d) we obtain that

$$\mathrm{LE}_{\leq}(d') < \mathrm{LE}_{\leq}(d) = \min_{\leq}\{\mathrm{LE}_{\leq}(u) \mid u \in U\}.$$

Therefore, $d' \notin U$ and so we find an element $r' \in \mathbb{W}$ and a family $(q'_{f'})_{f' \in F} \in \mathbb{W}^F$ such that

(a)′ $d' = \sum_{f' \in F} q'_{f'} f' + r'$;
(b)′ $\Phi(s') \notin \mathbb{P}\mathrm{LM}_{\leq}(f')$ for all $f' \in F$ and all $s' \in \mathrm{Supp}(r')$.
(c)′ $\mathrm{LE}_{\leq}(q'_{f'} f') \leq \mathrm{LE}_{\leq}(d')$ for all $f' \in F$ with $q'_{f'} \neq 0$.

Now, we set

$$r := r' \text{ and } q_f := \begin{cases} q'_{f'} & \text{if } f' \neq f, \\ q'_f + \frac{\mathrm{LC}_{\leq}(d)}{\mathrm{LC}_{\leq}(f)} e & \text{if } f = f'. \end{cases}$$

As

$$\mathrm{LE}_{\leq}(q'_{f'} f') \leq \mathrm{LE}_{\leq}(d') < \mathrm{LE}_{\leq}(d) \text{ and } \mathrm{LE}_{\leq}\big(\frac{\mathrm{LC}_{\leq}(d)}{\mathrm{LC}_{\leq}(f)} e\big) = \mathrm{LE}_{\leq}(e) \leq \mathrm{LE}_{\leq}(d),$$

we get

$$\mathrm{LE}_{\leq}(q_f F) = \mathrm{LE}_{\leq}\big((q'_f + \frac{\mathrm{LC}_{\leq}(d)}{\mathrm{LC}_{\leq}(f)} e) f\big) \leq \mathrm{LE}_{\leq}(d).$$

Now, it follows easily, that the requirements (a),(b) and (c) of our proposition are satisfied in the case (1).

So, let us assume that we are in the case (2). We set

$$d' := d - \mathrm{LD}_{\leq}(d).$$

If $d' = 0$ we have $d' = \mathrm{LD}_{\leq}(d)$ and it suffices to choose $q_f := 0$ for all $f \in F$ and $r = d$.

So, let $d' \neq 0$. Then, we have $\mathrm{LE}_{\leq}(d') < \mathrm{LE}_{\leq}(d)$ (see Definition, Reminder and Exercise 1.12.2 (D)(c)), so that again $d' \notin U$. But this means once more, that we get elements r' and $q'_{f'} \in \mathbb{W}$ (for all $f' \in F$) such that the above conditions (a)$'$, (b)$'$ and (c)$'$ are satisfied. Now, we set

$$r := r' + \mathrm{LD}_{\leq}(d) \text{ and } q_f := q'_f \text{ for all } f \in F.$$

As $\mathrm{supp}(r) \subseteq \mathrm{supp}(r') \cup \{\mathrm{LE}_{\leq}(d)\}$ and $\mathrm{LE}_{\leq}(q_f f) \leq \mathrm{LE}(d') \leq \mathrm{LE}_{\leq}(d)$ for all $f \in F$ with $q_f \neq 0$ the requirements (a),(b) and (c) are again satisfied for the suggested choice.

Now, we are ready to introduce the basic notion of this section: the concept of Gröbner basis.

Definition, Reminder and Exercise 1.12.6 (A) *(Monomial Ideals)* An ideal $I \subseteq \mathbb{P}$ is called a *monomial ideal* if there is a set $S \subset \mathbb{M} = \{\underline{Y}^{\underline{\nu}}\underline{Z}^{\underline{\mu}} \mid \underline{\nu}, \underline{\mu} \in \mathbb{N}_0^n\}$ such that

$$I = \sum_{s \in S} \mathbb{P}s.$$

Show that in this situation for all $m \in \mathbb{M} \setminus \{0\}$ we have

(a) If $m = \sum_{i=1}^{t} f_i s_i$ with $s_1, s_2, \ldots, s_t \in S$ and $f_1, f_2, \ldots, f_t \in \mathbb{P}$, then there is some $i \in \{1, 2, \ldots, t\}$ and some $n_i \in \mathrm{supp}(f_i)$ such that $m = n_i s_i$.
(b) $m \in I$ if and only if there are $n \in \mathbb{M}$ and some $s \in S$ such that $m = ns$.

(B) *(Leading Monomial Ideals)* Let $\leq \in \mathrm{AO}(\mathbb{E})$ and $T \subset \mathbb{W}$. Then, the ideal

$$\mathrm{LMI}_{\leq}(T) := \sum_{d \in T \setminus \{0\}} \mathbb{P}\mathrm{LM}_{\leq}(d)$$

is called the *leading monomial ideal of* T with respect to $\leq$.
Prove that for all $m \in \mathbb{M}$, we have the following statements.

(a) If $m = \sum_{i=1}^{s} f_i \mathrm{LM}_{\leq}(t_i)$ with $t_1, t_2, \ldots, t_s \in T$ and $f_1, f_2, \ldots, f_s \in \mathbb{P}$, then there is some $i \in \{1, 2, \ldots, s\}$ and some $n_i \in \mathrm{supp}(f_i)$ such that $t_i \neq 0$ and $m = n_i \mathrm{LM}_{\leq}(t_i)$.
(b) $m \in \mathrm{LMI}_{\leq}(T)$ if and only if there are elements $u \in \mathbb{E}$ and $t \in T$ such that $m = \mathrm{LM}_{\leq}(u)\mathrm{LM}_{\leq}(t)$.

(C) *(Gröbner Bases)* Let $\leq\in \mathrm{AO}(\mathbb{E})$ and let $L \subseteq \mathbb{W}$ be a left ideal. A *Gröbner basis* of L with respect to $\leq$ (or a $\leq$*-Gröbner basis* of L) is a subset $G \subseteq L$ such that

$$\#G < \infty \text{ and } \mathrm{LMI}_{\leq}(L) = \mathrm{LMI}_{\leq}(G).$$

Prove the following facts:

(a) If G is a $\leq$-Gröbner basis of L and $G \subseteq H \subseteq L$ with $\#H < \infty$, then H is a $\leq$-Gröbner basis of L.
(b) If G is a $\leq$-Gröbner basis of L, then for each $d \in L \setminus \{0\}$ there is some $u \in \mathbb{E}$ and some $g \in G \setminus \{0\}$ such that

$$\mathrm{LM}_{\leq}(d) = \mathrm{LM}_{\leq}(u)\mathrm{LM}_{\leq}(g) = \mathrm{LM}_{\leq}(ug).$$

(c) If G is a $\leq$-Gröbner basis of L, then for each $d \in L\setminus\{0\}$ there is some monomial $m = \underline{Y}^{\underline{\nu}}\underline{Z}^{\underline{\mu}} \in \mathbb{P}$ and some $g \in G \setminus \{0\}$ such that

$$\mathrm{LM}_{\leq}(d) = m\mathrm{LM}_{\leq}(g).$$

Now, we prove that Gröbner bases always exist, and that they deserve the name of "basis", as they generate the involved left ideal. Clearly, these statements correspond precisely to well known facts in multi-variate polynomial rings. After having established the announced existence and generating property of Gröbner bases, we shall add a few examples and exercises on the subject.

Proposition 1.12.7 (Existence and Generating Property of Gröbner Bases) *Let* $\leq\in \mathrm{AO(E)}$ *and let* $L \subseteq \mathbb{W}$ *be a left ideal. Then the following statements hold.*

(a) *L admits a $\leq$-Gröbner basis.*
(b) *If G is any $\leq$-Gröbner basis of L, then $L = \sum_{g\in G} \mathbb{W}g$.*

Proof

(a): This is clear as the ideal $\mathrm{LMI}_{\leq}(L)$ is generated by finitely many elements of the form $\mathrm{LM}_{\leq}(g)$ with $g \in L$.
(b): Let $G \subseteq L$ be a $\leq$-Gröbner basis of L and assume that $\sum_{g\in G} \mathbb{W}g \subsetneq L$. As $\leq\in \mathrm{WO}(\mathbb{E})$, we find some $e \in L \setminus \sum_{g\in G} \mathbb{W}g$ such that

$$\mathrm{LE}(e) = \min_{\leq}\{\mathrm{LE}_{\leq}(d) \mid d \in L \setminus \sum_{g\in G} \mathbb{W}g\}.$$

By Definition, Reminder and Exercise 1.12.6 (C)(b) we find some $u \in \mathbb{E}$ and some $g \in G$ such that

$$\mathrm{LM}_{\leq}(e) = \mathrm{LM}_{\leq}(u)\mathrm{LM}_{\leq}(g).$$

Setting

$$v := -\frac{\mathrm{LC}_{\leq}(e)}{\mathrm{LC}_{\leq}(g)} u$$

we now get on use of Proposition 1.12.4 (a) that

$$\begin{aligned}\mathrm{LT}_{\leq}(e) &= \mathrm{LC}_{\leq}(e)\mathrm{LM}_{\leq}(e) = \mathrm{LC}_{\leq}(e)\mathrm{LM}_{\leq}(u)\mathrm{LM}_{\leq}(g)\\ &= \mathrm{LC}_{\leq}(e)\mathrm{LT}_{\leq}(u)\frac{1}{\mathrm{LC}_{\leq}(g)}\mathrm{LT}_{\leq}(g) = \frac{\mathrm{LC}_{\leq}(e)}{\mathrm{LC}_{\leq}(g)}\mathrm{LT}_{\leq}(u)\mathrm{LT}_{\leq}(g)\\ &= -\mathrm{LT}_{\leq}(v)\mathrm{LT}_{\leq}(g) = -\mathrm{LT}_{\leq}(vg).\end{aligned}$$

As $e \notin \sum_{g\in G} \mathbb{W}g$ and $g \in G$, we have

$$e + vg \in L \setminus \sum_{g\in G} \mathbb{W}g.$$

In particular $e + vg \neq 0$. So by Definition, Reminder and Exercise 1.12.2 (D)(d) it follows that

$$\mathrm{LE}_{\leq}(e + vg) < \mathrm{LE}_{\leq}(e) = \min_{\leq}\{\mathrm{LE}_{\leq}(d) \mid d \in L \setminus \sum_{g\in G} \mathbb{W}g\}.$$

But this is a contradiction.

Now, we add the previously announced examples and exercises.

Examples and Exercises 1.12.8 (A) *(Leading Monomial Ideals)* Keep the above notations and hypotheses. Prove the following statements:

(a) Let $d \in \mathbb{W}\setminus\{0\}$ and $\leq \in \mathrm{AO}(\mathbb{E})$. Prove that $\mathrm{LMI}_{\leq}(\mathbb{W}d)$ is a principal ideal.
(b) Let $n = 1$, $X_1 =: X$ and $\partial_1 =: \partial$. Set $L := \mathbb{W}(X^2 - \partial) + \mathbb{W}(X\partial)$ and determine $\mathrm{LMI}_{\leq}(L)$ for $\leq := \leq_{\mathrm{lex}}$, $\leq_{\mathrm{deglex}}$ and $\leq := \leq_{\mathrm{degrevlex}}$.

(B) *(Gröbner Bases)* Keep the above notations and hypotheses. Prove the following statements:

(a) Let the notations be as in exercise (a) of part (A) and prove that $\{cd\}$ is a $\leq$-Gröbner basis of $\mathbb{W}d$ for all $c \in K \setminus \{0\}$, and that any singleton $\leq$-Gröbner bases of $\mathbb{W}d$ is of the above form.
(b) Let the notations and hypotheses be as in exercise (b) of part (A) and compute a $\leq$-Gröbner basis for $\leq := \leq_{\mathrm{lex}}$, $\leq_{\mathrm{deglex}}$ and $\leq := \leq_{\mathrm{degrevlex}}$

We now head for another basic result on Gröbner bases, which says that these bases enjoy a certain restriction property. This will be an important ingredient in our treatment of Universal Gröbner bases. We begin with the following preparations.

Notation 1.12.9 (A) For any set $S \subseteq \mathbb{W}$ we write (see also Definition, Reminder and Exercise 1.12.2 (D)):

$$\mathrm{supp}(S) := \bigcup_{s \in S} \mathrm{supp}(s) \text{ and } \mathrm{Supp}(S) := \bigcup_{s \in S} \mathrm{Supp}(s).$$

(B) Let $\leq \in \mathrm{TO}(\mathbb{E})$ (see Definition, Reminder and Exercise 1.12.2 (A)) and let $T \subset \mathbb{E}$. We write $\leq\restriction_T$ for the *restriction* of $\leq$ to T, thus—if we interpret binary relations on a set S as subsets of $S \times S$:

$$\leq\restriction_T \quad := \quad \leq \cap (T \times T), \text{ so that } : d \leq\restriction_T e \Leftrightarrow d \leq e \text{ for all } d, e \in T.$$

Proposition 1.12.10 (The Restriction Property of Gröbner Bases) *Let $L \subseteq \mathbb{W}$ be a left ideal. Let $\leq, \leq' \in \mathrm{AO}(\mathbb{E})$ and let G be a $\leq$-Gröbner basis of L. Assume that*

$$\leq\restriction_{\mathrm{Supp}(G)} \quad = \quad \leq'\restriction_{\mathrm{Supp}(G)} .$$

Then G is also a $\leq'$-Gröbner basis of L.

Proof Let $d \in L \setminus \{0\}$. We have to show that $\mathrm{LM}_{\leq'}(d) \in \mathrm{LMI}_{\leq'}(G)$. We may assume that $0 \notin G$. If we apply Proposition 1.12.5 to the ordering $\leq'$, we find an element r and a family $(q_g)_{g \in G} \in \mathbb{W}^G$ such that

(1) $d = \sum_{g \in G} q_g g + r$;
(2) $\Phi(s) \notin \mathbb{P}\mathrm{LM}_{\leq'}(g)$ for all $g \in G$ and all $s \in \mathrm{Supp}(r)$.
(3) $\mathrm{LE}_{\leq'}(q_g g) \leq' \mathrm{LE}_{\leq'}(d)$ for all $g \in G$ with $q_g \neq 0$.

Our immediate aim is to show that $r = 0$. Assume to the contrary that $r \neq 0$. As $r \in L$ and G is a $\leq$-Gröbner basis of L, we get $\mathrm{LM}_{\leq}(r) \in \mathrm{LMI}_{\leq}(G)$. So, there is some $g \in G$ such that $\mathrm{LM}_{\leq}(r) = m\mathrm{LM}_{\leq}(g)$ for some $m \in \mathbb{M}$ (see Definition, Reminder and Exercise 1.12.6 (C)(c)). As $\leq\restriction_{\mathrm{Supp}(G)} \; = \; \leq'\restriction_{\mathrm{Supp}(G)}$ it follows that

$$\Phi\big(\mathrm{LT}_{\leq}(r)\big) = \mathrm{LM}_{\leq}(r) \in \mathbb{P}\mathrm{LM}_{\leq'}(g).$$

As $\mathrm{LT}_{\leq}(r) \in \mathrm{Supp}(r)$, this contradicts the above condition (2). Therefore $r = 0$. But now, we may write

$$d = \sum_{g \in G^*} q_g g, \text{ whith } G^* := \{g \in G \mid q_g \neq 0\}.$$

By the above condition (3) we have $\mathrm{LE}_{\leq'}(q_g g) \leq' \mathrm{LE}_{\leq'}(d)$ for all $g \in G^*$. So, there is some $g \in G^*$ such that $\mathrm{LE}_{\leq'}(d) = \mathrm{LE}_{\leq'}(q_g g)$ (see Definition, Reminder and Exercise 1.12.2 (D)(d)), and hence $\mathrm{LM}_{\leq'}(d) = \mathrm{LM}_{\leq'}(q_g g)$. Thus, on use of Proposition 1.12.4 (b) we get indeed

$$\mathrm{LM}_{\leq'}(d) = \mathrm{LM}_{\leq'}(q_g)\mathrm{LM}_{\leq'}(g) \in \mathrm{LMI}_{\leq'}(G).$$

Now, we shall introduce the central concept of this section.

Definition 1.12.11 (Universal Gröbner Bases) Let $L \subseteq \mathbb{W}$ be a left ideal. A *universal Gröbner basis* of L is a (finite) subset $G \subset \mathbb{W}$ which is a $\leq$-Gröbner basis for all $\leq \in \mathrm{AO}(\mathbb{E})$.

Universal Gröbner bases have been studied by Sturmfels [41] in the polynomial ring $K[X_1, X_2, \ldots, X_n]$—and indeed this notion can be immediately extended to the Weyl algebra $\mathbb{W}$. Gröbner bases for left ideals in the Weyl algebra were introduced by Assi, Castro-Jiménez and Granger [3] and also by Saito et al. [38].

Clearly, our next aim should be to show, that universal Gröbner bases always exist. There are indeed various possible ways to prove this. Here, we shall do this by a topological approach which relies on an idea of Sikora [40], and which can be found in greater generality in Boldini's thesis [11]. We approach the subject by first introducing a natural metric on the set of total orderings of all elementary differential operators. Then, we make the reader prove in a series of exercises, that we get a complete metric space in this way.

Definition, Exercise and Convention 1.12.12 (A) (*The Natural Metric on the Set* $\mathrm{TO}(\mathbb{E})$) For all $i \in \mathbb{Z}$ we introduce the notation

$$\mathbb{E}_i := \{e \in \mathbb{E} \mid \deg(e) \leq i\} = \{\underline{X}^{\underline{\nu}}\underline{\partial}^{\underline{\mu}} \mid |\underline{\nu}| + |\underline{\mu}| \leq i\}.$$

We define a map

$$\mathrm{dist} : \mathrm{TO}(\mathbb{E}) \times \mathrm{TO}(\mathbb{E}) \longrightarrow \mathbb{R}, \text{ given by for all } \leq, \leq' \in \mathrm{TO}(\mathbb{E}) \text{ by}$$

$$\mathrm{dist}(\leq, \leq') := \begin{cases} 2^{-\sup\{r \in \mathbb{N}_0 \mid \leq\restriction_{\mathbb{E}_r} = \leq'\restriction_{\mathbb{E}_r}\}}, & \text{if } \leq \neq \leq', \\ 0, & \text{if } \leq = \leq'. \end{cases}$$

Prove that

(a) For all $\leq, \leq' \in \mathrm{TO}(\mathbb{E})$ and all $r \in \mathbb{N}_0$ we have

$$\mathrm{dist}(\leq, \leq') < \frac{1}{2^r} \text{ if and only if } \leq\restriction_{\mathbb{E}_{r+1}} = \leq'\restriction_{\mathbb{E}_{r+1}}.$$

(b) The map $\mathrm{dist} : \mathrm{TO}(\mathbb{E}) \times \mathrm{TO}(\mathbb{E}) \longrightarrow \mathbb{R}$ is a *metric* on $\mathrm{TO}(\mathbb{E})$.

From now on, we always endow $\mathrm{TO}(\mathbb{E})$ with this metric and the induced *Hausdorff topology*.

(B) (*Completeness of the Metric Space* $\mathrm{TO}(\mathbb{E})$) Let $(\leq_i)_{i \in \mathbb{N}_0}$ be a *Cauchy sequence* in $\mathrm{TO}(\mathbb{E})$. This means:

$$\text{For all } r \in \mathbb{N}_0 \text{ there is some } n(r) \in \mathbb{N}_0 \text{ such that } \mathrm{dist}(\leq_i, \leq_j) < \frac{1}{2^r} \text{ for all } i, j \geq n(r).$$

We introduce the binary relation $\leq \subseteq \mathbb{E} \times \mathbb{E}$ given for all $d, e \in \mathbb{E}$ by

$$d \leq e \text{ if and only if } d \leq_i e \text{ for all } i \gg 0.$$

Prove the following statements:

(a) If $r \in \mathbb{N}_0$, $d, e \in \mathbb{E}_{r+1}$, and $i, j \geq n(r)$, then $d \leq_i e$ if and only if $d \leq_j e$.
(b) If $r \in \mathbb{N}_0$, $d, e \in \mathbb{E}_{r+1}$, and $i \geq n(r)$, then $d \leq_i e$ if and only if $d \leq e$.
(c) $\leq \in \mathrm{TO}(\mathbb{E})$.
(d) If $r \in \mathbb{N}_0$, and $i \geq n(r)$, then $\mathrm{dist}(\leq_i, \leq) \leq \frac{1}{2^r}$.
(e) $\lim_{i \to \infty} \leq_i = \leq$.
(f) $\mathrm{TO}(\mathbb{E})$ is a *complete* metric space.

Now, we are ready to prove the basic ingredient of our existence proof for universal Gröbner bases.

Proposition 1.12.13 (Compactness of the Space of Total Orderings) *The space* $\mathrm{TO}(\mathbb{E})$ *is compact.*

Proof Let $(\leq_i)_{i \in \mathbb{N}_0}$ be a sequence in $\mathrm{TO}(\mathbb{E})$. It suffices to show, that $(\leq_i)_{i \in \mathbb{N}_0}$ has a convergent subsequence. Bearing in mind Definition, Exercise and Convention 1.12.12 (B)(f) (or (e)), it suffices to find a subsequence of $(\leq_i)_{i \in \mathbb{N}_0}$ which is a Cauchy sequence. Observe that all the sets $\mathbb{E}_r$ are finite. We want to construct a sequence $(\mathbb{S}_r)_{r \in \mathbb{N}_0}$ of infinite subsets $\mathbb{S}_r \subseteq \mathbb{N}_0$ such that for all $s \in \mathbb{N}_0$ we have

(1) $\mathbb{S}_{s+1} \subseteq \mathbb{S}_s$.
(2) $\leq_j \restriction_{\mathbb{E}_{s+1}} \; = \; \leq_k \restriction_{\mathbb{E}_{s+1}}$ for all $j, k \in \mathbb{S}_s$.

We construct the members $\mathbb{S}_r$ of the sequence $(\mathbb{S}_r)_{r \in \mathbb{N}_0}$ by induction r. As $\mathbb{E}_1$ is finite, we can find an infinite set $\mathbb{S}_0 \subseteq \mathbb{N}_0$ such that requirement (2) is satisfied with $s = 0$. Now, let $r > 0$ and assume that the sets $\mathbb{S}_0, \mathbb{S}_1, \ldots, \mathbb{S}_r$ are already defined such that requirement (1) holds for all $s < r$ and requirement (2) holds for all $s \leq r$. As $\mathbb{E}_{r+2}$ is finite, we find an infinite subset $\mathbb{S}_{r+1} \subseteq \mathbb{S}_r$ (which hence satisfies requirement (1) for $s = r$) such that requirement (2) is also satisfied with $s = r + 1$. This completes the step of induction and hence proves that a sequence $(\mathbb{S}_r)_{r \in \mathbb{N}_0}$ with the requested properties exists.
Now, we may choose a sequence $(i_k)_{k \in \mathbb{N}_0}$ in $\mathbb{N}_0$, such that

$$i_r < i_{r+1} \text{ and } i_r \in \mathbb{S}_r \text{ for all } r \in \mathbb{N}_0.$$

In particular it follows that

$$\leq_{i_j} \restriction_{\mathbb{E}_{r+1}} \; = \; \leq_{i_k} \restriction_{\mathbb{E}_{r+1}} \quad \text{for all } j, k \geq r$$

and hence (see Definition, Exercise and Convention 1.12.12 (A)(a))

$$\mathrm{dist}(\leq_{i_j}, \leq_{i_k}) < \frac{1}{2^r} \text{ for all } j, k \geq r.$$

So, the constructed subsequence $(\leq_{i_k})_{k\in\mathbb{N}_0}$ of our original sequence $(\leq_i)_{i\in\mathbb{N}_0}$ is indeed a Cauchy sequence.

What we need indeed to prove our main result, is the compactness of subspace of admissible orderings in the topological space of total orderings.

Proposition 1.12.14 (Compactness of the Space of Admissible Orderings) *The set* $\mathrm{AO}(\mathbb{E})$ *is a closed subset of* $\mathrm{TO}(\mathbb{E})$ *and hence compact.*

Proof Let $(\leq_i)_{i\in\mathbb{N}_0}$ be sequence in $\mathrm{AO}(\mathbb{E})$, which converges in $\mathrm{TO}(\mathbb{E})$ and let

$$\lim_{i\to\infty} \leq_i \quad = \quad \leq .$$

We aim to show, that $\leq \in \mathrm{AO}(\mathbb{E})$. According to Definition, Reminder and Exercise 1.12.2 (C), we must show, that for all $\underline{\lambda}, \underline{\lambda}', \underline{\kappa}, \underline{\kappa}', \underline{\nu}, \underline{\mu} \in \mathbb{N}_0^n$ the following statements hold.

(1) $1 \leq \underline{X}^{\underline{\nu}}\underline{\partial}^{\underline{\mu}}$.
(2) If $\underline{X}^{\underline{\lambda}}\underline{\partial}^{\underline{\kappa}} \leq \underline{X}^{\underline{\lambda}'}\underline{\partial}^{\underline{\kappa}'}$ then $\underline{X}^{\underline{\lambda}+\underline{\nu}}\underline{\partial}^{\underline{\kappa}+\underline{\mu}} \leq \underline{X}^{\underline{\lambda}'+\underline{\nu}}\underline{\partial}^{\underline{\kappa}'+\underline{\mu}}$.

So, fix $\underline{\lambda}, \underline{\lambda}', \underline{\kappa}, \underline{\kappa}', \underline{\nu}, \underline{\mu} \in \mathbb{N}_0^n$. Then we find some $r \in \mathbb{N}_0$ such that all the elementary differential operators which occur in (1) and (2) belong to $\mathbb{E}_{r+1}$. Now, we find some $i \in \mathbb{N}_0$ such that $\mathrm{dist}(\leq_i, \leq) < \frac{1}{2^r}$, hence such that $\leq\restriction_{\mathbb{E}_{r+1}} \; = \; \leq_i\restriction_{\mathbb{E}_{r+1}}$. As $\leq_i \in \mathrm{AO}(\mathbb{E})$ the required inequalities hold for $\leq_i$. But then, by the coincidence of $\leq$ and $\leq_i$ on $\mathbb{E}_{r+1}$, they hold also for $\leq$.

Now, after having established the following auxiliary result, we are ready to prove the announced main result.

Lemma 1.12.15 *Let* $L \subset \mathbb{W}$ *be a left ideal and let* $G \subseteq L$ *be a finite subset. Then, the set*

$$\mathbb{U}_L(G) := \{\leq \in \mathrm{AO}(\mathbb{E}) \mid G \textit{ is a } \leq -\textit{ Gröbner basis of } L\}$$

is open in $\mathrm{AO}(\mathbb{E})$.

Proof We may assume that $\mathbb{U}_L(G)$ is not empty and choose $\leq \in \mathbb{U}_L(G)$. We find some $r \in \mathbb{N}_0$ with $\mathrm{supp}(G) \subseteq \mathbb{E}_{r+1}$. Let $\leq' \in \mathrm{AO}(\mathbb{E})$ such that $\mathrm{dist}(\leq, \leq') < \frac{1}{2^r}$. So, we obtain that $\leq\restriction_{\mathbb{E}_{r+1}} \; = \; \leq'\restriction_{\mathbb{E}_{r+1}}$ and hence in particular that $\leq\restriction_{\mathrm{Supp}(G)} \; = \; \leq'\restriction_{\mathrm{Supp}(G)}$. By Proposition 1.12.10 it follows that G is a $\leq'$-Gröbner basis of L and hence that $\leq' \in \mathbb{U}_L(G)$. But this means, that the open neighborhood

$$\{\leq' \in \mathrm{AO}(\mathbb{E}) \mid \mathrm{dist}(\leq', \leq) < \frac{1}{2^r}\}$$

of $\leq$ belongs to $\mathbb{U}_L(G)$.

Theorem 1.12.16 (Existence of Universal Gröbner Bases) *Each left ideal L of $\mathbb{W}$ admits a universal Gröbner basis.*

Proof Let $L \subseteq \mathbb{W}$ be a left ideal. For each $\leq \in \mathrm{AO}(\mathrm{E})$ we choose a $\leq$-Gröbner basis $G_{\leq}$ of L. In the notations of Lemma 1.12.15 we have $\leq \in \mathbb{U}_L(G_{\leq})$. So, by this same Lemma the family

$$\big(\mathbb{U}_L(G_{\leq})\big)_{\leq \in \mathrm{AO}(\mathbb{E})}$$

is an open covering of $\mathrm{AO}(\mathbb{E})$. By Proposition 1.12.14 we thus find finitely many elements

$$\leq_1, \leq_2, \ldots, \leq_r \in \mathrm{AO}(\mathbb{E})$$

such that

$$\mathrm{AO}(\mathbb{E}) = \bigcup_{i=1}^{r} \mathbb{U}_L(G_{\leq_i}).$$

Let $\leq \in \mathrm{AO}(\mathbb{E})$. Then $\leq \in \mathbb{U}_L(G_{\leq_i})$ for some $i \in \{1, 2, \ldots, r\}$. Therefore $G_{\leq_i}$ is a $\leq$-Gröbner basis of L. So $\bigcup_{i=1}^{r} G_{\leq_i}$ is a Gröbner basis of L for all $\leq \in \mathrm{AO}(\mathbb{E})$.

As a first application of the previous existence result we get the following finiteness result.

Corollary 1.12.17 (Finiteness of the Set of Leading Monomial Ideals) *Let $L \subseteq \mathbb{W}$ be a left ideal. Then the set*

$$\{\mathrm{LMI}_{\leq}(L) \mid \leq \in \mathrm{AO}(\mathbb{E})\}$$

of all leading monomial ideals of L with respect to admissible orderings of $\mathbb{E}$ is finite.

Proof Let $G \subseteq L$ be a universal Gröbner basis of L. Then we have

$$\{\mathrm{LMI}_{\leq}(L) \mid \leq \in \mathrm{AO}(\mathbb{E})\} = \{\mathrm{LMI}_{\leq}(G) \mid \leq \in \mathrm{AO}(\mathbb{E})\}.$$

Therefore

$$\begin{aligned} \#\{\mathrm{LMI}_{\leq}(L) \mid \leq \in \mathrm{AO}(\mathbb{E})\} &\leq \#\{\sum_{h \in H} \mathbb{P}\Phi(h) \mid H \subseteq \mathrm{supp}(G)\} \\ &\leq \#\{H \subseteq \mathrm{supp}(G)\} = 2^{\#\mathrm{supp}(G)}. \end{aligned}$$

1.13 Weighted Orderings

This section is devoted to the study of admissible orderings which are compatible with a given weight and the related notion of weighted (admissible) ordering. Such weighted orderings were first studied by Assi, Castro-Jiménez and Granger [3] and by Saito et al. [38].
In relation to these weighted orderings, we shall introduce the fundamental notion of symbol of a differential operator with respect to a given weight. We will see, that these symbols, which are indeed polynomials, behave again multiplicatively. Moreover, we shall see that the symbols of all members of a Gröbner basis of a given left ideal generate the so-called induced ideal of the given left ideal. Our ultimate goal is to prove, that the number of characteristic varieties of given D-module with respect to all weights is finite. Moreover, we shall prove a certain stability result for characteristic varieties found in Boldini's thesis [11], which is published in [12].

Notation 1.13.1 (A) As previously, we fix a positive integer n, a field K of characteristic 0 and consider the standard Weyl algebra

$$\mathbb{W} := \mathbb{W}(K, n) = K[X_1, X_2, \ldots, X_n, \partial_1, \partial_2, \ldots, \partial_n],$$

the polynomial ring

$$\mathbb{P} := K[Y_1, Y_2, \ldots, Y_n, Z_1, Z_2, \ldots, Z_n]$$

in the indeterminates $Y_1, Y_2, \ldots, Y_n, Z_1, Z_2, \ldots, Z_n$ with coefficients in the field K and the isomorphism of K-vector spaces

$$\Phi : \mathbb{W} \xrightarrow{\cong} \mathbb{P}, \quad \underline{X}^{\underline{\nu}}\underline{\partial}^{\underline{\mu}} \mapsto \underline{Y}^{\underline{\nu}}\underline{Z}^{\underline{\mu}} \text{ for all } \underline{\nu}, \underline{\mu} \in \mathbb{N}_0^n.$$

(B) We also write

$$\Omega := \{(\underline{v}, \underline{w}) \in \mathbb{N}_0^n \times \mathbb{N}_0^n \mid (v_i, w_i) \neq (0, 0) \text{ for all } i = 1, 2, \ldots, n\} \subset \mathbb{N}_0^n \times \mathbb{N}_0^n$$

for the set of all weights. If

$$\underline{\omega} = (\underline{v}, \underline{w}) \in \Omega$$

we also use the suffix $\underline{\omega}$ instead of the suffix $\underline{vw}$ in all the previously introduced notations. So we write for example

$$\mathbb{W}_{\bullet}^{\underline{\omega}} := \mathbb{W}_{\bullet}^{\underline{vw}}, \quad \deg^{\underline{\omega}}(d) := \deg^{\underline{vw}}(d), \quad \mathbb{P}^{\underline{\omega}} := \mathbb{P}^{\underline{vw}}, \quad \ldots$$

Observe, that

$$\underline{\omega} + \underline{\alpha} \in \Omega \text{ and } s\underline{\omega} \in \Omega \text{ for all } \underline{\omega}, \underline{\alpha} \in \Omega \text{ and all } s \in \mathbb{N},$$

where the arithmetic operations are performed in $\mathbb{N}_0^{2n}$.

Now, we introduce the concept of admissible orderings which are compatible with a given weight.

Definition and Exercise 1.13.2 (A) *(Weight Compatible Orderings)* We fix a weight and an admissible ordering of the set $\mathbb{E}$ of elementary differential operators in $\mathbb{W}$ (see Definition, Reminder and Exercise 1.12.2 (C)):

$$\underline{\omega} = (\underline{v}, \underline{w}) \in \Omega \text{ and } \leq \in \mathrm{AO}(\mathbb{E}).$$

We say that $\leq$ is *compatible* with the weight $\underline{\omega} = (\underline{v}, \underline{w}) \in \Omega$ (or $\underline{\omega}$*-compatible*), if for all $d, e \in \mathbb{E}$ we have:

$$\text{If } \deg^{\underline{\omega}}(d) < \deg^{\underline{\omega}}(e), \text{ then } d < e.$$

So, $\leq$ is compatible with $\underline{\omega} = (\underline{v}, \underline{w})$ if and only if for all $\underline{\nu}, \underline{\mu}, \underline{\nu}', \underline{\mu}' \in \mathbb{N}_0^n$ we have the following implication:

$$\text{If } \underline{\nu}\underline{v} + \underline{\mu}\underline{w} < \underline{\nu}'\underline{v} + \underline{\mu}'\underline{w}, \text{ then } \underline{X}^{\underline{\nu}}\underline{\partial}^{\underline{\mu}} < \underline{X}^{\underline{\nu}'}\underline{\partial}^{\underline{\mu}'}.$$

We set

$$\mathrm{AO}^{\underline{\omega}}(\mathbb{E}) = \mathrm{AO}^{\underline{v}\underline{w}}(\mathbb{E}) := \{\leq \in \mathrm{AO}(\mathbb{E}) \mid \ \leq \text{ is compatible with } \underline{\omega} = (\underline{v}, \underline{w})\}.$$

(B) *(Weighted Admissible Orderings)* Keep the notations and hypotheses of part (A). We define a new binary relation

$$\leq^{\underline{\omega}} = \leq^{\underline{v}\underline{w}} \subseteq \mathbb{E} \times \mathbb{E}$$

on $\mathbb{E}$, by setting, for all $d, e \in \mathbb{E}$:

$$d \leq^{\underline{\omega}} e \text{ if } \begin{cases} \text{either} & \deg^{\underline{\omega}}(d) < \deg^{\underline{\omega}}(e) \\ \text{or else} & \deg^{\underline{\omega}}(d) = \deg^{\underline{\omega}}(e) \text{ and } d < e. \end{cases}$$

Prove that for each weight $\underline{\omega} = (\underline{v}, \underline{w}) \in \Omega$ and each $\leq \in \mathrm{AO}(\mathbb{E})$ the following statements hold.

(a) $\leq^{\underline{\omega}} \in \mathrm{AO}^{\underline{\omega}}(\mathbb{E})$.
(b) $(\leq^{\underline{\omega}})^{\underline{\omega}} = \leq^{\underline{\omega}}$.
(c) $\leq \in \mathrm{AO}^{\underline{\omega}}(\mathbb{E})$ if and only if $\leq = \leq^{\underline{\omega}}$.

The admissible ordering $\leq^{\underline{\omega}} \in \mathrm{AO}(\mathbb{E})$ is called the $\underline{\omega}$*-weighted ordering associated to* $\leq$.

Another important concept, which was already mentioned in the introduction to this section, is the notion of symbol of a differential operator. We now will introduce this notion after a few preparatory steps.

Definition and Exercise 1.13.3 (A) Let $\underline{\omega} = (\underline{v}, \underline{w}) \in \Omega$, let $i \in \mathbb{N}_0$ and let

$$d = \sum_{(\underline{\nu},\underline{\mu})\in \mathrm{supp}(d)} c^{(d)}_{\underline{\nu}\underline{\mu}} \underline{X}^{\underline{\nu}} \underline{\partial}^{\underline{\mu}} \in \mathbb{W} \quad \text{with } c^{(d)}_{\underline{\nu}\underline{\mu}} \in K \setminus \{0\} \text{ for all } (\underline{\nu}, \underline{\mu}) \in \mathrm{supp}(d).$$

We set

$$\mathrm{supp}^{\underline{\omega}}_i(d) := \{(\underline{\nu}, \underline{\mu}) \in \mathrm{supp}(d) \mid \underline{\nu}\underline{v} + \underline{\mu}\underline{w} = i\}.$$

and

$$d^{\underline{\omega}}_i = d^{\underline{v}\underline{w}}_i := \sum_{(\underline{\nu},\underline{\nu})\in \mathrm{supp}^{\underline{\omega}}_i(d)} c^{(d)}_{\underline{\nu}\underline{\mu}} \underline{X}^{\underline{\nu}} \underline{\partial}^{\underline{\mu}}.$$

Prove that for all $d, e \in \mathbb{W}$, all $i, j \in \mathbb{N}_0$ and for all weights $\underline{\omega} = (\underline{v}, \underline{w}) \in \Omega$ the following statements hold:

(a) If $i > \deg^{\underline{\omega}}(d)$, then $d^{\underline{\omega}}_i = 0$.
(b) $d^{\underline{\omega}}_i = (d^{\underline{\omega}}_i)^{\underline{\omega}}_i$.
(c) $(d + e)^{\underline{\omega}}_i = d^{\underline{\omega}}_i + e^{\underline{\omega}}_i$.
(d) If $d, e \neq 0$, $i := \deg^{\underline{\omega}}(d)$ and $j := \deg^{\underline{\omega}}(e)$, then

$$\mathrm{supp}^{\underline{\omega}}_{i+j}(de) = \{(\underline{\nu}+\underline{\nu}', \underline{\mu}+\underline{\mu}') \mid (\underline{\nu}, \underline{\mu}) \in \mathrm{supp}^{\underline{\omega}}_i(d) \text{ and } (\underline{\nu}', \underline{\mu}') \in \mathrm{supp}^{\underline{\omega}}_j(e)\}.$$

(e) If $d, e \neq 0$, $i := \deg^{\underline{\omega}}(d)$ and $j := \deg^{\underline{\omega}}(e)$, then

$$(de)^{\underline{\omega}}_{i+j} = \sum_{(\underline{\nu},\underline{\mu})\in \mathrm{supp}^{\underline{\omega}}_i(d), (\underline{\nu}',\underline{\mu}')\in \mathrm{supp}^{\underline{\omega}}_j(e)} c^{(d)}_{\underline{\nu}\underline{\mu}} c^{(e)}_{\underline{\nu}'\underline{\mu}'} \underline{X}^{\underline{\nu}+\underline{\nu}'} \underline{\partial}^{\underline{\mu}+\underline{\mu}'}.$$

(B) Keep the notations and hypotheses of part (A). We set

$$\sigma^{\underline{\omega}}_i(d) := \Phi\left(d^{\underline{\omega}}_i\right) = \sum_{(\underline{\nu},\underline{\nu})\in \mathrm{supp}^{\underline{\omega}}_i(d)} c^{(d)}_{\underline{\nu}\underline{\mu}} \underline{Y}^{\underline{\nu}} \underline{Z}^{\underline{\mu}}.$$

Prove on use of statements (a)–(e) of part (A) that for all $d, e \in \mathbb{W}$, all $i, j \in \mathbb{N}_0$ and for all weights $\underline{\omega} = (\underline{v}, \underline{w}) \in \Omega$ the following statements hold:

(a) $\sigma^{\underline{\omega}}_i(d) := \sigma^{\underline{\omega}}_i(d^{\underline{\omega}}_i)$.
(b) If $i > \deg^{\underline{\omega}}(d)$, then $\sigma^{\underline{\omega}}_i(d) = 0$.
(c) $\sigma^{\underline{\omega}}_i(d) = \sigma^{\underline{\omega}}_i(d^{\underline{\omega}}_i)$.
(d) $\sigma^{\underline{\omega}}_i(d + e) = \sigma^{\underline{\omega}}_i(d) + \sigma^{\underline{\omega}}_i(e)$.

(C) *(The Symbol of a Differential operator with Respect to a Weight)* Keep the notations of part (A), (B). We define the $\underline{\omega} = (\underline{v}, \underline{w})$-*symbol* of the differential operator $d \in \mathbb{W}$ by

$$\sigma^{\underline{\omega}}(d) := \begin{cases} 0 & \text{if } d = 0, \\ \sigma^{\underline{\omega}}_{\deg^{\underline{\omega}}(d)}(d) & \text{if } d \neq 0. \end{cases}$$

Prove that for all $d, e \in \mathbb{W} \setminus \{0\}$ the following statements hold.

(a) $\sigma^{\underline{\omega}}(d) = \Phi(d^{\underline{\omega}}_{\deg^{\underline{\omega}}(d)}) = \sigma^{\underline{\omega}}\big(d^{\underline{\omega}}_{\deg^{\underline{\omega}}(d)}(d)\big)$.

(b) $\sigma^{\underline{\omega}}(d + e) = \begin{cases} \sigma^{\underline{\omega}}(d) + \sigma^{\underline{\omega}}(e) & \text{if } \deg^{\underline{\omega}}(d) = \deg^{\underline{\omega}}(e) = \deg^{\underline{\omega}}(d + e) \\ \sigma^{\underline{\omega}}(d) & \text{if } \deg^{\underline{\omega}}(d) > \deg^{\underline{\omega}}(e). \end{cases}$

First, we now prove that symbols behave well with respect to products of differential operators.

Proposition 1.13.4 (Multiplicativity of Symbols) *Let $\underline{\omega} = (\underline{v}, \underline{w}) \in \Omega$ and let $d, e \in \mathbb{W}$. Then*

$$\sigma^{\underline{\omega}}(de) = \sigma^{\underline{\omega}}(d)\sigma^{\underline{\omega}}(e).$$

Proof If $d = 0$ or $e = 0$, our claim is obvious. So, let $d, e \neq 0$. We write $i := \deg^{\underline{\omega}}(d)$ and $j := \deg^{\underline{\omega}}(e)$. Observe that $\deg^{\underline{\omega}}(de) = i + j$. So, by Definition and Exercise 1.13.3 (A)(e) we have

$$\begin{aligned}
\sigma^{\underline{\omega}}(de) &= \sigma^{\underline{\omega}}_{i+j}(de) = \Phi\big((de)^{\underline{\omega}}_{i+j}\big) \\
&= \Phi\Big(\sum_{(\underline{\nu},\underline{\mu}) \in \operatorname{supp}^{\underline{\omega}}_i(d), (\underline{\nu}',\underline{\mu}') \in \operatorname{supp}^{\underline{\omega}}_j(e)} c^{(d)}_{\underline{\nu}\underline{\mu}} c^{(e)}_{\underline{\nu}'\underline{\mu}'} \underline{X}^{\underline{\nu}+\underline{\nu}'} \underline{\partial}^{\underline{\mu}+\underline{\mu}'}\Big) \\
&= \sum_{(\underline{\nu},\underline{\mu}) \in \operatorname{supp}^{\underline{\omega}}_i(d), (\underline{\nu}',\underline{\mu}') \in \operatorname{supp}^{\underline{\omega}}_j(e)} c^{(d)}_{\underline{\nu}\underline{\mu}} c^{(e)}_{\underline{\nu}'\underline{\mu}'} \underline{Y}^{\underline{\nu}+\underline{\nu}'} \underline{Z}^{\underline{\mu}+\underline{\mu}'} \\
&= \Big(\sum_{(\underline{\nu},\underline{\mu}) \in \operatorname{supp}^{\underline{\omega}}_i(d)} c^{(d)}_{\underline{\nu}\underline{\mu}} \underline{Y}^{\underline{\nu}} \underline{Z}^{\underline{\mu}}\Big)\Big(\sum_{(\underline{\nu}',\underline{\mu}') \in \operatorname{supp}^{\underline{\omega}}_j(e)} c^{(e)}_{\underline{\nu}'\underline{\mu}'} \underline{Y}^{\underline{\nu}'} \underline{Z}^{\underline{\mu}'}\Big) \\
&= \Phi(d^{\underline{\omega}}_i)\Phi(e^{\underline{\omega}}_j) = \sigma^{\underline{\omega}}_i(d)\sigma^{\underline{\omega}}_j(e) = \sigma^{\underline{\omega}}(d)\sigma^{\underline{\omega}}(e).
\end{aligned}$$

In Definition and Remark 1.11.5 we have seen, that each left ideal L of the standard Weyl algebra $\mathbb{W}$ induces a graded ideal in the associated graded ring with respect to a given weight. These induced ideals will play a crucial role in our future considerations. We just revisit now these ideals.

Reminder, Definition and Exercise 1.13.5 (A) *(Induced Graded Ideals)* Let $L \subset \mathbb{W}$ be a left ideal, let $\underline{\omega} = (\underline{v}, \underline{w}) \in \Omega$ be a weight and let us consider the $\underline{\omega}$-graded ideal (see Definition and Remark 1.11.5)

$$\mathbb{G}^{\underline{\omega}}(L) := \bigoplus_{i\in\mathbb{Z}} \big((L \cap \mathbb{W}_i^{\underline{\omega}}) + \mathbb{W}_{i-1}^{\underline{\omega}}\big)/\mathbb{W}_{i-1}^{\underline{\omega}} \cong \bigoplus_{i\in\mathbb{Z}} L_i^{\underline{\omega}}/L_{i-1}^{\underline{\omega}} = \mathrm{Gr}_{L_\bullet^{\underline{\omega}}}(L) \subseteq \mathbb{G}^{\underline{\omega}}(\mathbb{W}),$$

where

$$L_\bullet^{\underline{\omega}} = L \cap \mathbb{W}_\bullet^{\underline{\omega}} := \big(L \cap \mathbb{W}_i^{\underline{\omega}}\big)_{i\in\mathbb{N}_0}$$

is the filtration induced on L by the weighted filtration $\mathbb{W}_\bullet^{\underline{\omega}}$. We now consider the $\underline{\omega}$-graded ideal of $\mathbb{P}^{\underline{\omega}} = \mathbb{P}$ given by

$$\overline{\mathbb{G}}^{\underline{\omega}}(L) := (\eta^{\underline{\omega}})^{-1}\big(\mathbb{G}^{\underline{\omega}}(L)\big),$$

where

$$\eta^{\underline{vw}} = \eta^{\underline{\omega}} : \mathbb{P} = \mathbb{P}^{\underline{\omega}} \xrightarrow{\cong} \mathbb{G}^{\underline{\omega}}.$$

is the canonical isomorphism of graded rings of Theorem 1.9.4. We call $\overline{\mathbb{G}}^{\underline{\omega}}(L)$ the ($\underline{\omega}$-graded) *ideal induced by* L in $\mathbb{P}$.

(B) Let the notations and hypotheses be as part (A). Fix $i \in \mathbb{N}_0$ and consider the i-th $\underline{\omega}$-graded part

$$\overline{\mathbb{G}}^{\underline{\omega}}(L)_i = \overline{\mathbb{G}}^{\underline{\omega}}(L) \cap \mathbb{P}_i^{\underline{\omega}} = (\eta^{\underline{\omega}})^{-1}\big(\mathbb{G}_i^{\underline{\omega}}\big)$$

of the ideal $\overline{\mathbb{G}}^{\underline{\omega}}(L) \subseteq \mathbb{P}$. Prove the following statements:

(a) Let $d \in L$ with $\deg^{\underline{\omega}}(d) = i$ and let $\overline{d} := d + \mathbb{W}_{i-1}^{\underline{\omega}} \in \mathbb{G}^{\underline{\omega}}(L)_i$. Then it holds

$$(\eta^{\underline{\omega}})^{-1}(\overline{d}) = \Phi(d_i^{\underline{\omega}}) = \sigma^{\underline{\omega}}(d) \in \overline{\mathbb{G}}^{\underline{\omega}}(L)_i.$$

(b) Each element $h \in \mathbb{G}^{\underline{\omega}}(L)_i \setminus \{0\}$ can be written as

$$h = \sigma^{\underline{\omega}}(d), \text{ with } d \in L \text{ and } \deg^{\underline{\omega}}(d) = i.$$

(C) *(The Induced Exact Sequence Associated to a Left Ideal with Respect to a Weight)* Keep the above notations and hypotheses. Prove the following statements:

(a) There is a short exact sequence of graded $\mathbb{P}^{\underline{\omega}}$-modules

$$0 \longrightarrow \overline{\mathbb{G}}^{\underline{\omega}}(L) \longrightarrow \mathbb{G}^{\underline{\omega}} \longrightarrow \mathrm{Gr}_{\mathbb{W}_\bullet^{\underline{\omega}} K\overline{1}}(\mathbb{W}/L) \longrightarrow 0,$$

where $\overline{1} := 1 + L \in \mathbb{W}/L$ and $\mathbb{W}_\bullet^{\underline{\omega}} K\overline{1}$ is the $\underline{\omega}$ -filtration induced on the cyclic D-module $\mathbb{W}/L$ by its subspace $K\overline{1}$.

(b) $\mathrm{Ann}_{\mathbb{P}}\big(\mathrm{Gr}_{\mathbb{W}^{\underline{\omega}}_{\bullet}K\overline{1}}(\mathbb{W}/L)\big) = \overline{\mathbb{G}}^{\underline{\omega}}(L)$.
(c) $\mathbb{V}^{\underline{\omega}}(\mathbb{W}/L) = \mathrm{Var}\big(\overline{\mathbb{G}}^{\underline{\omega}}(L)\big)$.

We call this sequence the *short exact sequence associated to the left ideal L with respect to the weight $\underline{\omega}$.*

Now, we are ready to formulate and to prove a result which we already announced in the introduction to this section. It relates the symbols of the members of a Gröbner bases of a left ideal with the induced ideal with respect to a given weight.

Proposition 1.13.6 (Generation of the Induced Ideal by the Symbols of a Gröbner Basis) *Let $\underline{\omega} \in \Omega$, let $L \subseteq \mathbb{W}$ be a left ideal, let $\leq \in$ AO($\mathbb{E}$) and let G be a $\leq^{\underline{\omega}}$-Gröbner basis of L. Then it holds*

(a) $\overline{\mathbb{G}}^{\underline{\omega}}(L) = \sum_{g\in G} \mathbb{P}\sigma^{\underline{\omega}}(g)$.
(b) *For each $h \in \overline{\mathbb{G}}^{\underline{\omega}}(L) \setminus \{0\}$ there is some $g \in G \setminus \{0\}$ and some monomial $m = \underline{Y}^{\underline{\nu}}\underline{Z}^{\underline{\mu}} \in \mathbb{P}$ such that*

$$\mathrm{LM}_{\leq}\big(\Phi^{-1}(h)\big) = m\mathrm{LM}_{\leq}\big(\Phi^{-1}(\sigma^{\underline{\omega}}(g))\big).$$

Proof (a): As the ideal $\overline{\mathbb{G}}^{\underline{\omega}}(L) \subseteq \mathbb{P}^{\underline{\omega}}$ is graded, it suffices to show, that for each $i \in \mathbb{N}_0$ and each $h \in \overline{\mathbb{G}}^{\underline{\omega}}(L)_i \setminus \{0\}$ we have $h \in \sum_{g\in G} \mathbb{P}\sigma^{\underline{\omega}}(g)$. So, fix $i \in \mathbb{N}_0$ and assume that $h \notin \sum_{g\in G} \mathbb{P}\sigma^{\underline{\omega}}(g)$ for some $h \in \overline{\mathbb{G}}^{\underline{\omega}}(L)_i \setminus \{0\}$. Then, by Reminder, Definition and Exercise 1.13.5 (B)(b), the set

$$\mathfrak{S} := \{e \in L \mid \deg^{\underline{\omega}}(e) = i \text{ and } \sigma^{\underline{\omega}}(e) \notin \sum_{g\in G} \mathbb{P}\sigma^{\underline{\omega}}(g)\}$$

is not empty. Choose $d \in \mathfrak{S}$ such that

$$\mathrm{LE}_{\leq^{\underline{\omega}}}(d) = \min_{\leq^{\underline{\omega}}}\{\mathrm{LE}_{\leq^{\underline{\omega}}}(e) \mid e \in \mathfrak{S}\}.$$

As G is a $\leq^{\underline{\omega}}$-Gröbner basis of L we find some $g \in G$ and some $u \in \mathbb{E}$ such that $\mathrm{LM}_{\leq^{\underline{\omega}}}(d) = \mathrm{LM}_{\leq^{\underline{\omega}}}(ug)$ (see Definition, Reminder and Exercise 1.12.6 (C)(b)). With

$$v := \frac{\mathrm{LC}_{\leq^{\underline{\omega}}}(d)}{\mathrm{LC}_{\leq^{\underline{\omega}}}(g)}u$$

it follows that $\mathrm{LE}_{\leq^{\underline{\omega}}}(d) = \mathrm{LE}_{\leq^{\underline{\omega}}}(vg)$, hence

$$\mathrm{LD}_{\leq^{\underline{\omega}}}(d) = \mathrm{LC}_{\leq^{\underline{\omega}}}(d)\mathrm{LE}_{\leq^{\underline{\omega}}}(d) = \mathrm{LC}_{\leq^{\underline{\omega}}}(d)\mathrm{LE}_{\leq^{\underline{\omega}}}(ug) = \mathrm{LD}_{\leq^{\underline{\omega}}}(vg)$$

and $\deg^{\underline{\omega}}(vg) = i$. So, by Definition, Reminder and Exercise 1.12.2 (D)(d) we may conclude that either

(1) $\deg^{\underline{\omega}}(d - vg) < i$, or else
(2) $\deg^{\underline{\omega}}(d - vg) = i$ and $\mathrm{LE}_{\leq^{\underline{\omega}}}(d - vg) < \mathrm{LE}_{\leq^{\underline{\omega}}}(d)$.

In the case (1) we have (see Definition and Exercise 1.13.3 (C)(b) and Proposition 1.13.4)

$$\sigma^{\underline{\omega}}(d) = \sigma^{\underline{\omega}}(d - (d - vg)) = \sigma^{\underline{\omega}}(vg)) = \sigma^{\underline{\omega}}(v)\sigma^{\underline{\omega}}(g) \in \sum_{g \in G} \mathbb{P}\sigma^{\underline{\omega}}(g)$$

and hence get a contradiction.
So, assume that we are in the case (2). As $d - vg \in L$ it follows by our choice of d, that $\sigma^{\underline{\omega}}(d - vg) \in \sum_{g \in G} \mathbb{P}\sigma^{\underline{\omega}}(g)$. Observe that we have

$$i = \deg^{\underline{\omega}}(d - vg) = \deg^{\underline{\omega}}(vg) = \deg^{\underline{\omega}}(d) = \deg^{\underline{\omega}}\big((d - vg) + vg\big).$$

So, by Definition and Exercise 1.13.3 (C)(b) and by Proposition 1.13.4 we have

$$\begin{aligned} \sigma^{\underline{\omega}}(d) &= \sigma^{\underline{\omega}}\big((d - vg) + vg\big) = \sigma^{\underline{\omega}}(d - vg) + \sigma^{\underline{\omega}}(vg) \\ &= \sigma^{\underline{\omega}}(d - vg) + \sigma^{\underline{\omega}}(v)\sigma^{\underline{\omega}}(g) \in \sum_{g \in G} \mathbb{P}\sigma^{\underline{\omega}}(g), \end{aligned}$$

and this is again a contradiction.

(b): We find some $i \in \mathbb{N}_0$ such that $\mathrm{LM}_{\leq}\big(\Phi^{-1}(h)\big) = \mathrm{LM}_{\leq}\big(\Phi^{-1}(h_i^{\underline{\omega}}(h))\big)$. As the ideal $\overline{\mathbb{G}}^{\underline{\omega}}(L) \subseteq \mathbb{P}^{\underline{\omega}}$ is graded, we have $h_i^{\underline{\omega}}(h) \in \overline{\mathbb{G}}^{\underline{\omega}}(L)$. So we may assume, that $h \in \overline{\mathbb{G}}^{\underline{\omega}}(L)_i \setminus \{0\}$. Now, by Reminder, Definition and Exercise 1.13.5 (B), we find some $d \in L$ with $\deg^{\underline{\omega}}(d) = i$ and $\Phi^{-1}(h) = d_i^{\underline{\omega}}$, whence

$$\mathrm{LM}_{\leq}\big(\Phi^{-1}(h)\big) = \mathrm{LM}_{\leq}(d_i^{\underline{\omega}}) = \mathrm{LM}_{\leq^{\underline{\omega}}}(d).$$

As G is a $\leq^{\underline{\omega}}$-Gröbner basis of L, we find some $g \in G \setminus \{0\}$ with $\deg^{\underline{\omega}}(g) = j$ and some monomial $m = \underline{Y}^{\underline{\nu}}\underline{Z}^{\underline{\mu}} \in \mathbb{P}$ such that (see Definition, Reminder and Exercise 1.12.6 (C)(c) and also Definition and Exercise 1.13.3 (C)(a))

$$\mathrm{LM}_{\leq^{\underline{\omega}}}(d) = m\mathrm{LM}_{\leq^{\underline{\omega}}}(g) = m\mathrm{LM}_{\leq}(g_j^{\underline{\omega}}) = m\mathrm{LM}_{\leq}\big(\Phi^{-1}(\sigma_j^{\underline{\omega}}(g))\big),$$

and so we get our claim.

Now, we are ready to prove our first basic finiteness result. It says that the set of all induced ideals of a given left ideal in the Weyl algebra is finite.

Corollary 1.13.7 (Finiteness of the Set of Induced Ideals) *Let $L \subseteq \mathbb{W}$ be a left ideal. Then, the following statements hold:*

(a) $\#\{\overline{\mathbb{G}}^{\underline{\omega}}(L) \mid \underline{\omega} \in \Omega\} < \infty$.
(b) $\#\{\mathbb{V}^{\underline{\omega}}(\mathbb{W}/L) \mid \underline{\omega} \in \Omega\} < \infty$.

Proof (a): Let G be an universal Gröbner basis of L. Then, by Proposition 1.13.6, for each $\underline{\omega} \in \Omega$ we have $\overline{\mathbb{G}}^{\underline{\omega}}(L) = \sum_{g \in G} \mathbb{P}\sigma^{\underline{\omega}}(g)$. For each $g \in G$ we write

$$g = \sum_{(\underline{\nu},\underline{\mu}) \in \operatorname{supp}(g)} c^{(g)}_{\underline{\nu}\underline{\mu}} \underline{X}^{\underline{\nu}} \underline{\partial}^{\underline{\mu}}.$$

Then, for each $\underline{\omega} \in \Omega$ we have

$$\sigma^{\underline{\omega}}(g) = \Phi(g^{\underline{\omega}}_{\deg_{\underline{\omega}}(g)}) = \sum_{(\underline{\nu},\underline{\mu}) \in \operatorname{supp}^{\underline{\omega}}_{\deg^{\underline{\omega}}(g)}(g)} c^{(g)}_{\underline{\nu}\underline{\mu}} \underline{Y}^{\underline{\nu}} \underline{Z}^{\underline{\mu}}.$$

Therefore

$$\#\{\sigma^{\underline{\omega}}(g) \mid \underline{\omega} \in \Omega\} \leq \#\{H \subseteq \operatorname{supp}(g)\} = 2^{\#\operatorname{supp}(g)}.$$

It follows that

$$\#\{\overline{\mathbb{G}}^{\underline{\omega}}(L) = \sum_{g \in G} \mathbb{P}\sigma^{\underline{\omega}}(g) \mid \underline{\omega} \in \Omega\} \leq \#\{\big(\sigma^{\underline{\omega}}(g)\big)_{g \in G} \in \mathbb{P}^G \mid \underline{\omega} \in \Omega\} \leq$$
$$\leq \prod_{g \in G} 2^{\#\operatorname{supp}(g)} = 2^{\#\operatorname{supp}(G)}.$$

(b): This follows immediately from statement (a) on use of Reminder, Definition and Exercise 1.13.5 (C)(c).

The second statement of the previous result says that a given cyclic D-module has only finitely many characteristic varieties, if $\underline{\omega}$ runs through all weights. Our first main theorem says, that this finiteness statement holds indeed for arbitrary D-modules. To prove this, we first have to investigate the behavior of characteristic varieties in short exact sequences of D-modules. This needs some preparations.

Exercise and Definition 1.13.8 (A) Let $\underline{\omega} \in \Omega$ and let

$$0 \longrightarrow Q \xrightarrow{\iota} U \xrightarrow{\pi} P \longrightarrow 0$$

be an exact sequence of D-modules. Let $V \subseteq U$ be a finitely generated K-vector subspace such that $U = \mathbb{W}V$. We endow Q with the filtration

$$Q_\bullet := \big(\iota^{-1}(\mathbb{W}^{\underline{\omega}}_i V)\big)_{i \in \mathbb{N}_0}.$$

Prove the following statements:

(a) For each $i \in \mathbb{N}_0$ there is a K-linear map

$$\bar{\iota}_i : Q_i/Q_{i-1} \longrightarrow \mathbb{W}^{\underline{\omega}}_i V/\mathbb{W}^{\underline{\omega}}_{i-1} V, \quad q + Q_{i-1} \mapsto \iota(q) + \mathbb{W}^{\underline{\omega}}_{i-1} V.$$

(b) For each $i \in \mathbb{N}_0$ there is a K-linear map

$$\overline{\pi}_i : \mathbb{W}_i^{\underline{\omega}} V/\mathbb{W}_{i-1}^{\underline{\omega}} V \longrightarrow \mathbb{W}_i^{\underline{\omega}} \pi(V)/\mathbb{W}_{i-1}^{\underline{\omega}} \pi(V), \quad q+\mathbb{W}_{i-1}^{\underline{\omega}} V \mapsto \pi(q)+\mathbb{W}_{i-1}^{\underline{\omega}} \pi(V).$$

(c) For each $i \in \mathbb{N}_0$ it holds

$$\pi^{-1}\big(\mathbb{W}_{i-1}^{\underline{\omega}} \pi(V)\big) = \iota(Q) + \mathbb{W}_{i-1}^{\underline{\omega}} V.$$

(d) For each $i \in \mathbb{N}_0$ there is a short exact sequence of K-vector spaces

$$0 \longrightarrow Q_i/Q_{i-1} \xrightarrow{\overline{\iota}_i} \mathbb{W}_i^{\underline{\omega}} V/\mathbb{W}_{i-1}^{\underline{\omega}} V \xrightarrow{\overline{\pi}_i} \mathbb{W}_i^{\underline{\omega}} \pi(V)/\mathbb{W}_{i-1}^{\underline{\omega}} \pi(V) \longrightarrow 0.$$

(B) *(The Graded Exact Sequence associated to a Short Exact Sequence of D-Modules)* Keep the hypotheses and notations of part (A). Prove the following statements:

(a) For each $i \in \mathbb{N}_0$ there is a short exact sequence of K-vector spaces

$$0 \longrightarrow \mathrm{Gr}_{Q_\bullet}(Q)_i \xrightarrow{\overline{\iota}_i} \mathrm{Gr}_{\mathbb{W}_\bullet^{\underline{\omega}} V}(U)_i \xrightarrow{\overline{\pi}_i} \mathrm{Gr}_{\mathbb{W}_\bullet^{\underline{\omega}} \pi(V)}(P)_i \longrightarrow 0.$$

(b) There is an exact sequence of graded $\mathbb{P}^{\underline{\omega}}$-modules

$$0 \longrightarrow \mathrm{Gr}_{Q_\bullet}(Q) \xrightarrow{\overline{\iota}} \mathrm{Gr}_{\mathbb{W}_\bullet^{\underline{\omega}} V}(U) \xrightarrow{\overline{\pi}} \mathrm{Gr}_{\mathbb{W}_\bullet^{\underline{\omega}} \pi(V)}(P) \longrightarrow 0,$$

with $\overline{\iota} := \bigoplus_{i\in\mathbb{N}_0} \overline{\iota}_i$ and $\overline{\pi} := \bigoplus_{i\in\mathbb{N}_0} \overline{\pi}_i$.

The exact sequence of statement (b) is called the *exact sequence induced by the exact sequence* $0 \to Q \xrightarrow{\iota} U \xrightarrow{\pi} P \to 0$ and the generating vector space V of U.

(C) Keep the previous notations and hypotheses. Prove the following statements:

(a) For each finitely generated K-vector subspace $T \subseteq Q$ with $Q = \mathbb{W}T$ and $V \subseteq \iota(T)$, the two filtrations $Q_\bullet$ and $\mathbb{W}_\bullet^{\underline{\omega}} T$ of Q are equivalent.
(b) $\mathrm{Var}\big(\mathrm{Ann}_{\mathbb{P}}(\mathrm{Gr}_{Q_\bullet}(Q))\big) = \mathbb{V}^{\underline{\omega}}(Q)$.

Now, we can prove the crucial result, needed to extend the previous finiteness statement for characteristic varieties from cyclic to arbitrary D-modules.

Proposition 1.13.9 (Additivity of Characteristic Varieties) *Let $\underline{\omega} \in \Omega$ and let*

$$0 \longrightarrow Q \xrightarrow{\iota} U \xrightarrow{\pi} P \longrightarrow 0$$

be an exact sequence of D-modules. Then it holds

$$\mathbb{V}^{\underline{\omega}}(U) = \mathbb{V}^{\underline{\omega}}(Q) \cup \mathbb{V}^{\underline{\omega}}(P).$$

Proof We fix a finitely generated K-vector subspace $V \subseteq U$ with $\mathbb{W}V = U$ and consider the corresponding induced short exact sequence (see Exercise and Definition 1.13.8 (B))

$$0 \longrightarrow \mathrm{Gr}_{Q_\bullet}(Q) \xrightarrow{\bar{\iota}} \mathrm{Gr}_{\mathbb{W}_\bullet^{\underline{\omega}}V}(U) \xrightarrow{\bar{\pi}} \mathrm{Gr}_{\mathbb{W}_\bullet^{\underline{\omega}}\pi(V)}(P) \longrightarrow 0.$$

On use of Exercise and Definition 1.13.8 (C)(b) we obtain

$$\begin{aligned}\mathbb{V}^{\underline{\omega}}(U) &= \mathrm{Var}\big(\mathrm{Ann}_{\mathbb{P}}(\mathrm{Gr}_{\mathbb{W}_\bullet^{\underline{\omega}}V}(U))\big)\\ &= \mathrm{Var}\big(\mathrm{Ann}_{\mathbb{P}}(\mathrm{Gr}_{Q_\bullet}(Q))\big) \cup \mathrm{Var}\big(\mathrm{Ann}_{\mathbb{P}}(\mathrm{Gr}_{\mathbb{W}_\bullet^{\underline{\omega}}\pi(V)}(P))\big) = \mathbb{V}^{\underline{\omega}}(Q) \cup \mathbb{V}^{\underline{\omega}}(P).\end{aligned}$$

Now, we are ready to prove the announced first main theorem of this section.

Theorem 1.13.10 (Finiteness of the Set of Characteristic Varieties) *Let U be a D-module. Then*

$$\#\{\mathbb{V}^{\underline{\omega}}(U) \mid \underline{\omega} \in \Omega\} < \infty.$$

Proof We proceed by induction on the number r of generators of U. If $r = 1$ we have $U \cong \mathbb{W}/L$ for some left ideal $L \subseteq \mathbb{W}$. In this case, we may conclude by Corollary 1.13.7 (b). So, let $r > 1$. Then, we find a short exact of D-modules

$$0 \longrightarrow Q \xrightarrow{\iota} U \xrightarrow{\pi} P \longrightarrow 0$$

such that Q and P are generated by less than r elements. By induction, we have

$$\#\{\mathbb{V}^{\underline{\omega}}(Q) \mid \underline{\omega} \in \Omega\} < \infty \text{ and } \#\{\mathbb{V}^{\underline{\omega}}(P) \mid \underline{\omega} \in \Omega\} < \infty.$$

By Proposition 1.13.9 we also have

$$\{\mathbb{V}^{\underline{\omega}}(U) \mid \underline{\omega} \in \Omega\} = \{\mathbb{V}^{\underline{\omega}}(Q) \cup \mathbb{V}^{\underline{\omega}}(P) \mid \underline{\omega} \in \Omega\},$$

hence

$$\#\{\mathbb{V}^{\underline{\omega}}(U) \mid \underline{\omega} \in \Omega\} \leq \#\{\mathbb{V}^{\underline{\omega}}(Q) \mid \underline{\omega} \in \Omega\} + \#\{\mathbb{V}^{\underline{\omega}}(P) \mid \underline{\omega} \in \Omega\} < \infty.$$

As already announced in the introduction to this section, our ultimate goal is to establish a certain stability result for characteristic varieties of a given D-module. To pave the way for this, we perform a number of preparatory considerations, which are the subject of the exercises to come.

Definition and Exercise 1.13.11 (A) *(Leading Forms)* We consider the polynomial ring $\mathbb{P}$. Let

$$f = \sum_{(\underline{\nu},\underline{\mu})\in \mathrm{supp}(f)} c_{\underline{\nu}\underline{\mu}}^{(f)} \underline{Y}^{\underline{\nu}} \underline{Z}^{\underline{\mu}} \in \mathbb{P} \quad \text{with } c_{\underline{\nu}\underline{\mu}}^{(f)} \in K \setminus \{0\} \text{ for all } (\underline{\nu}, \underline{\mu}) \in \mathrm{supp}(f).$$

We set

$$\mathrm{supp}_i^{\underline{\omega}}(f) := \{(\underline{\nu}, \underline{\mu}) \in \mathrm{supp}(f) \mid \underline{\nu}\underline{v} + \underline{\mu}\underline{w} = i\}$$

and consider the $i - th$ *homogeneous component* of f with respect to ω, thus the polynomial

$$f_i^{\underline{\omega}} = f_i^{\underline{v}\underline{w}} := \sum_{(\underline{\nu}, \underline{\nu}) \in \mathrm{supp}_i^{\underline{\omega}}(f)} c_{\underline{\nu}\underline{\mu}}^{(f)} \underline{Y}^{\underline{\nu}} \underline{Z}^{\underline{\mu}}.$$

The *leading form* of f with respect to the weight $\underline{\omega}$ is defined by

$$\mathrm{LF}^{\underline{\omega}}(f) := \begin{cases} 0 & \text{if } f = 0, \\ f_{\deg^{\underline{\omega}}(f)}^{\underline{\omega}} & \text{if } f \neq 0. \end{cases}$$

Prove that for all $f, g \in \mathbb{P}$, all $i, j \in \mathbb{N}_0$ and for all weights $\underline{\omega} = (\underline{v}, \underline{w}) \in \Omega$ the following statements hold:

(a) If $i > \deg^{\underline{\omega}}(f)$, then $f_i^{\underline{\omega}} = 0$.
(b) $f_i^{\underline{\omega}} = f_i^{\underline{\omega}}(f_i^{\underline{\omega}})$.
(c) $(f + g)_i^{\underline{\omega}} = f_i^{\underline{\omega}} + g_i^{\underline{\omega}}$.
(d) $(fg)_i^{\underline{\omega}} = \sum_{j+k=i} f_j^{\underline{\omega}} g_k^{\underline{\omega}}$.
(e) $\mathrm{LF}^{\underline{\omega}}(fg) = \mathrm{LF}^{\underline{\omega}}(f)\mathrm{LF}^{\underline{\omega}}(g)$.
(f) $\mathrm{LF}(f) = f$ if and only if f is homogeneous with respect to the $\underline{\omega}$-grading of $\mathbb{P}$.
(g) If $d \in \mathbb{W}$, then $\sigma^{\underline{\omega}}(d) = \mathrm{LF}^{\underline{\omega}}\big(\Phi(d)\big)$.

(B) *(Leading Form Ideals)* Keep the notations and hypotheses of part (A). If $S \subset \mathbb{P}$ is any subset, we define the *leading form ideal* of S with respect to $\underline{\omega}$ by

$$\mathrm{LFI}^{\underline{\omega}}(S) := \sum_{f \in S} \mathbb{P}\mathrm{LF}^{\underline{\omega}}(f).$$

Let $S \subseteq T \subseteq \mathbb{P}$ and $\leq \in \mathrm{AO}(\mathbb{E})$. Prove the following statements:

(a) $\mathrm{LFI}^{\underline{\omega}}(S) \subseteq \mathrm{LFI}^{\underline{\omega}}(T)$.
(b) If for each $t \in T \setminus \{0\}$ there is some monomial $m = \underline{Y}^{\underline{\nu}}\underline{Z}^{\underline{\mu}} \in \mathbb{M} \subset \mathbb{P}$ and some $s \in S$ such that $\mathrm{LM}_{\leq^{\underline{\omega}}}\big(\Phi^{-1}(t)\big) = m\mathrm{LM}_{\leq^{\underline{\omega}}}\big(\Phi^{-1}(s)\big)$, then $\mathrm{LFI}^{\underline{\omega}}(S) = \mathrm{LFI}^{\underline{\omega}}(T)$.
(c) For each ideal $I \subseteq \mathbb{P}$ it holds

$$\sqrt{\mathrm{LFI}^{\underline{\omega}}(I)} = \sqrt{\mathrm{LFI}^{\underline{\omega}}(\sqrt{I})}.$$

(d) If $I, J \subseteq \mathbb{P}$ are ideals, then

(1) $\mathrm{LFI}^{\underline{\omega}}(I \cap J) \subseteq \mathrm{LFI}^{\underline{\omega}}(I) \cap \mathrm{LFI}^{\underline{\omega}}(I)$ and $\mathrm{LFI}^{\underline{\omega}}(I)\mathrm{LFI}^{\underline{\omega}}(J) \subseteq \mathrm{LFI}^{\underline{\omega}}(IJ)$;
(2) $\sqrt{\mathrm{LFI}^{\underline{\omega}}(I \cap J)} = \sqrt{\mathrm{LFI}^{\underline{\omega}}(I) \cap \mathrm{LFI}^{\underline{\omega}}(J)} = \sqrt{\mathrm{LFI}^{\underline{\omega}}(I)} \cap \sqrt{\mathrm{LFI}^{\underline{\omega}}(J)}$.

The announced Stability Theorem for Characteristic Varieties we are heading for, concerns the behavior of characteristic varieties under certain changes of the involved weights. To prepare this new type of considerations, we suggest the following exercise.

Exercise 1.13.12 (A) Prove that for all $d \in \mathbb{W}$, all $i, j \in \mathbb{N}_0$, all $s \in \mathbb{N}$ and for all weights $\underline{\alpha} = (\underline{a}, \underline{b}), \underline{\omega} = (\underline{v}, \underline{w}) \in \Omega$ the following statements hold (For the unexplained notations see Definition and Exercise 1.13.3):

(a) $\text{supp}\big([d_i^{\underline{\omega}}]_j^{\underline{\alpha}}\big) = \text{supp}_i^{\underline{\omega}}(d) \bigcap \text{supp}_j^{\underline{\alpha}}(d)$.
(b) $\text{supp}\big([d_i^{\underline{\omega}}]_j^{\underline{\alpha}}\big) \subseteq \text{supp}_{j+si}^{\underline{\alpha}+s\underline{\omega}}(d)$.
(c) If $i \geq \deg^{\underline{\omega}}(d)$, $j \geq \deg^{\underline{\alpha}}(d_i^{\underline{\omega}})$ and $s > \deg^{\underline{\alpha}}(d) - j$, then the inclusion of statement (b) becomes an equality.
(d) If $i \geq \deg^{\underline{\omega}}(d)$, $j \geq \deg^{\underline{\alpha}}\big(d_i^{\underline{\omega}}\big)$ and $s > \deg^{\underline{\alpha}}(d) - j$, then

$$[d_i^{\underline{\omega}}]_j^{\underline{\alpha}} = d_{j+si}^{\underline{\alpha}+s\underline{\omega}}.$$

(B) Prove on use of statements (a)–(d) of part (A) that for all $d \in \mathbb{W}$, all $i, j \in \mathbb{N}_0$, all $s \in \mathbb{N}$ and for all weights $\underline{\omega} = (\underline{v}, \underline{w}), \underline{\alpha} = (\underline{a}, \underline{b}) \in \Omega$ the following statements hold:

(a) $\sigma_j^{\underline{\alpha}}(d_i^{\underline{\omega}}) = \sum_{(\underline{\nu},\underline{\mu})\in\text{supp}_i^{\underline{\omega}}(d)\cap\text{supp}_j^{\underline{\alpha}}(d)} c_{\underline{\nu}\underline{\mu}}^{(d)} \underline{Y}^{\underline{\nu}} \underline{Z}^{\underline{\mu}} = \sigma_i^{\underline{\omega}}(d_j^{\underline{\alpha}})$.
(b) If $i \geq \deg^{\underline{\omega}}(d)$, $j \geq \deg^{\underline{\alpha}}(d_i^{\underline{\omega}})$ and $s > \deg^{\underline{\alpha}}(d) - j$, then

$$[\sigma_i^{\underline{\omega}}(d)]_j^{\underline{\alpha}} = \sigma_{j+si}^{\underline{\alpha}+s\underline{\omega}}(d).$$

The next two auxiliary results are of fairly technical nature. But they will play a crucial role in the proof of our Stability Theorem.

Lemma 1.13.13 *Let $\underline{\alpha}, \underline{\omega} \in \Omega$, let $d \in \mathbb{W} \setminus \{0\}$ and let $s \in \mathbb{N}$ such that*

$$s > \deg^{\underline{\alpha}}(d) - \deg^{\underline{\alpha}}\big(\sigma^{\underline{\omega}}(d)\big).$$

Then, the following statements hold:

(a) $\deg^{\underline{\alpha}+s\underline{\omega}}(d) = \deg^{\underline{\alpha}}\big(\sigma^{\underline{\omega}}(d)\big) + s\deg^{\underline{\omega}}(d)$.
(b) $\text{LF}^{\underline{\alpha}}\big(\sigma^{\underline{\omega}}(d)\big) = \sigma^{\underline{\alpha}+s\underline{\omega}}(d)$.

Proof We write

$$i := \deg^{\underline{\omega}}(d) \text{ and } j := \deg^{\underline{\alpha}}\big(\sigma^{\underline{\omega}}(d)\big).$$

Observe, that $\sigma^{\underline{\omega}}(d) = \sigma_i^{\underline{\omega}}(d) = \Phi(d_i^{\underline{\omega}})$, so that

$$j = \deg^{\underline{\alpha}}\big(\sigma^{\underline{\omega}}(d)\big) = \deg^{\underline{\alpha}}(d_i^{\underline{\omega}}) \text{ and also } s > \deg^{\underline{\alpha}}(d) - j.$$

Now, by Exercise 1.13.12 (B)(b) we obtain

$$\mathrm{LF}^{\underline{\alpha}}\big(\sigma^{\underline{\omega}}(d)\big) = [\sigma_i^{\underline{\omega}}(d)]_j^{\underline{\alpha}} = \sigma_{j+si}^{\underline{\alpha}+s\underline{\omega}}(d).$$

It remains to show that

$$j + si = \deg^{\underline{\alpha}+s\underline{\omega}}(d).$$

As $\mathrm{LF}^{\underline{\alpha}}\big(\sigma^{\underline{\omega}}(d)\big) \neq 0$ we have $\sigma_{j+si}^{\underline{\alpha}+s\underline{\omega}}(d) \neq 0$ and hence $j + si \leq \deg^{\underline{\alpha}+s\underline{\omega}}(d)$ (see Definition and Exercise 1.13.3 (B)(b)).
Assume that $j + si > \deg^{\underline{\alpha}+s\underline{\omega}}(d)$. Then, we may write $\deg^{\underline{\alpha}+s\underline{\omega}}(d) = k + si$, with $k > j$. It follows, that $s > \deg^{\underline{\alpha}}(d) - k$. On application of Exercise 1.13.12 (B)(b) we get that

$$[\sigma_i^{\underline{\omega}}(d)]_k^{\underline{\alpha}} = \sigma_{k+si}^{\underline{\alpha}+s\underline{\omega}}(d) = \sigma^{\underline{\alpha}+s\underline{\omega}}(d) \neq 0.$$

As $k > j = \deg^{\underline{\alpha}}\big(\sigma^{\underline{\omega}}(d)\big)$ we have $[\sigma_i^{\underline{\omega}}(d)]_k^{\underline{\alpha}} = 0$ (see Definition and Exercise 1.13.11 (A)(a)). This contradiction completes our proof.

Lemma 1.13.14 *Let $L \subseteq \mathbb{W}$ be a left ideal, let $\underline{\alpha}, \underline{\omega} \in \Omega$, let $\leq \in \mathrm{AO}(\mathbb{E})$ and let G be a $(\leq^{\underline{\alpha}})^{\underline{\omega}}$-Gröbner basis of L. Then*

$$\mathrm{LFI}^{\underline{\alpha}}\big(\mathbb{G}^{\underline{\omega}}(L)\big) = \mathrm{LFI}^{\underline{\alpha}}\big(\{\sigma^{\underline{\omega}}(g) \mid g \in G\}\big).$$

Proof By Reminder, Definition and Exercise 1.13.5 (B)(a) we have

$$S := \{\sigma^{\underline{\omega}}(g) \mid g \in G \setminus \{0\}\} \subseteq \overline{\mathbb{G}}^{\underline{\omega}}(L) =: T$$

If we apply Proposition 1.13.6 (b) with $\leq^{\underline{\alpha}}$ instead of $\leq$, we see that for all $t \in T$ there is some monomial $m = \underline{Y}^{\underline{\nu}}\underline{Z}^{\underline{\mu}} \in \mathbb{M} \subset \mathbb{P}$ and some $s \in S$ such that $\mathrm{LM}_{\leq^{\underline{\alpha}}}\big(\Phi^{-1}(t)\big) = m\mathrm{LM}_{\leq^{\underline{\alpha}}}\big(\Phi^{-1}(s)\big)$. By Definition and Exercise 1.13.11 (B)(b) it follows that

$$\mathrm{LFI}^{\underline{\alpha}}\big(\overline{\mathbb{G}}^{\underline{\omega}}(L)\big) = \mathrm{LFI}^{\underline{\alpha}}(S) = \mathrm{LFI}^{\underline{\alpha}}(T) = \mathrm{LFI}^{\underline{\alpha}}\big(\{\sigma^{\underline{\omega}}(g) \mid g \in G\}\big).$$

Now, we are ready to formulate and to prove the announced stability result.

Theorem 1.13.15 (Stability of Induced Graded Ideals, Boldini [11, 12]) *Let $L \subseteq \mathbb{W}$ be a left ideal and let $\underline{\alpha} \in \Omega$. Then, there exists an integer $\overline{s} = \overline{s}(\underline{\alpha}, L) \in \mathbb{N}_0$ such that for all $s \in \mathbb{N}$ with $s > \overline{s}$ and all $\underline{\omega} \in \Omega$ we have*

$$\mathrm{LFI}^{\underline{\alpha}}\big(\overline{\mathbb{G}}^{\underline{\omega}}(L)\big) = \overline{\mathbb{G}}^{\underline{\alpha}+s\underline{\omega}}(L).$$

Proof Let G be a universal Gröbner basis of L. Then, by Lemma 1.13.14, for each $\underline{\omega} \in \Omega$ we have

$$\mathrm{LFI}^{\underline{\alpha}}\big(\overline{\mathbb{G}}^{\underline{\omega}}(L)\big) = \mathrm{LFI}^{\underline{\alpha}}\big(\{\sigma^{\underline{\omega}}(g) \mid g \in G\}\big) = \sum_{g \in G} \mathbb{P}\mathrm{LF}^{\underline{\alpha}}\big(\sigma^{\underline{\omega}}(g)\big).$$

Now, we set

$$\overline{s} := \max\{\deg^{\underline{\alpha}}(g) \mid g \in G \setminus \{0\}\}.$$

By Lemma 1.13.13 it follows that $\mathrm{LF}^{\underline{\alpha}}\big(\sigma^{\underline{\omega}}(g)\big) = \sigma^{\underline{\alpha}+s\underline{\omega}}(g)$ for all $s \in \mathbb{N}$ with $s > \overline{s}$, all $\underline{\omega} \in \Omega$ and all $g \in G \setminus \{0\}$. So, for all $s \in \mathbb{N}$ with $s > \overline{s}$ and all $\underline{\omega} \in \Omega$ we have

$$\mathrm{LFI}^{\underline{\alpha}}\big(\overline{\mathbb{G}}^{\underline{\omega}}(L)\big) = \sum_{g \in G} \mathbb{P}\sigma^{\underline{\alpha}+s\underline{\omega}}(g).$$

If we apply Proposition 1.13.6 (a) with $\underline{\alpha} + s\underline{\omega}$ instead of $\underline{\omega}$ we also get

$$\overline{\mathbb{G}}^{\underline{\alpha}+s\underline{\omega}}(L) = \sum_{g \in G} \mathbb{P}\sigma^{\underline{\alpha}+s\underline{\omega}}(g)$$

for all $s \in \mathbb{N}$ with $s > \overline{s}$ and all $\underline{\omega} \in \Omega$. This completes our proof.

Notation 1.13.16 If $\mathfrak{Z} \subseteq \mathrm{Spec}(\mathbb{P})$ is a closed set we denote the *vanishing ideal* of $\mathfrak{Z}$ by $I_{\mathfrak{Z}}$, thus:

$$I_{\mathfrak{Z}} := \bigcap_{\mathfrak{p} \in \mathfrak{Z}} \mathfrak{p} = \sqrt{J}, \text{ for all ideals } J \subseteq \mathbb{P} \text{ with } \mathfrak{Z} = \mathrm{Var}(J).$$

Theorem 1.13.17 (Stability of Characteristic Varieties, Boldini [11, 12]) *Let U be a D-module, and let $\underline{\alpha} \in \Omega$. Then, there exists an integer $\overline{s} = \overline{s}(\underline{\alpha}, U) \in \mathbb{N}_0$ such that for all $s \in \mathbb{N}$ with $s > \overline{s}$ and all $\underline{\omega} \in \Omega$ we have*

$$\mathrm{Var}\big(\mathrm{LFI}^{\underline{\alpha}}\big(I_{\mathbb{V}^{\underline{\omega}}(U)}\big)\big) = \mathbb{V}^{\underline{\alpha}+s\underline{\omega}}(U).$$

Proof We proceed by induction on the number r of generators of U. First, let $r = 1$. Then we have $U \cong \mathbb{W}/L$ for some left ideal $L \subseteq \mathbb{W}$. By Theorem 1.13.15 we find some $\overline{s} \in \mathbb{N}_0$ such that for all $s \in \mathbb{N}$ with $s > \overline{s}$ and all $\underline{\omega} \in \Omega$ we have

$$\mathrm{LFI}^{\underline{\alpha}}\big(\overline{\mathbb{G}}^{\underline{\omega}}(L)\big) = \overline{\mathbb{G}}^{\underline{\alpha}+s\underline{\omega}}(L).$$

By Reminder, Definition and Exercise 1.13.5 (C)(c) we have

$$\mathbb{V}^{\underline{\alpha}+s\underline{\omega}}(U) = \mathrm{Var}\big(\overline{\mathbb{G}}^{\underline{\alpha}+s\underline{\omega}}(L)\big) \text{ and } I_{\mathbb{V}^{\underline{\omega}}(U)} = \sqrt{\overline{\mathbb{G}}^{\underline{\omega}}(L)}.$$

By Definition and Exercise 1.13.11 (B)(c) we thus get

$$\sqrt{\mathrm{LFI}^{\underline{\alpha}}\big(I_{\mathbb{V}^{\underline{\omega}}(U)}\big)} = \sqrt{\mathrm{LFI}^{\underline{\alpha}}\big(\sqrt{\overline{\mathbb{G}}^{\underline{\omega}}(L)}\big)} = \sqrt{\mathrm{LFI}^{\underline{\alpha}}\big(\overline{\mathbb{G}}^{\underline{\omega}}(L)\big)},$$

so that indeed—for all $s \in \mathbb{N}$ with $s > \overline{s}$ and all $\underline{\omega} \in \Omega$—we have

$$\mathrm{Var}\big(\mathrm{LFI}^{\underline{\alpha}}\big(I_{\mathbb{V}^{\underline{\omega}}(U)}\big)\big) = \mathrm{Var}\big(\mathrm{LFI}^{\underline{\alpha}}\big(\overline{\mathbb{G}}^{\underline{\omega}}(L)\big) = \mathrm{Var}\big(\overline{\mathbb{G}}^{\underline{\alpha}+s\underline{\omega}}(L)\big) = \mathbb{V}^{\underline{\alpha}+s\underline{\omega}}(U).$$

Now, let $r > 1$. Then, we find a short exact of D-modules

$$0 \longrightarrow Q \xrightarrow{\iota} U \xrightarrow{\pi} P \longrightarrow 0$$

such that Q and P are generated by less than r elements. By induction, we thus find a number $\overline{s} \in \mathbb{N}_0$, such that for all $\underline{\omega} \in \Omega$ and all $s \in \mathbb{N}$ with $s > \overline{s}$ it holds

$$\mathrm{Var}\big(\mathrm{LFI}^{\underline{\alpha}}(I_{\mathbb{V}^{\underline{\omega}}(Q)})\big) = \mathbb{V}^{\underline{\alpha}+s\underline{\omega}}(Q) \text{ and } \mathrm{Var}\big(\mathrm{LFI}^{\underline{\alpha}}(I_{\mathbb{V}^{\underline{\omega}}(P)})\big) = \mathbb{V}^{\underline{\alpha}+s\underline{\omega}}(P).$$

By Proposition 1.13.9 we have

$$\mathbb{V}^{\underline{\alpha}+s\underline{\omega}}(U) = \mathbb{V}^{\underline{\alpha}+s\underline{\omega}}(Q) \cup \mathbb{V}^{\underline{\alpha}+s\underline{\omega}}(P)$$

and hence, moreover

$$I_{\mathbb{V}^{\underline{\omega}}(U)} = I_{\mathbb{V}^{\underline{\omega}}(Q) \cup \mathbb{V}^{\underline{\omega}}(Q)} = I_{\mathbb{V}^{\underline{\omega}}(Q)} \cap I_{\mathbb{V}^{\underline{\omega}}(P)}.$$

By Definition and Exercise 1.13.11 (B)(d)(2) it follows from the last equality that

$$\sqrt{\mathrm{LFI}^{\underline{\alpha}}\big(I_{\mathbb{V}^{\underline{\omega}}(U)}\big)} = \sqrt{\mathrm{LFI}^{\underline{\alpha}}\big(I_{\mathbb{V}^{\underline{\omega}}(Q)}\big)} \cap \sqrt{\mathrm{LFI}^{\underline{\alpha}}\big(I_{\mathbb{V}^{\underline{\omega}}(P)}\big)}.$$

Therefore

$$\mathrm{Var}\big(\mathrm{LFI}^{\underline{\alpha}}(I_{\mathbb{V}^{\underline{\omega}}(U)})\big) = \mathrm{Var}\big(\mathrm{LFI}^{\underline{\alpha}}(I_{\mathbb{V}^{\underline{\omega}}(Q)})\big) \cup \mathrm{Var}\big(\mathrm{LFI}^{\underline{\alpha}}(I_{\mathbb{V}^{\underline{\omega}}(P)})\big).$$

It follows, that

$$\mathrm{Var}\big(\mathrm{LFI}^{\underline{\alpha}}(I_{\mathbb{V}^{\underline{\omega}}(U)})\big) = \mathbb{V}^{\underline{\alpha}+s\underline{\omega}}(Q) \cup \mathbb{V}^{\underline{\alpha}+s\underline{\omega}}(P) = \mathbb{V}^{\underline{\alpha}+s\underline{\omega}}(U)$$

for all $\underline{\omega} \in \Omega$ and all $s \in \mathbb{N}$ with $s > \overline{s}$. This completes the step of induction and hence proves our claim.

To formulate our Stability Theorem in a more geometric manner, we introduce the following notion.

Definition 1.13.18 (The Critical Cone) Let $\mathfrak{Z} \subseteq \mathrm{Spec}(\mathbb{P})$ be a closed set. Then, the *critical cone* of $\mathfrak{Z}$ is defined as

$$\mathrm{CCone}(\mathfrak{Z}) := \mathrm{Var}\big(\mathrm{LFI}^{\underline{1}}(I_{\mathfrak{Z}})\big),$$

where $\underline{1} = (\underline{1}, \underline{1}) \in \Omega$ denotes the *standard weight.*

On use of the introduced terminology, we now can define our Stability Theorem as follows.

Corollary 1.13.19 (Affine Deformation of Characteristic Varieties to Critical Cones, Boldini [11, 12]) *Let U be a D-module. Then, there is an integer $\overline{s} = \overline{s}(U) \in \mathbb{N}_0$ such that for all $\underline{\omega} \in \Omega$ and all $s \in \mathbb{N}$ with $s > \overline{s}$ it holds*

$$\mathbb{V}^{\underline{1}+s\underline{\omega}}(U) = \mathrm{CCone}\big(\mathbb{V}^{\underline{\omega}}(U)\big).$$

Proof This is immediate by Theorem 1.13.17.

1.14 Standard Degree and Hilbert Polynomials

In this section, we give an outlook to the relation between D-modules and Castelnuovo-Mumford regularity, which we mentioned in the introduction. We shall consider a situation, which is exclusively related to the standard degree filtration $\mathbb{W}_\bullet = \mathbb{W}_\bullet^{\mathrm{deg}} = \mathbb{W}_\bullet^{\underline{11}}$ of the underlying Weyl algebra $\mathbb{W}$. Having in mind to approach the bounding result for the degree of defining equations of characteristic varieties mentioned in the introduction, we shall restrict ourselves to consider D-modules U endowed with filtrations $V\mathbb{W}_\bullet$ induced by a finite-dimensional generating vector space V of U.

Preliminary Remark 1.14.1 (A) Let $n \in \mathbb{N}$, let K be a field of characteristic 0 and consider the standard Weyl algebra $\mathbb{W} = \mathbb{W}(K, n) = K[X_1, X_2, \ldots, X_n, \partial_1, \partial_2, \ldots, \partial_n]$. Moreover let $\mathscr{A}$ be a ring of smooth functions in $X_1, X_2, \ldots, X_n$ over K (see Remark and Definition 1.11.11 (A)). One concern of Analysis is to study whole *families of differential equations. So for fixed $r, s \in \mathbb{N}$ one chooses a family $\mathbb{F} \subseteq \mathbb{W}^{s\times r}$ of matrices of differential operators. Then one studies all systems of equations (see Remark and Definition 1.11.11 (B))*

$$\mathscr{D}\begin{pmatrix} f_1 \\ f_2 \\ \vdots \\ f_r \end{pmatrix} = \begin{pmatrix} 0 \\ 0 \\ \vdots \\ 0 \end{pmatrix}, \text{ with } \mathscr{D} \in \mathbb{F}.$$

(B) Let the notations and hypotheses by as in part (A). One aspect of the above approach is to study the behavior of the characteristic varieties $\mathbb{V}^{\deg}(\mathscr{D}) := \mathbb{V}_{\mathbb{W}_\bullet^{\deg}}(U_{\mathscr{D}})$ with respect to the degree filtration (see Definition and Remark 1.8.6 and Definition and Remark 1.11.2 (D)) of the D-module $U_{\mathscr{D}}$ defined by the matrix $\mathscr{D}$ (see Remark and Definition 1.11.11 (C)) if this latter runs through the family $\mathbb{F}$. The goal of this section is to prove that the degree of hypersurfaces which cut out set-theoretically the characteristic variety $\mathbb{V}^{\deg}(\mathscr{D})$ is bounded, if $\mathscr{D}$ runs through appropriate families $\mathbb{F}$.

Below, we recall a few notions from Commutative Algebra.

Reminder, Definition and Exercise 1.14.2 (Hilbert Functions, Hilbert Polynomials and Hilbert Coefficients for Modules Over Very Well Filtered Algebras) (A) Let K be a field and let $R = \bigoplus_{i\in\mathbb{N}_0} R_i$ be a homogeneous Noetherian K-algebra (see Conventions, Reminders and Notations 1.1.1 (I) for this notion), so that $R_0 = K$ and $R = K[x_1, x_2, \ldots, x_r]$ with finitely many elements $x_1, x_2, \ldots, x_r \in R_1$. Moreover, let $M = \bigoplus_{i\in\mathbb{Z}} M_i$ be a finitely generated graded R-module. Then we denote the *Hilbert function* of M by h_M, so that $h_M(i) := \dim_K(M_i)$ for all $i \in \mathbb{Z}$. We denote by $P_M(X)$ the *Hilbert polynomial* of M, so that $h_M(i) = P_M(i)$ for all $i \gg 0$. Keep in mind that $\dim(M) = \dim\big(R/\mathrm{Ann}_R(M)\big)$ and

$$\deg\big(P_M(X)\big) = \begin{cases} \dim(M) - 1, & \text{if } \dim(M) > 0 \\ -\infty, & \text{if } \dim(M) \le 0. \end{cases}.$$

The Hilbert polynomial $P_M(X)$ has a *binomial presentation*:

$$P_M(X) = \sum_{k=0}^{\dim(M)-1} (-1)^k e_k(M) \binom{X + \dim(M) - k - 1}{\dim(M) - k - 1} \quad \big(e_k(M) \in \mathbb{Z}, e_0(M) \ge 0\big).$$

The integer $e_k(M)$ is called the k-th *Hilbert coefficient* of M. If $\dim(M) > 0$, $e_0(M) > 0$ is called the *multiplicity* of M. Finally let us also introduce the *postulation number* of M, thus the number $\mathrm{pstln}(M) := \sup\{i \in \mathbb{Z} \mid h_M(i) \neq P_M(i)\}$.

(B) Now, let $(A, A_\bullet)$ be a very well filtered K-algebra (see Definition and Remark 1.3.4 (A)). Let U be a finitely generated (left) A-module. Chose a vector space $V \subseteq U$ of finite dimension such that $AV = U$. Then, the graded $\mathrm{Gr}_{A_\bullet}(A)$-module $\mathrm{Gr}_{A_\bullet V}(U)$ is generated by finitely many homogeneous elements of degree 0 (see Exercise and Definition 1.10.5 (B)(c)). So, by part (A) this graded module admits a Hilbert function $h_{U,A_\bullet V} := h_{\mathrm{Gr}_{A_\bullet V}(U)}$ with $h_{U,A_\bullet V}(i) := \dim_K\big(\mathrm{Gr}_{A_\bullet V}(U)_i\big)$ for all $i \in \mathbb{Z}$, the *Hilbert function* of U with respect to the filtration induced by V. Moreover, by part (A), the module $\mathrm{Gr}_{A_\bullet V}(U)$ admits a Hilbert polynomial, thus a polynomial $P_{U,A_\bullet V}(X) := P_{\mathrm{Gr}_{A_\bullet V}(U)}(X) \in \mathbb{Q}[X]$ with $h_{U,A_\bullet V}(i) = P_{U,A_\bullet V}(i)$ for all $i \gg 0$. We call this polynomial the *Hilbert polynomial* of U with respect to the filtration induced by V. Keep in mind that according to part (A) we have $d_{A_\bullet}(U) := \dim\big(\mathrm{Gr}_{A_\bullet V}(U)\big) = \dim\big(\mathbb{V}_{A_\bullet}(U)\big)$. Moreover the polynomial $P_{U,A_\bullet V}(X)$ has a binomial presentation:

$$P_{U,A_\bullet V}(X) = \sum_{k=0}^{d_{A_\bullet}(U)-1} (-1)^k e_k\big(U, A_\bullet V\big)\binom{X + d_{A_\bullet}(U) - k - 1}{d_{A_\bullet}(U) - k - 1} \quad \big(e_k(U, A_\bullet V) \in \mathbb{Z}\big).$$

The integer $e_k(U, A_\bullet V)$ is called the k-th *Hilbert coefficient* of U with respect to the filtration induced by V. Finally, keep in mind, that by part (A) we have $e_0(U, A_\bullet V) > 0$ if $d_{A_\bullet}(U) > 0$. In this situation the number $e_0(U, A_\bullet V)$ is called the *multiplicity* of U with respect to the filtration induced by V. For the sake of completeness, we set $e_0(U, A_\bullet V) := 0$ if $d_{A_\bullet}(U) \leq 0$. Finally, according to part (A) we define the *postulation number* of U with respect to the filtration induced by V:

$$\mathrm{pstln}_{U,A_\bullet V}(U) := \mathrm{pstln}(\mathrm{Gr}_{A_\bullet V}(U)) := \sup\{i \in \mathbb{Z} \mid h_{U,A_\bullet V}(i) \neq P_{U,A_\bullet V}(i)\}.$$

(C) Keep the notations and hypotheses of part (B) and assume that $d_{A_\bullet}(U) > 0$. Prove the following claims.

(a) There is a polynomial $Q_{U,A_\bullet V}(X) \in \mathbb{Q}[X]$ such that:

(1) $\deg\big(Q_{U,A_\bullet V}(X)\big) = d_{A_\bullet}(U)$,
(2) $\Delta\big(Q_{U,A_\bullet V}(X)\big) := Q_{U,A_\bullet V}(X) - Q_{U,A_\bullet V}(X-1) = P_{U,A_\bullet V}(X)$ and
(3) $\dim_K(A_i V) = Q_{U,A_\bullet V}(i)$ for all $i \gg 0$.
(4) For each $t \in \mathbb{Z}$ the polynomial $Q_{U,A_\bullet V}(X+t) \in \mathbb{Q}[X]$ has leading term

$$\frac{e_0(U, A_\bullet V)}{d_{A_\bullet}(U)!} X^{d_{A_\bullet}(U)}.$$

(Hint: Observe that for all $i \in \mathbb{N}$ we have $\dim_K(A_i V) = \sum_{j=0}^{i} \dim_K \big(\mathrm{Gr}_{A_\bullet V}(U)_j\big) = \sum_{j=0}^{i} h_{U,A_\bullet}(j)\big)$.)

(b) The multiplicity $e_{A_\bullet}(U) := e_0(U, A_\bullet V)$ is the same for each finite dimensional K-subspace $V \subseteq U$ with $AV = U$.
(Hint: Let $V^{(1)}, V^{(2)} \subset U$ be two finite dimensional K-subspaces such that $AV^{(1)} = AV^{(2)} = U$. Use Exercise and Definition 1.10.5 (C)(a) and Definition and Remark 1.10.1 (C)(a) to find some $r \in \mathbb{N}_0$ such that for all $i \in \mathbb{Z}$ it holds $A_{i-r}V^{(1)} \subseteq A_i V^{(2)} \subseteq A_{i+r}V^{(1)}$. Then apply (a).)

(D) Let $A = \mathbb{W} = K[X_1, X_2, \ldots, X_n, \partial_1, \partial_2, \ldots, \partial_n]$ and let $A_\bullet = \mathbb{W}_\bullet = \mathbb{W}_\bullet^{\underline{11}}$ be the standard degree filtration of $\mathbb{W}$ (see Definition and Remark 1.8.6). Let $U = K[X_1, X_2, \ldots, X_n]$ be the D-module of Example 1.11.9. Compute the two polynomials $P_{U,A_\bullet K}(X)$ and $Q_{U,A_\bullet K}(X)$.

The next Exercise and Remark intends to present the *Bernstein Inequality* and the related notion of *holonomic D-module*. For those readers, who aim to learn more about these important subjects, we recommend to consult one of [7–9, 24, 37] or [38].

Exercise and Remark 1.14.3 (A) Endow the Weyl algebra

$$\mathbb{W} = \mathbb{W}(K, n) = K[X_1, X_2, \ldots, X_n, \partial_1, \partial_2, \ldots, \partial_n]$$

with its standard degree filtration $\mathbb{W}_\bullet := \mathbb{W}_\bullet^{\text{deg}}$ (see Definition and Remark 1.8.6). If $d \in \mathbb{W}$ write $\deg(d)$ for the standard degree $\deg^{\underline{1}\underline{1}}(d)$ of d. Use Exercise 1.6.4 (D) to prove the following statement:

If $d \in \mathbb{W} \setminus K$, then there is some $i \in \{1, 2, \ldots, n\}$ such that $\deg\big([X_i, d]\big) = \deg(d) - 1$ or else $\deg\big([\partial_i, d]\big) = \deg(d) - 1$.

(B) *(The Bernstein Monomorphisms)* Keep the notations of part (A) and let U be a non-zero D-module over the Weyl algebra $\mathbb{W}$. Let $V \subseteq U$ be a K-vector space of finite dimension and endow U with the induced filtration $U_\bullet := \mathbb{W}_\bullet V$ (see Exercise and Definition 1.10.5 (A),(B) and Definition and Remark 1.11.2 (D)). Let $k \in \mathbb{N}_0$, let $d \in \mathbb{W}$ with $\deg(d) = k$ and let $i \in \{1, 2, \ldots, n\}$. Prove the following statement

(a) If $k > 0$ and $dU_k = 0$, then $[X_i, d]U_{k-1} = [\partial_i, d]U_{k-1} = 0$.

Use part (A) and statement (B)(a) to prove the following claim by induction on k:

(b) For each $k \in \mathbb{N}_0$ there is a K-linear injective map $\phi_k : \mathbb{W}_k \longrightarrow \text{Hom}_K(U_k, U_{2k})$, given by $\phi_k(d)(u) := du$, for all $d \in \mathbb{W}_k$ and all $u \in U_k$.

(Hint: The existence of the linear map ϕ_k is easy to verify. The injectivity of ϕ_0 is obvious. If $k > 0$ and ϕ_k is not injective, part (A) and statement (B)(b) imply that ϕ_{k-1} is not injective.)

(C) *(The Bernstein Inequality)* Keep the previous notations. Use statement (B)(b) to prove

(a) For all $k \in \mathbb{N}_0$ it holds $\binom{k+2n}{2n} \leq \dim_K(U_k)\dim_K(U_{2k})$.

(Hint: Determine $\dim_K(\mathbb{W}_k)$ for all $k \in \mathbb{N}_0$ and keep in mind that for any two K-vector spaces S, T of finite dimension one has $\dim\big(\text{Hom}_K(S, T)\big) = \dim_K(S)\dim_K(T)$.)
Use statement (a) and Reminder, Definition and Exercise 1.14.2 (C)(a) to prove Bernstein's Inequality:

(b) If $U \neq 0$, then $d_{\mathbb{W}_\bullet}(U) = d_{\mathbb{W}_\bullet^{1,1}}(U) \geq n$.

(D) *(Holonomic D-Modules)* Keep the above notations. It is immediate from the definition, that one always has the inequality $d_{\mathbb{W}_\bullet}(U) \leq 2n$. The D-module U is called *holonomic* if $d_{\mathbb{W}_\bullet}(U) \leq n$, hence if $U = 0$ or else (by Bernsteins' Inequality) $U \neq 0$ and $d_{\mathbb{W}_\bullet}(U) = n$. Holonomic D-modules are of particular interest and play a crucial role in many applications of D-modules. The result of

Reminder, Definition and Exercise 1.14.2 (D) shows that the (simple!) D-module $U = K[X_1, X_2, \ldots, X_n]$ be the D-module of Example 1.11.9 is holonomic.
Use Proposition 1.13.9 to prove the following result:

(a) If $0 \longrightarrow Q \longrightarrow U \longrightarrow P \longrightarrow 0$ is an exact sequence of D-modules, then U is holonomic if and only Q and P are holonomic.

Accepting without proof the fact that all simple D-modules are holonomic, one can prove by statement (a) that a D-module U is holonomic if and only if it is of finite length, hence if and only if it admits a finite ascending chain $0 = U_0 \subsetneq U_1 \subsetneq \cdots \subsetneq U_{l-1} \subsetneq U_l = U$ of submodules, such that U_i/U_{i-1} is simple for all $i = 1, \ldots, l$.

We now recall some basics facts on Local Cohomology Theory. As a reference we suggest [18].

Reminder 1.14.4 (Local Cohomology Modules) (A) Let R be a commutative Noetherian ring and let $\mathfrak{a} \subset R$ be an ideal. The $\mathfrak{a}$-*torsion submodule* of an R-module M is given by

$$\Gamma_{\mathfrak{a}}(M) := \bigcup_{n \in \mathbb{N}_0} (0 :_M \mathfrak{a}^n) \cong \lim_{\underset{n}{\rightarrow}} \operatorname{Hom}_R(R/\mathfrak{a}^n, M).$$

Observe, that the assignment $M \mapsto \Gamma_{\mathfrak{a}}(M)$ gives rise to a covariant left-exact functor of R-modules (indeed a sub-functor of the identity functor)—called the $\mathfrak{a}$-*torsion functor*—so that for each short exact sequence of R-modules $0 \longrightarrow N \longrightarrow M \longrightarrow P \longrightarrow 0$ we naturally have an exact sequence $0 \longrightarrow \Gamma_{\mathfrak{a}}(N) \longrightarrow \Gamma_{\mathfrak{a}}(M) \longrightarrow \Gamma_{\mathfrak{a}}(P)$.
If $i \in \mathbb{N}_0$, the i-*th local cohomology functor* $H^i_{\mathfrak{a}}(\bullet)$ with respect to the ideal $\mathfrak{a}$ can be defined as the i-th right derived functor $\mathscr{R}^i \Gamma_{\mathfrak{a}}(\bullet)$ of the $\mathfrak{a}$-torsion functor, so that for each R-module M one has:

$$H^i_{\mathfrak{a}}(M) = \mathscr{R}^i \Gamma_{\mathfrak{a}}(M) \cong \lim_{\underset{n}{\rightarrow}} \operatorname{Ext}^i_R(R/\mathfrak{a}^n, M).$$

For each short exact sequence of R-modules $0 \longrightarrow N \longrightarrow M \longrightarrow P \longrightarrow 0$ there is a natural exact sequence of R-modules

$$\begin{array}{l} 0 \longrightarrow H^0_{\mathfrak{a}}(N) \longrightarrow H^0_{\mathfrak{a}}(M) \longrightarrow H^0_{\mathfrak{a}}(P) \longrightarrow H^1_{\mathfrak{a}}(N) \longrightarrow H^1_{\mathfrak{a}}(M) \longrightarrow H^1_{\mathfrak{a}}(P) \longrightarrow \\ \longrightarrow H^2_{\mathfrak{a}}(N) \longrightarrow H^2_{\mathfrak{a}}(M) \longrightarrow H^2_{\mathfrak{a}}(P) \longrightarrow H^3_{\mathfrak{a}}(N) \longrightarrow H^3_{\mathfrak{a}}(M) \longrightarrow H^3_{\mathfrak{a}}(P) \cdots, \end{array}$$

the *cohomology sequence* associated to the given short exact sequence. In particular, local cohomology commutes with finite direct sums.
Moreover, we have

(a) If $\sqrt{\mathfrak{a}} = \sqrt{\sum_{i=1}^r Rx_i}$ for some elements $x_1, x_2, \ldots, x_r \in R$, then $H^i_{\mathfrak{a}}(M) = 0$ for all $i > r$ and all R-modules M.
(b) $H^i_{\mathfrak{a}}(M) = 0$ for all $i > \dim(M)$ and all (finitely generated) R-modules M.

(B) *(Graded Local Cohomology)* Assume from now on, that the ring R of part (A) is (positively) graded and that the ideal $\mathfrak{a} \subseteq R$ is graded, so that

$$R = \bigoplus_{j \in \mathbb{N}_0} R_j \text{ and } \mathfrak{a} = \bigoplus_{j \in \mathbb{N}_0} \mathfrak{a}_j, \text{ with } \mathfrak{a}_j = \mathfrak{a} \cap R_j \quad (\forall j \in \mathbb{N}_0).$$

If $M = \bigoplus_{k \in \mathbb{Z}} M_k$ is a graded R-module then, for each $i \in \mathbb{N}_0$, the local cohomology module of M with respect to $\mathfrak{a}$ carries a *natural grading*:

$$H^i_{\mathfrak{a}}(M) = \bigoplus_{j \in \mathbb{Z}} H^i_{\mathfrak{a}}(M)_j.$$

Moreover, if $h : M \longrightarrow N$ is a homomorphism of graded R-modules, then the induced homomorphism in cohomology $H^i_{\mathfrak{a}}(M) \longrightarrow H^i_{\mathfrak{a}}(N)$ is a homomorphism of graded R modules. If $0 \longrightarrow N \longrightarrow M \longrightarrow P \longrightarrow 0$ is an exact sequence of graded R-modules, then so is its associated cohomology sequence (see part (A)).

(C) *(Graded Local Cohomology with Respect to the irrelevant Ideal)* Let $R = \bigoplus_{j \in \mathbb{N}_0} R_j$ be as in part (B). The *irrelevant ideal* of R is defined by

$$R_+ := \bigoplus_{j \in \mathbb{N}} R_j.$$

The graded components of local cohomology modules of finitely generated graded R modules with respect to the irrelevant ideal R_+ behave particularly well, namely:

(a) Let $i \in \mathbb{N}_0$ and let $M = \bigoplus_{j \in \mathbb{Z}} M_j$ be a finitely generated graded R-module. Then:

(1) $H^i_{R_+}(M)_j$ is a finitely generated R_0-module for all $j \in \mathbb{Z}$.
(2) $H^i_{R_+}(M)_j = 0$ for all $j \gg 0$.

Bearing in mind what we just said in Part (C), we no can introduce the cohomological invariant which plays the crucial rôle in this section: Castelnuovo-Mumford regularity. As a reference we suggest Chapter 17 of [18].

Reminder, Remark and Exercise 1.14.5 (Castelnuovo-Mumford Regularity) (A) Keep the notations and hypotheses of Reminder, Definition and Exercise 1.14.2(A) and of Reminder 1.14.4. For each finitely generated graded module $M = \bigoplus_{j \in \mathbb{Z}} M_j$ over the homogeneous Noetherian K-algebra $R = \bigoplus_{j \in \mathbb{N}_0} R_j = K[x_1, x_2, \ldots, x_r]$ and for each $k \in \mathbb{N}_0$ by Reminder 1.14.4 (A)(a),(b) and (C)(a)(2) we now can define the *Castelnuovo-Mumford regularity at and above level* k of M by

$$\operatorname{reg}^k(M) := \sup\{a_i(M) + i \mid i \geq k\} = \max\{a_i(M) + i \mid i = k, k+1, \ldots, \dim(M)\}$$

with

$$a_i(M) := \sup\{j \in \mathbb{Z} \mid H^i_{R_+}(M)_j \neq 0\} \text{ for all } i \in \mathbb{N}_0,$$

where $H^i_{R_+}(M)_j$ denotes the j-th graded component of the i-th local cohomology module $H^i_{R_+}(M) = \bigoplus_{k\in\mathbb{Z}} H^i_{R_+}(M)_k$ of M with respect to the irrelevant ideal $R_+ := \bigoplus_{j\in\mathbb{N}} R_j = \sum_{m=1}^r Rx_m$ (see Reminder 1.14.4 (B),(C)).
Keep in mind that the *Castelnuovo-Mumford regularity* of M is defined by

$$\operatorname{reg}(M) := \operatorname{reg}^0(M) = \sup\{a_i(M) + i \mid i \in \mathbb{N}_0\} = \max\{a_i(M) + i \mid i = 0, 1, \ldots, \dim(M)\}$$

and keep in mind the fact that

$$\operatorname{reg}^1(M) = \operatorname{reg}(M/\Gamma_{R_+}(M)) \text{ and } P_{M/\Gamma_{R_+}(M)}(X) = P_M(X).$$

(B) Keep the notations and hypotheses of part (A). Let

$$\operatorname{gendeg}(M) := \inf\{m \in \mathbb{Z} \mid M = \sum_{k\leq m} RM_k\} \quad \big(\leq \operatorname{reg}(M)\big)$$

denote the *generating degree* of M. Keep in mind, that the ideal $\operatorname{Ann}_R(M) \subseteq R$ is homogeneous. Use the previous inequality to prove the following claims:

(a) If $b \in \mathbb{Z}$ such that $\operatorname{reg}\big(\operatorname{Ann}_R(M)\big) \leq b$, there are elements

$$f_1, f_2, \ldots, f_s \in \operatorname{Ann}_R(M) \cap \big(\bigcup_{i\leq b} R_i\big) \text{ with } \operatorname{Var}\big(\operatorname{Ann}_R(M)\big) = \bigcap_{i=1}^{s} \operatorname{Var}(f_i).$$

(C) We recall a few basic facts on Castelnuovo-Mumford regularity.

(a) If $r \in \mathbb{N}$ and $R = K[T_1, T_2, \ldots, T_r]$ is a polynomial ring over the field K, then $\operatorname{reg}(R) = \operatorname{reg}\big(K[T_1, T_2, \ldots, T_r]\big) = 0$.
(b) If $0 \longrightarrow N \longrightarrow M \longrightarrow P \longrightarrow 0$ is a short exact of finitely generated graded R -modules, then we have the equality $\operatorname{reg}(N) \leq \max\{\operatorname{reg}(M), \operatorname{reg}(P) + 1\}$.
(c) If $r \in \mathbb{N}$ and if $M^{(1)}, M^{(2)}, \ldots, M^{(r)}$ are finitely generated graded R-modules, then we have the equality $\operatorname{reg}\big(\bigoplus_{i=1}^r M^{(i)}\big) = \max\{\operatorname{reg}(M^{(i)}) \mid i = 1, 2, \ldots, r\}$.

(D) We mention the following bounding result (see Corollary 17.4.2 of [18]):

(a) Let $R = \bigoplus_{j\in\mathbb{N}_0} R_j$ be a Noetherian homogeneous ring (see Conventions, Reminders and Notations 1.1.1 (I) for this notion) such that R_0 is Artinian and local. Let $W = \bigoplus_{j\in\mathbb{Z}} W_j$ be a finitely generated graded R-module and let $P \in \mathbb{Q}[X]\setminus\{0\}$. Then, there is an integer G such that for each R-homomorphism $f : W \longrightarrow M$ of finitely generated graded R-modules, which is surjective in all large degrees and such that $P_M = P$, we have $\operatorname{reg}^1(M) \leq G$.

Use the bounding result of statement (a) to prove the following result.

(b) There is a function $\overline{B} : \mathbb{N}_0^2 \times \mathbb{Q}[X] \longrightarrow \mathbb{Z}$ such that for each choice of $r, t \in \mathbb{N}$, for each field K, for each homogeneous Noetherian K-algebra $R = \bigoplus_{i \in \mathbb{N}_0} R_i$ with $h_R(1) \leq t$ and each finitely generated graded R-module $M = \bigoplus_{i \in \mathbb{Z}} M_i$ with $M = RM_0$ and $h_M(0) \leq r$ we have

$$\operatorname{reg}^1(M) \leq \overline{B}\big(t, r, P_M\big).$$

Another bounding result, which we shall use later is (see Corollary 6.2 of [17]):

(c) Let $R = K[T_1, T_2, \dots, T_r]$ be a polynomial ring over the field K, furnished with its standard grading. Let $f : W \longrightarrow V$ be a homomorphism of finitely generated graded R-modules such that $V \neq 0$ is generated by μ homogeneous elements of degree 0. Then

$$\operatorname{reg}\big(\operatorname{Im}(f)\big) \leq \big[\max\{\operatorname{gendeg}(W), \operatorname{reg}(V) + 1\} + \mu + 1\big]^{2^{r-1}}.$$

We now prove a special case of Theorem 3.10 of [16].

Proposition 1.14.6 *Let $r \in \mathbb{N}$, let $R := K[T_1, T_2, \dots, T_r]$ be the polynomial ring over the field K and let $M = \bigoplus_{n \in \mathbb{N}_0} M_n$ be finitely generated graded R-module with $M = RM_0$. Then*

$$\operatorname{reg}\big(\operatorname{Ann}_R(M)\big) \leq \big[\operatorname{reg}(M) + h_M(0)^2 + 2\big]^{2^{r-1}} + 1.$$

Proof Observe first, that we have an exact sequence of graded R-modules

$$0 \longrightarrow \operatorname{Ann}_R(M) \longrightarrow R \xrightarrow{\varepsilon} \operatorname{Hom}_R(M, M), \text{ with } x \mapsto \varepsilon(x) := x\operatorname{Id}_M, \text{ for all } x \in R.$$

Moreover, there is an epimorphism of graded R-modules

$$\pi : R^{h_M(0)} \longrightarrow M \longrightarrow 0.$$

So, with $g := \operatorname{Hom}_R(\pi, \operatorname{Id}_M)$ we get an induced monomorphism of graded R-modules

$$0 \longrightarrow \operatorname{Hom}_R(M, M) \xrightarrow{g} \operatorname{Hom}_R\big(R^{h_M(0)}, M\big) \cong M^{h_M(0)}.$$

So, we get a composition map

$$f := g \circ \varepsilon : R \longrightarrow M^{h_M(0)} =: V, \text{ with } \operatorname{Im}(f) = \operatorname{Im}(\varepsilon) \cong R/\operatorname{Ann}_R(M).$$

Now, observe that $\operatorname{gendeg}(R) = 0$ (see Reminder, Remark and Exercise 1.14.5 (C)(a)), $\operatorname{reg}(V) = \operatorname{reg}(M)$ (see Reminder, Remark and Exercise 1.14.5 (C)(c)) and that V is generated by $h_M(0)^2$ homogeneous elements of degree 0. So, by Reminder, Remark and Exercise 1.14.5 (D)(c) we obtain

$$\operatorname{reg}\big(R/\operatorname{Ann}_R(M)\big) = \operatorname{reg}\big(\operatorname{Im}(f)\big) \le \big[\operatorname{reg}(M) + h_M(0)^2 + 2\big]^{2^{r-1}}.$$

On application of Reminder, Remark and Exercise 1.14.5 (C) (b) to the short exact sequence of graded R-modules

$$0 \longrightarrow \operatorname{Ann}_R(M) \longrightarrow R \longrightarrow R/\operatorname{Ann}_R(M) \longrightarrow 0$$

and keeping in mind that $\operatorname{reg}(R) = 0$, we thus get indeed our claim.

Exercise 1.14.7 Let the notations and hypotheses be as in Proposition 1.14.6. Show that

(a) $\operatorname{reg}\big(\operatorname{Ann}_R(M/\Gamma_{R_+}(M))\big) \le \big[\operatorname{reg}^1(M) + h_M(0)^2 + 2\big]^{2^{r-1}} + 1.$

(b) $\operatorname{Var}\big(\operatorname{Ann}_R(M/\Gamma_{R_+}(M))\big) = \begin{cases} \operatorname{Var}\big(\operatorname{Ann}_R(M)\big), & \text{if } \dim_R(M) > 0 \\ \emptyset, & \text{if } \dim_R(M) = 0. \end{cases}$

Notation, Remark and Exercise 1.14.8 (A) Let $\overline{B} : \mathbb{N}_0^2 \times \mathbb{Q}[X] \longrightarrow \mathbb{Z}$ be the bounding function introduced in Reminder, Remark and Exercise 1.14.5 (D)(b). We define a new function

$$F : \mathbb{N}^2 \times \mathbb{Q}[X] \longrightarrow \mathbb{Z} \text{ by } F(t,r,P) := \big[\overline{B}(t,r,P) + r^2 + 2\big]^{2^{r-1}} + 1 \quad \big(t,r \in \mathbb{N},\ P \in \mathbb{Q}[X]\big).$$

(B) Let the notations as in part (A). Use Proposition 1.14.6, Reminder, Remark and Exercise 1.14.5 (B) and Exercise 1.14.7 to show that for each field K, for each choice of $r, t \in \mathbb{N}$, for each polynomial ring $R = K[T_1, T_2, \ldots, T_t]$ and for each finitely generated graded R-module $M = \bigoplus_{n \in \mathbb{N}_0} M_n$ with $M = RM_0$, $h_M(0) \le r$ and $P_M = P$, we have the following statements:

(a) $\operatorname{reg}\big(\operatorname{Ann}_R(M/\Gamma_{R_+}(M))\big) \le F(t,r,P)$.

(b) There are homogeneous polynomials $f_1, f_2, \ldots, f_s \in \operatorname{Ann}_R\big(M/\Gamma_{R_+}(M)\big)$ with

   (1) $\deg(f_i) \le F(t,r,P)$ for all $i = 1, 2, \ldots, s$.

   (2) $\operatorname{Var}\big(\operatorname{Ann}_R(M)\big) = \operatorname{Var}(f_1, f_2, \ldots, f_s) = \bigcap_{i=1}^{s} \operatorname{Var}(f_i)$.

No, we are ready to prove the main result of this section.

Theorem 1.14.9 (Boundedness of the Degrees of Defining Equations of Characteristic Varieties, Compare [16]) *Let $n \in \mathbb{N}$, let K be a field of characteristic* 0, *let U be a D-module over the standard Weyl algebra*

$$\mathbb{W} = \mathbb{W}(K,n) = K[X_1, X_2, \ldots, X_n \partial_1, \partial_2, \ldots, \partial_n]$$

and let $V \subseteq U$ *be a* K*-subspace with* $\dim_K(V) \leq r < \infty$ *and* $U = \mathbb{W}V$*. Moreover, let*

$$F : \mathbb{N}^2 \times \mathbb{Q}[X] \longrightarrow \mathbb{Z}$$

be the bounding function defined in Notation, Remark and Exercise 1.14.8 (A). Keep in mind that the degree filtration $\mathbb{W}_\bullet^{\deg}$ *of* $\mathbb{W}$ *(see Definition and Remark 1.8.6) is very good (see Corollary 1.8.7 (a)) and let*

$$P_{U,\mathbb{W}_\bullet^{\deg} V} \in \mathbb{Q}[X]$$

be the Hilbert polynomial of U *induced by* V *with respect to the degree filtration* $\mathbb{W}_\bullet^{\deg}$ *(see Reminder, Definition and Exercise 1.14.2 (B)).*

Then, there are homogeneous polynomials

$$f_1, f_2, \ldots, f_s \in \mathbb{P} = K[Y_1, Y_2, \ldots, Y_n, Z_1, Z_2, \ldots, Z_n]$$

such that

(a) $\deg(f_i) \leq F\big(2n, r, P_{U,\mathbb{W}_\bullet^{\deg}}\big)$.
(b) $\mathbb{V}_{\mathbb{W}_\bullet^{\deg}}(U) = \mathrm{Var}(f_1, f_2, \ldots, f_s) = \bigcap_{i=1}^{s} \mathrm{Var}(f_i)$.

Proof Observe that (see Definition and Remark 1.11.2)

$$\mathbb{V}_{\mathbb{W}_\bullet^{\deg}}(U) = \mathrm{Var}\big(\mathrm{Ann}_{\mathbb{P}}(\mathrm{Gr}_{\mathbb{W}_\bullet^{\deg} V}(U)\big).$$

Now, we may conclude by Notation, Remark and Exercise 1.14.8 (B)(b), applied to the graded $\mathbb{P}$-module $\mathrm{Gr}_{\mathbb{W}_\bullet^{\deg} V}(U)$ and bearing in mind that—by Exercise and Definition 1.10.5 (B)(c)—this latter graded module is generated in degree 0.

Conclusive Remark 1.14.10 (A) Keep the above notations. To explain the meaning of this result, we fix $r, s \in \mathbb{N}$ and we fix a polynomial $P \in \mathbb{Q}[X]$. For any matrix

$$\mathscr{D} = \begin{pmatrix} d_{11} & d_{12} & \ldots & d_{1r} \\ d_{21} & d_{22} & \ldots & d_{2r} \\ \vdots & \vdots & & \vdots \\ d_{s1} & d_{s2} & \ldots & d_{sr} \end{pmatrix} \in \mathbb{W}^{s \times r}$$

of polynomial partial differential operators we consider the induced epimorphism of D-modules

$$\mathbb{W}^r \xrightarrow{\pi_{\mathscr{D}}} U_{\mathscr{D}} \longrightarrow 0,$$

consider the K-subspace

$$K^r = (\mathbb{W}_0^{\deg})^r \subset \mathbb{W}^r$$

and set

$$V_{\mathscr{D}} := \pi_{\mathscr{D}}(K^r).$$

Then, referring to our Preliminary Remark 1.14.1 we consider the family of systems of differential equations

$$\mathbb{F} = \mathbb{F}^P := \{\mathscr{D} \in \mathbb{W}^{s \times r} \mid P_{U_{\mathscr{D}}, \mathbb{W}_\bullet^{\deg} V_{\mathscr{D}}} = P\}$$

whose *canonical Hilbert polynomial* $P_{U_{\mathscr{D}}, \mathbb{W}_\bullet^{\deg} V_{\mathscr{D}}}$ equals P. As an immediate application of Theorem 1.14.9 we can say

The degree of hypersurfaces which cut out set-theoretically the characteristic variety $\mathbb{V}^{\deg}(\mathscr{D})$ is bounded, if $\mathscr{D}$ runs through the family $\mathbb{F}^P$.

Clearly, our results give much more, as they bound the invariant

$$\mathrm{reg}\big(\mathrm{Ann}_{\mathbb{P}}\big[\mathrm{Gr}_{\mathbb{W}_\bullet^{\deg} V_{\mathscr{D}}}(U_{\mathscr{D}})/\Gamma_{\mathbb{P}_+}(\mathrm{Gr}_{\mathbb{W}_\bullet^{\deg} V_{\mathscr{D}}}(U_{\mathscr{D}})\big]\big)$$

along the class $\mathbb{F}^P$.

(B) Our motivation to prove Theorem 1.14.9 was a question arising in relation with the PhD thesis [5], namely: Does the Hilbert function (with respect to an appropriate filtration) of a D-module U over a standard Weyl algebra $\mathbb{W}$ bound the degrees of polynomials which cut out set-theoretically the characteristic variety of U? This leads to the question, whether the Hilbert function h_M of a graded module M which is generated over the polynomial ring $K[X_1, X_2, \ldots, X_r]$ by finitely many elements of degree 0 bounds the (Castelnuovo-Mumford) regularity $\mathrm{reg}(\mathrm{Ann}_R(M))$ of the annihilator $\mathrm{Ann}_R(M)$ of M. This latter question was answered affirmatively in the Master thesis [39] and lead to the article [16].
Theorem 1.14.9 above actually improves what has been shown in [16] and in Theorem 14.6 of [15]. There it is shown, that the degrees of the polynomials $f_1, f_2, \ldots, f_s \in \mathbb{P}$ which occur in Theorem 1.14.9 are bounded in terms of n and the Hilbert function $h_{U,A_\bullet V} = h_{\mathrm{Gr}_{A_\bullet V}(U)}$ (see Reminder, Definition and Exercise 1.14.2 (B)). More precisely, in these previous results, the degrees in question are bounded in terms of n, $h_{U,A_\bullet V}(0)$ and the postulation number (see Reminder, Definition and Exercise 1.14.2 (A))

$$\mathrm{pstln}_{A_\bullet V}(U) := \sup\{i \in \mathbb{Z} \mid h_{U,A_\bullet V}(i) \neq P_{U,A_\bullet V}(0)\} = \mathrm{pstln}\big(\mathrm{Gl}_{A_\bullet V}(U)$$

of U with respect to the filtration $A_\bullet V$. Theorem 1.14.9 shows, that the postulation number $\mathrm{pstln}_{A_\bullet V}(U)$ is not needed to bound the degrees we are interested in.

(C) We thank the referee for having pointed out to us, that Aschenbrenner and Leykin [2] have proved a result, which is closely related to Theorem 1.14.9 and which furnishes a bound on the degree of the elements of Gröbner bases of a left ideal $I \subseteq \mathbb{W}$ of our Weyl algebra. More precisely, if $\underline{\omega} \in \Omega$ (see Notation 1.13.1), if $d \in \mathbb{N}$ and if I is generated by elements whose $\underline{\omega}$-weighted degree $\deg^{\underline{\omega}}(\bullet)$ does not exceed d, then I admits a $\leq^{\underline{\omega}}$-Gröbner basis consisting of elements whose $\underline{\omega}$-weighted degree does not exceed the bound $2\big(\frac{d^2}{2} + d\big)^{2^{2n-1}}$.
As the Castelnuovo-Mumford regularity $\operatorname{reg}(\mathfrak{a})$ of a graded ideal in the polynomial ring $\mathfrak{a} \subseteq K[X_1, x_2, \ldots, X_n]$ over a field K is an upper bound for the degree of the polynomials occurring in some Gröbner basis of $\mathfrak{a}$, the mentioned result in [2] corresponds to the "classical" regularity bound $\operatorname{reg}(\mathfrak{a}) \leq \big(2\operatorname{gendeg}(\mathfrak{a})\big)^{2^{n-2}}$ for graded ideals in the polynomial ring (see [23, 27, 28], but also [17] and [22]). Via Gröbner bases and Macaulay's Theorem for Hilbert Functions (see [26], for example), this latter regularity bound on its turn, is also related to the module theoretic form of Mumford's regularity bound ([18], Corollary 17.4.2), we were using as an important tool in the proof of Theorem 1.14.9 (see Reminder, Remark and Exercise 1.14.5 (D)(a)).

References

1. W. Adams, P. Loustaunau, *An Introduction to Gröbner Bases*. Graduate Studies in Mathematics, vol. 3 (American Mathematical Society, Providence, 1994)
2. M. Aschenbrenner, A. Leykin, Degree bounds for Gröbner bases in algebras of solvable type. J. Pure Appl. Algebra **213**(8), 1578–1605 (2009)
3. A. Assi, F.J. Castro-Jiménez, J.M. Granger, The Gröbner fan of an A_n-module. J. Pure Appl. Algebra **150**(1), 27–39 (2000)
4. J. Ayoub, *Introduction to Algebraic D-Modules*. Lecture Notes (University of Zürich, Zürich, 2009). (www.math.uzh.ch → Professoren → J.Ayoub → Vorlesungen und Seminare)
5. M. Bächtold, Fold-type solution singularities and characteristic varieties of non-linear PDEs, PhD Dissertation, Institute of Mathematics, University of Zürich (2009)
6. T. Becker, V. Weispfennig, *Gröbner Bases* (Springer, New York, 1993).
7. J.N. Bernstein, Modules over a ring of differential operators. study of the fundamental solutions of equations with constant coefficients. Funkttsional'nyi Analiz Ego Prilozheniya **5**(2), 1–16 (1971). Translated from the Russian
8. J.-E. Björk, *Rings of Differential Operators*, vol. 21. North-Holland Mathematical Library (North-Holland, Amsterdam, 1979)
9. J.-E. Björk, *Analytic $\mathscr{D}$-Modules and Applications*. Mathematics and its Applications (Kluwer Academic Publishers, Dordrecht, 1993)
10. R. Boldini, Modules over Weyl algebras. Diploma thesis, Institute of Mathematics, University of Zürich (2007)
11. R. Boldini, Finiteness of leading monomial ideals and critical cones of characteristic varieties. PhD Dissertation, Institute of Mathematics, University of Zürich (2012)
12. R. Boldini, Critical cones of characteristic varieties. Trans. Am. Math. Soc. **365**(1), 143–160 (2013)
13. A. Borel, et al., *Algebraic D-Modules*. Perspectives in Mathematics, vol. 2 (Academic, London, 1987)

14. J. Briançon, P. Maisonobe, Idéaux de germes d'opérateurs différentiels à une variable. Enseignement Mathématique **30**(1–2), 7–38 (1984)
15. M. Brodmann, Notes on Weyl algebras and d-modules, in *Four Lecture Notes on Commutative Algebra*, ed. by N.T. Cuong, L.T. Hoa, N.V. Trung. Lecture Notes Series of the Vietnam Institute for Advanced Study in Mathematics, Hanoi, vol. 4 (2015), pp. 7–96
17. M. Brodmann, T. Götsch, Bounds for the castelnuovo-mumford regularity. J. Commutative Algebra **1**(2), 197–225 (2009)
18. M. Brodmann, R.Y. Sharp, *Local Cohomology – An Algebraic Introduction with Geometric Applications*. Cambridge Studies in Advanced Mathematics, vol. 136, 2nd edn. (Cambridge University Press, Cambridge, 2013)
16. M. Brodmann, C.H. Linh, M.-H. Seiler, Castelnuovo-mumford regularity of annihilators, ext and tor modules, in *Commutative Algebra*, ed. by I. Peeva. Expository Papers Dedicated to David Eisenbud on the Occasion of his 65th Birthday (Springer, New York, 2013), pp. 207–236.
19. B. Buchberger, F. Winkler, *Gröbner Bases and Applications*. London Mathematical Society Lecture Notes Series, vol. 251 (Cambridge University Press, Cambridge, 1998)
20. J.L. Bueso, J. Gómez-Torrecillas, A. Verschoren, *Algorithmic Methods in Non-Commutative Algebra: Applications to Quantum Groups*. Mathematical Modelling: Theory and Applications, vol. 17 (Kluwer Academic Publishers, Dordrecht, 2003)
21. F.J. Castro-Jiménez, Théorè me de division pour les opérateurs différentiels et calcul des multiplictés. Thèse de 3ème cycle, Paris VII (1984)
23. G. Caviglia, E. Sbarra, Characteristic free bounds for the Castelnuovo-Mumford regularity. Compositio Mathematica **141**, 1365–1373 (2005)
22. M. Chardin, A.L. Fall, U. Nagel, Bounds for the Castelnuovo-Mumford regularity of modules. Math. Z. **258**, 69–80 (2008)
24. S.C. Coutinho, *A Primer of Algebraic D-Modules*. LMS Student Texts, vol. 33 (Cambridge University Press, Cambridge, 1995)
25. D. Cox, J. Little, D. O'Shea, *Ideals, Varieties and Algorithms* (Springer, New York, 1992)
26. D. Fröberg, *An Introduction to Gröbner Bases* (Wiley, Chichester, 1997)
27. A. Galligo, Théorème de division et stabilité en géometrie analytique locale. Annales de l'Institut Fourier **29**, 107–184 (1979)
28. M. Giusti, Some effectivity problems in polynomial ideal theory, in *Eurosam 84 (Cambridge, 1984)*. Springer Lecture Notes in Computational Sciences, vol. 174 (Springer, Berlin, 1984), pp. 159–171
29. M. Kashiwara, *D-Modules and Microlocal Calculus*. Translation of Mathematical Monographs, vol. 217 (American Mathematical Society, Providence, 2003)
30. M. Kreuzer, L. Robbiano, *Computational Commutative Algebra 1* (Springer, Berlin, 2000)
31. V. Levandovskyy, Non-commutative computer algebra for polynomial algebras: Gröbner bases, applications and implementations. PhD thesis, University of Kaiserslautern (2005)
32. H. Li, F. van Oystaeyen, *Zariski Filtrations*. *K*-Monographs in Mathematics, vol. 2 (Kluwer Academic Publishers, Dordrecht, 1996)
33. G. Lyubeznik, Finiteness properties of local cohomology modules (an application of *D*-modules to commutative algebra). Invent. Math. **113**, 14–55 (1993)
34. G. Lyubeznik, A partial survey of local cohomology, in *Local Cohomology and its Applications*, ed. by G. Lyubeznik. M. Dekker Lecture Notes in Pure and Applied Mathematics, vol. 226 (CRC Press, Boca Raton, 2002), pp. 121–154
35. J.C. McConnell, J.C. Robson, *Noncommutative Noetherian Rings*. AMS Graduate Studies in Mathematics, vol. 30 (American Mathematical Society, Providence, 2001)
36. B. Mishra, *Algorithmic Algebra* (Springer, New York, 1993)
37. F. Pham, *Singularités des Systèmes Différentiels de Gauss-Manin*. Progress in Mathematics (Birkhäuser, Basel, 1979)
38. M. Saito, B. Sturmfels, N. Takayama, *Gröbner Deformations of Hypergeometric Differential Equations*. Algorithms and Computation in Mathematics, vol. 6 (Springer, Berlin, 2000)

39. M.-H. Seiler, Castelnuovo-Mumford regularity of annihilators. Diploma thesis, Institute of Mathematics, University of Zürich (2010)
40. A.S. Sikora, Topology on the Space of Orderings of Groups. Bull. Lond. Math. Soc. **36**, 519–526 (2004)
41. B. Sturmfels, *Gröbner Bases and Convex Polytopes*. University Lecture Notes Series, vol. 8 (American Mathematical Society, Providence, 1996)
42. W. Vasconcelos, *Computational Methods in Commutative Algebra and Algebraic Geometry*. Algorithms and Computation in Mathematics, vol. 2 (Springer, Berlin, 1998)
43. V. Weispfennig, *Constructing Universal Gröbner Bases*. Lecture Notes in Computer Science, vol. 356 (Springer, Berlin, 1989)

Chapter 2
Inverse Systems of Local Rings

Juan Elias

Abstract Matlis duality and the particular case of Macaulay correspondence provide a dictionary between the Artin algebras and their inverse systems. Inspired in a result of Emsalem we translate the problem of classification of Artin algebras to a problem of linear system of equations on the inverse systems.

The main purpose of these notes is to use this result to classify Artin Gorenstein algebras with Hilbert function $\{1, 3, 3, 1\}$, level algebras and compressed algebras. The main results presented in these notes were obtained in collaboration with M.E. Rossi.

2.1 Introduction

These notes are based on a series of lectures given by the author at the Vietnam Institute for Advanced Study in Mathematics, Hanoi, during the period February 8–March 7, 2014. The aim of these three lectures was to present some recent results on the classification of Artin Gorenstein and level algebras by using the inverse system of Macaulay. These notes are not a review on the known results of Macaulay's inverse systems. See [12, 20–23] and [11] for further details on inverse systems.

Let $R = \mathbf{k}[[x_1, \dots x_n]]$ be the ring of the formal series and let $S = \mathbf{k}[y_1, \dots, y_n]$ be a polynomial ring. Macaulay established a one-to-one correspondence between the Gorenstein Artin algebras $A = R/I$ and cyclic submodules $\langle F \rangle$ of the polynomial ring S. This correspondence is a particular case of Matlis duality because the injective hull of $\mathbf{k}$ as R-module is isomorphic to S. The structure of S as R-module is defined, depending on the characteristic of the residue field $\mathbf{k}$, by derivation or by contraction. Macaulay's correspondence establish a dictionary between the algebraic-geometric properties of Artin Gorenstein algebras A and the algebraic properties of its inverse system F or the geometric properties of the variety

J. Elias (✉)
Department de Matemàtiques i Informàtica, Universitat de Barcelona, Barcelona, Spain
e-mail: elias@ub.edu

N. Tu CUONG et al. (eds.), *Commutative Algebra and its Interactions to Algebraic Geometry*, Lecture Notes in Mathematics 2210,
https://doi.org/10.1007/978-3-319-75565-6_2

defined by $F \equiv 0$. See [13] for the extension to higher dimensions of Macaulay's correspondence.

In the second chapter we review the main results on injective modules. We prove the existence on the injective hull of a ring and we prove Matlis' duality for a complete ring. The main references used in this chapter are: [28] and [27].

In the third chapter we study Macaulay's correspondence that is a particular case of Matlis' duality. In the main result of this we prove that S is the injective hull of the residue field of the R-module $\mathbf{k}$. From this result and Matlis' duality we deduce Macaulay's correspondence. We end the chapter computing the Hilbert function of a quotient $A = R/I$ in terms of its inverse system. The main references used in this chapter are: [18, 20–23] and [25].

The fourth chapter is devoted to give a quick introduction to Artin Gorenstein, level and compressed algebras. We only quote the results needed to achieve the main goal of these notes. The main references used in this chapter are: [20] and [21].

The fifth chapter is the core of these notes. We present the main results obtained in collaboration with M.E. Rossi on the classification of Artinian Gorenstein algebras, level algebras and compressed algebras, [12] and [11]. After a short review of the classification of Artin algebra we show the difficulty of the problem of the classification of Artin algebras recalling some results obtained in collaboration with Valla, [14] and [15].

Inspired in a result of Emsalem, [16], we translate the problem of classification of Artin algebras to a problem of linear systems of equations. The study of these systems of equations permits to establish the main result of this paper, Theorem 2.5.10. We end the chapter by giving a complete analytic classification of Artin Gorenstein algebras with Hilbert function $\{1, 3, 3, 1\}$ by using the Weierstrass form of an elliptic plane curve. The main references used in this chapter are: [12, 20] and [11].

In Sect. 2.6 we consider the problem of computing the Betti numbers of an ideal I by considering only its inverse system without computing the ideal I. The main open problem is to characterize the complete intersection ideals in terms of their inverse systems. In this chapter we focus the study on the computation of the last Betti number (i.e. the Cohen-Macaulay type) and the first Betti number (i.e. the minimal number of generators)

In the last chapter we show that some results of the chapter four cannot be generalized and we present several explicit computations of the minimal number of generators of some families of Artin Gorenstein and level algebras.

In these notes we omit reviewing some recent interesting results on the rationality of the Poincaré series of an Artin Gorenstein algebra, on the smoothability of the Artinian algebras, and the applications of these results to the study of the geometric properties of Hilbert schemes, see for instance [4, 5] and their reference's list.

The examples of this paper are done by using the Singular library [7, 8], and Mathematica®.

2.2 Injective Modules: Matlis' Duality

Given a commutative ring R we denote by R_mod, resp. $R\_mod.Noeth$, $R\_mod.Artin$, the category of R-modules, resp. category of Noetherian R-modules, Artinian R-modules.

Definition 2.2.1 (Injective Module) Let R be a commutative ring and let E be an R-module. E is injective if and only if $\mathrm{Hom}_R(\cdot, E)$ is an exact functor.

Since for all R-module E the contravariant functor $\mathrm{Hom}_R(\cdot, E)$ is right exact, we have that E is injective if and only for all injective morphism $h : M \longrightarrow N$ and for all morphism $f : M \longrightarrow E$, where M and N are R-modules, there exists a morphism $g : N \longrightarrow E$ making the following diagram commutative:

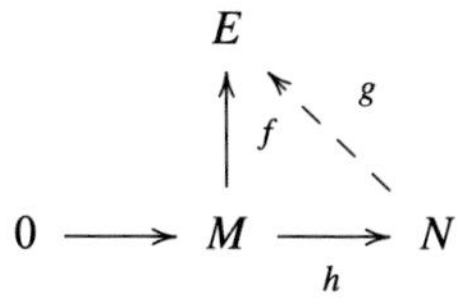

In the following result we collect some basic properties of injective modules.

Proposition 2.2.2

(i) If a R-module E is injective, then every short exact sequence splits:

$$0 \longrightarrow E \longrightarrow M \longrightarrow N \longrightarrow 0$$

(ii) If an injective module E is a submodule of a module M, then E is a direct summand of M, in other words, there is a complement S with $M = S \oplus E$.
(iii) If $(E_j)_{\in J}$ is a family of injective R-modules, then $\prod_{j \in J} E_j$ is also an injective module.
(iv) Every direct summand of an injective R-module is injective.
(v) A finite direct sum of injective R-modules is injective.

Now that we have showed some of the properties of the injective modules, we need to find an easier way to check the injectivity of a module. This criterion is the following:

Proposition 2.2.3 (Baer's Criterion) *A R-module E is injective if and only if every homomorphism $f : I \to E$, where I is an ideal of R, can be extended to R.*

Proof First, if E is injective, then, as I is a submodule of R, the existence of an extension g of f is just a a straight consequence of the injectivity of E.

Consider that we have the following diagram, where M is a submodule of a R-module N:

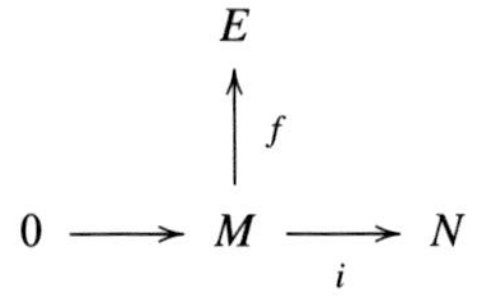

We may assume that M is a submodule of N. Let us consider the set $X = \{(M', g') | M \subset M' \subset N,\ g'|_M = f\}$. Note that $X \neq \emptyset$ because $(M, f) \in X$. Now we put a partial order in X, $(M', g') \preceq (M'', g'')$, which means that $M' \subset M''$ and g'' extends g'. It is easy to see that any chain in X has an upper bound in X (just take the union). By Zorn's Lemma we have that there is a maximal element (M_0, g_0) of X. If $M_0 = N$ we are done, so we can assume that there is some $b \in N$ that is not in M_0. Define $I = \{r \in R : r.b \in M_0\}$, which is clearly an ideal of R. Now define $h : I \to E$ by $h(r) = g_0(r.b)$. By hypothesis, there is a map h^* extending h. Finally define $M_1 = M_0 + \langle b \rangle$ and $g_1 : M_1 \to E$ by

$$g_1(a_0 + br) = g_0(a_0) + r \cdot h^*(1),$$

where $a_0 \in M_0$ and $r \in R$. Notice that if $a_0 + r.b = a_0' + r'.b$ then $(r - r')b = a_0' - a_0 \in M_0$ and $(r - r') \in I$. Therefore, $g_0((r - r')b)$ and $h(r - r')$ are defined and we have:

$$g_0(a_0' - a_0) = g_0((r - r')b) = h(r - r') = h^*(r - r') = (r - r') \cdot h^*(1).$$

Thus, $g_0(a_0') - g_0(a_0) = r \cdot h^*(1) - r' \cdot h^*(1)$ and this shows that $g_0(a_0') + r' \cdot h^*(1) = g_0(a_0) + r \cdot h^*(1)$.

Clearly, $g_1(a_0) = g_0(a_0)$ for all $a_0 \in M_0$, so that the map g_1 extends g_0. We conclude that $(M_0, g_0) \preceq (M_1, g_1)$ and $M_0 \neq M_1$, contradicting the maximality of (M_0, g_0). Therefore, $M_0 = N$, the map g_0 is a lifting of f and then E is injective.

Proposition 2.2.4 *If R is a Noetherian ring and $(E_j)_{j \in J}$ is a family of injective R-modules, then $\bigoplus_{j \in J} E_j$ is an injective R-module.*

Proof By the Baer criterion, it suffices to complete the diagram

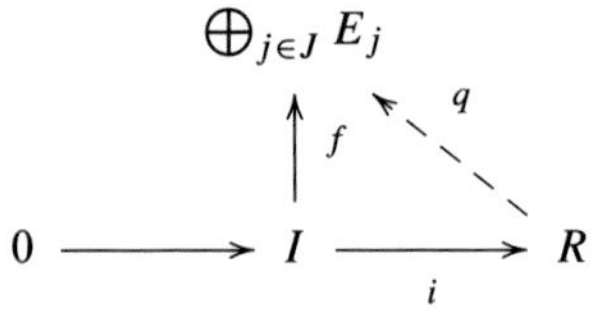

where I is an ideal of R. If $x \in \bigoplus_j E_j$, then $x = (e_j)$, where $e_j \in E_j$. Since R is noetherian, I is finitely generated. There exists a finite set S such that $\mathrm{Im}(f) \subset \oplus_{s \in S} E_s$. But we already know that the finite direct sums are injective. Hence, there is a homomorphism $g' : R \to \oplus_{s \in S} E_s$. Finally, composing g' with the inclusion of $\oplus_{s \in S} E_s$ into $\oplus_{j \in J} E_j$ completes the given diagram.

Next step is to show that any R-module is a sub-module of an injective module, for this end we have to recall the basics of divisible modules.

Definition 2.2.5 (Divisible Modules) Let M be an R-module over a ring R and let $r \in R \backslash Z(R)$ and $m \in M$. We say that m is divisible by r if there is some $m' \in M$ with $m = rm'$. In general, we say that M is a divisible module if for all $r \in R \backslash Z(R)$ and for all $m \in M$ we have that m is divisible by r.

Proposition 2.2.6 *Every injective module E is divisible.*

Proof Assume that E is injective. Let $e \in E$ and $a \in R \backslash Z(R)$, we must find $x \in E$ with $e = ax$. Define $f : (a) \to E$ by $f(ra) = rm$. Observe that this map is well defined because a is not a zero divisor. Since E is injective we have the following diagram:

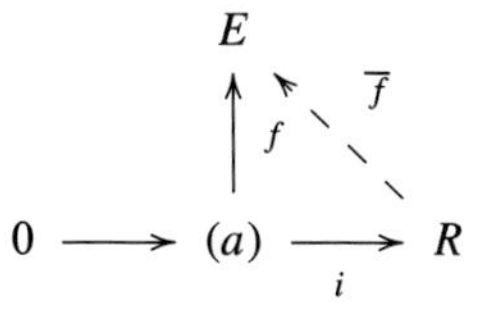

where $\overline{f}$ extends f. In particular, $m = f(a) = \overline{f}(a) = a\overline{f}(1)$. So, the x that we need is $x = \overline{f}(1)$.

Proposition 2.2.7 *Let R be a principal ideal domain and M an R-module. Then we have that M is divisible if and only if M is injective.*

Proof We are going to use Baer's criterion. Assume that $f : I \to E$ is a homomorphism where I is a non zero ideal. By hypothesis, $I = (a)$ for some non zero $a \in I$. Since E is divisible, there is some $e \in E$ with $f(a) = ae$. Define $h : R \to E$ by $h(s) = se$. It is easy to check that h is a homomorphism, moreover, it extends f. That is, if $s = ra \in I$, we have that $h(s) = h(ra) = rae = rf(a) = f(ra)$. Therefore, by Baer's criterion, E is injective.

Lemma 2.2.8 *Let R be a ring. Then:*

(i) For all G abelian groups, $\mathrm{Hom}_{\mathbb{Z}}(R, G)$ *is an R-module.*
(ii) If G is injective as a $\mathbb{Z}$-module, then $\mathrm{Hom}_{\mathbb{Z}}(R, G)$ *is R-injective.*

Proof

(i) This statement is clear, because the addition is as usual, and with the multiplication by elements of R, we define $(rf)(x) = f(rx)$ if $r \in R$ and $f \in \mathrm{Hom}_{\mathbb{Z}}(R, G)$.
(ii) If we have a monomorphism $g : M_1 \to M_2$ and a homomorphism $f : M_1 \to \mathrm{Hom}_{\mathbb{Z}}(R, G)$, we have to find an extension from M_2 to $\mathrm{Hom}_{\mathbb{Z}}(R, G)$. But if we have that f, we can also define a homomorphism f' between M_1 and G in the following way, $f'(m_1) = (f(m_1))(1)$. Is an homomorphism because f is also an homomorphism. So, as G is injective, we can find an extension of f', namely $\overline{f'}$. With this map, we can define the extension we wanted $\overline{f}(m_2) : R \to G$

where $\overline{f}(m_2)(r) = \overline{f'}(rm_2)$. The way that we constructed the map assure us that is an homomorphism and that extends f.

Theorem 2.2.9 *Let R be a ring and M an R-module. Then there exists an R-injective module E and a monomorphism $f : M \to E$. In other words, any module M can be embedded as a submodule of an injective module.*

Proof Since M is a $\mathbb{Z}$-module we have that $M \cong \mathbb{Z}^{(I)}/H$ for a suitable subgroup H of $\mathbb{Z}^{(I)}$. Notice that $\mathbb{Z}^{(I)} \subset \mathbb{Q}^{(I)}$ as abelian groups, so $M \subset G = \mathbb{Q}^{(I)}/H$. But as $\mathbb{Q}$ is divisible, we have that also G is divisible. Hence $M \hookrightarrow G$, where G is an injective abelian group. So from the last Lemma we deduce that $\mathrm{Hom}_{\mathbb{Z}}(R, G)$ is an R-injective module. Then we have the exact sequence of R-modules

$$0 \longrightarrow \mathrm{Hom}_{\mathbb{Z}}(R, M) \longrightarrow E = \mathrm{Hom}_{\mathbb{Z}}(R, G).$$

Next step is to embed M in E; it is enough to show that the linear map $f : M \to \mathrm{Hom}_{\mathbb{Z}}(R, M)$, defined by $f(m)(r) = rm$ if $r \in R$, is injective. If $f(m)(r) = 0$ for all $r \in R$, we have that $f(m)(1) = m = 0$.

Definition 2.2.10 (Proper Essential Extensions) Let R be a ring and let $N \subset M$ be R-modules. We say that M is an essential extension of N if for any non-zero submodule U of M one has $U \cap N \neq 0$. An essential extension M of N is called proper if $N \neq M$.

Proposition 2.2.11 *Let R be a ring.*

(i) An R-module N is injective if and only if it has no proper essential extensions.
(ii) Let $N \subset M$ be an essential extension. Let E be an injective module containing N. Then there exists a monomorphism $\phi : M \longrightarrow E$ extending the inclusion $N \subset M$.

Proof

(i) Let's assume that N is injective and let $N \subset M$ be an essential extension. Since N is injective, N is a direct summand of M, Proposition 2.2.2. Let S be the complement of N in M, Proposition 2.2.2. Then $N \cap S = 0$ and so, the extension $N \subset M$ is essential, so $S = 0$ and $N = M$. Conversely, suppose that N has no proper essential extensions. Let E be an injective module containing N, Theorem 2.2.9. Let us consider the set of submodules $M \subset E$ such that $M \cap N = 0$. This set is not empty $0 \in X$ and it is inductively ordered. By Zorn's Lemma there is a maximal element $L \in X$, so $N \cong N + L/L \subset E/L$. This extension is essential. Let K be an R-module $L \subset K \subset E$ such that $K/L \cap (N + L)/L = 0$. Hence $K \cap (N + L) = 0$, so $K \cap N = 0$. From the maximality of L we deduce $K = L$. Since N has no proper essential extensions we obtain $E = N + L$. On the other hand we have $L \cap N = 0$, so $E = N \oplus L$. From Proposition 2.2.2 we get that N is injective.

(ii) Since E is injective there exists a homomorphism $\phi : M \longrightarrow E$ extending the inclusion $N \subset M$. If $\ker(\phi) \neq 0$ then $\ker(\phi) \cap M \neq 0$ because the extension

$N \subset M$ is essential. Let $0 \neq x \in \ker(\phi) \cap M$ then we get a contradiction: $x = \phi(x) = 0$.

Definition 2.2.12 Let be R a ring and M an R-module. An injective hull of M is an injective module $E_R(M)$ such that $M \subset E_R(M)$ is an essential extension.

Proposition 2.2.13 *Let R be a ring and let M be an R-module.*

(i) M admits an injective hull. Moreover, if $M \subset I$ and I is injective, then a maximal essential extension of M in I is an injective hull of M.

(ii) Let E be an injective hull of M, let I be an injective R-module, and $\alpha : M \to I$ a monomorphism. Then there exists a monomorphism $\varphi : E \to I$ such that the following diagram is commutative, where i is the inclusion:

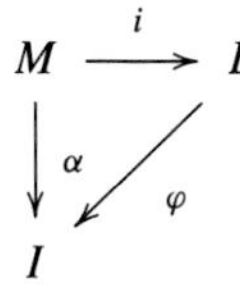

In other words, the injective hulls of M are the "minimal" injective modules in which M can be embedded.

(iii) If E and E' are injective hulls of M, then there exists an isomorphism $\varphi : E \to E'$ such that the following diagram commutes:

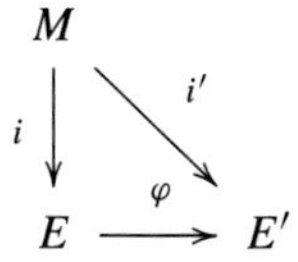

Proof

(i) We know by Theorem 2.2.9 that we can embed M into an injective module I. Now consider $\mathcal{S}$ to be the set of all essential extensions N with $M \subset N \subset I$. Applying Zorn's Lemma to this set yields to a maximal essential extension $M \subset E$ such that $E \subset I$. We claim that E has no proper essential extensions and because of Proposition 2.2.11 we can say that E will be injective and therefore it will be the injective hull we are looking for. Assume that E has a proper essential extension E'. Since I is injective, there exists $\psi : E' \to I$ extending the inclusion $E \subset I$. Suppose $\ker \psi = 0$; then $\mathrm{Im}\psi \subset I$ is an essential extension of M (in I) properly containing E, which contradicts the fact that E is maximal. On the other hand, since ψ extends the inclusion $E \subset I$ we have $E \cap \ker \psi = 0$. But this contradicts with the essentiality of the extension $E \subset E'$. And then we have the result we were looking for.

(ii) Since I is injective, α can be extended to an homomorphism $\varphi : E \to I$. We have that $\varphi|_M = \alpha$, and so $M \cap \ker \varphi = \ker \alpha = 0$. Thus, since the extension $M \subset E$ is essential, we even have $\ker \phi = 0$ and therefore φ is a monomorphism.

(iii) By (ii) there is a monomorphism $\phi : E \to E'$ such that $\phi|_M$ equals the inclusion $M \subset E'$. Then, as $\mathrm{Im}\phi \cong E$ because of the injectivity, $\mathrm{Im}\phi$ is also injective and hence a direct summand of E'. However, since the extension $M \subset E'$ is essential, ϕ is exhaustive because there can't be direct summands different than the total. Therefore, ϕ is an isomorphism.

Remark We can use this proposition to build an injective resolution, $E_R^*(M)$ of a module M. We let $E^0(M) = E_R(M)$ and denote the embedding by ∂^{-1}. Now suppose that the injective resolution has been constructed till the i-th step:

$$0 \longrightarrow E^0(M) \xrightarrow{\partial^0} E^1(M) \xrightarrow{\partial^1} \dots \longrightarrow E^{i-1}(M) \xrightarrow{\partial^{i-1}} E^i(M)$$

We define then $E^{i+1} = E_R(\mathrm{Coker}\, \partial^{i-1})$, and ∂^i is defined as the inclusion.

Definition 2.2.14 Let $(R, \mathfrak{m}, \mathbf{k})$ be a local ring. Given an R-module M the Matlis dual of M is $M^\vee = \mathrm{Hom}_R(M, E_R(\mathbf{k}))$. We write $(-)^\vee = \mathrm{Hom}_R(-, E_R(\mathbf{k}))$, which is a contravariant exact functor from the category *R_mod* into itself.

Proposition 2.2.15 *Let $(R, \mathfrak{m}, \mathbf{k})$ be a local ring. Then $(-)^\vee$ is a faithful functor. Furthermore, if M is a R-module of finite length, then $\ell_R(M^\vee) = \ell_R(M)$. If R is in addition an Artin ring then $\ell_R(E_R(\mathbf{k})) = \ell_R(R) < \infty$.*

Proof We have to show that if M is a nonzero R-module then $M^\vee$ is nonzero. Let's take a non-zero cyclic submodule $R/\mathfrak{a}$ of M. Since $\mathfrak{a} \subset \mathfrak{m}$ we have the maps

$$M \hookleftarrow R/\mathfrak{a} \twoheadrightarrow R/\mathfrak{m} \cong \mathbf{k}.$$

Notice that $\mathbf{k}^\vee = \mathrm{Hom}_R(\mathbf{k}, E_R(\mathbf{k})) \cong \mathbf{k}$. Applying the functor $(-)^\vee$ to this diagram we get

$$M^\vee \twoheadrightarrow (R/\mathfrak{a})^\vee \hookleftarrow \mathbf{k}^\vee \cong \mathbf{k},$$

implying that $M^\vee$ is nonzero.

Let M be a finite length R-module, we use induction on $\ell(M)$ to prove $\ell_R(M) = \ell_R(M^\vee)$. If $\ell_R(M) = 1$, then M is a simple R-module and thus $M \cong R/\mathfrak{m} = \mathbf{k}$. Thus $\ell_R(M^\vee) \cong \ell_R(\mathbf{k}) = 1$. For the general case, pick a simple submodule $S \subset M$. We apply $(-)^\vee$ to the short exact sequence:

$$0 \longrightarrow S \longrightarrow M \longrightarrow M/S \longrightarrow 0$$

Since $S \cong \mathbf{k}$, we have $\ell(S^\vee) = 1$. Now, by induction, $\ell_R((M/S)^\vee) = \ell_R(M/S) = \ell_R(M) - 1$. We conclude then $\ell_R(M^\vee) = \ell_R(M)$.

Let us assume that R is Artin, so $\ell_R(R) < \infty$. From the first part we get $\ell_R(E_R(\mathbf{k})) = \ell_R(R) < \infty$.

Proposition 2.2.16 *Let R be a ring, $\mathfrak{a}$ an ideal of R and M a R-module annihilated by $\mathfrak{a}$. Then, if $E = E_R(M)$:*

$$E_{R/\mathfrak{a}}(M) = \{e \in E \ : \ \mathfrak{a}e = 0\} = (0 :_E \mathfrak{a})$$

Proof Both M and $(0 :_E \mathfrak{a})$ are annihilated by $\mathfrak{a}$ and thus can be thought as $R/\mathfrak{a}$-modules. Clearly $M \subset (0 :_E \mathfrak{a}) \subset E$. Since all $R/\mathfrak{a}$-submodule of $(0 :_E \mathfrak{a})$ is also a R-submodule of E, necessarily $(0 :_E \mathfrak{a})$ is an essential extension on M. So now we need to check that $(0 :_E \mathfrak{a})$ is injective. So let us consider a diagram of $R/\mathfrak{a}$-modules:

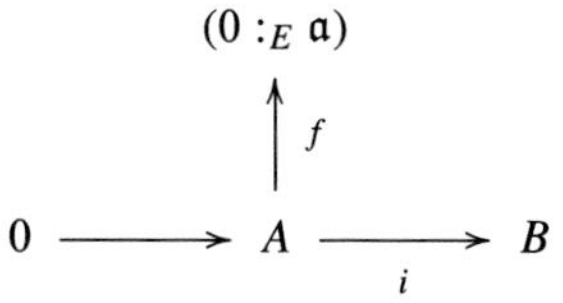

We have to prove that there is $g : B \longrightarrow 0 :_E \mathfrak{a})$ such that $f = g \circ i$. But as we can think these modules as R-modules, we can replace $(0 :_E \mathfrak{a})$ by E and, since E is injective, we can extend the diagram and make the diagram commutative. But this commutativity implies that $\mathrm{Im}(g) \subset (0 :_E \mathfrak{a})$ and therefore the original diagram also commutes.

Corollary 2.2.17 *Let $(R, \mathfrak{m}, \mathbf{k})$ be a local ring and $E = E_R(\mathbf{k})$. Let $\mathfrak{a}$ be an ideal of R. Then:*

(i) $E_{R/\mathfrak{a}}(\mathbf{k}) = (0 :_E \mathfrak{a})$
(ii) $E = \bigcup_{t \geq 1} E_{R/\mathfrak{m}^t}(\mathbf{k})$

Now it's time to prove some technical results with the assumption that we need, the completeness of the Noetherian local ring.

Lemma 2.2.18 *Let $(R, \mathfrak{m}, \mathbf{k})$ be a complete Noetherian local ring and $E = E_R(\mathbf{k})$. Then:*

(i) $R^\vee \cong E$ and $E^\vee \cong R$.
(ii) For every R-module M the natural map $M \to M^{\vee\vee}$ induce isomorphisms $R \to R^{\vee\vee}$ and $E \to E^{\vee\vee}$.

Proof

(i) It is well known that $R^\vee = Hom_R(R, E) \cong E$. Now let's prove $E^\vee \cong R$. Assume first that R is Artinian. Consider the map $\theta : R \to E^\vee = \mathrm{Hom}_R(E, E)$ which sends an element $r \in R$ to the homothety defined by r. Since $\ell(R) = \ell(E^\vee)$, Proposition 2.2.15, we only need to prove that θ is injective. Suppose that $rE = 0$. Then, by the last Corollary, $E_{R/(r)}(\mathbf{k}) = (0 :_E (r)) = E$, and, by the same argument, $\ell(E) = \ell(R/(r))$. This implies that $\ell(R) = \ell(R/(r))$, then $r = 0$.

Assume now that R is Noetherian and complete. We consider the map $\theta : R \to E^\vee = \mathrm{Hom}_R(E, E)$ as above, we will prove that θ is an isomorphism. Let's write $R_t = R/\mathfrak{m}^t$ for each t. By the last corollary $E_t := E_{R_t}(\mathbf{k}) = (0 :_E \mathfrak{m}^t)$. Let $\varphi \in \mathrm{Hom}_R(E, E) = E^\vee$. It is clear that $\varphi(E_t) \subset E_t$ and thus $\varphi \in \mathrm{Hom}_{R_t}(E_t, E_t)$. Since R_t is Artinian we have φ is a homothety defined by an element $r_t \in R_t$. The fact $E_t \subset E_{t+1}$ implies that $r_t - r_{t+1} \in \mathfrak{m}^t$ for all $t \geq 1$. In consequence, $r = (r_t)_t \in \hat{R} = R$ and $r_t = r + \mathfrak{m}^t$ for all $t \geq 1$. We claim that φ is given by multiplication by r. This follows from the fact that $E = \cup_t E_t$ and that $\varphi(e) = r_t e$ for all $e \in E_t$. Moreover, r is uniquely determined by φ, and we conclude that θ is bijective.

(ii) We consider the natural homomorphism $\gamma : M \to M^{\vee\vee} = \mathrm{Hom}_R(\mathrm{Hom}_R(M, E), E)$ given by $\gamma(m)(\varphi) = \varphi(m)$. Fisrt we prove that $\gamma : R \to R^{\vee\vee}$ is an isomorphism. This map is the composition of the two isomorphisms given in part (i) $R \cong E^\vee \cong (R^\vee)^\vee$. In fact, if $r \in R$, the map $R \cong E^\vee$ sends r to multiplication by r, $h_r : E \to E$. Now the map $E^\vee \cong (R^\vee)^\vee$ sends h_r to α_r defined by $\alpha_r(\varphi) = h_r(\varphi(1)) = \varphi(r)$, so $\alpha_r = \gamma(r)$. The case of E is analogous to this one.

Proposition 2.2.19 *Let $(R, \mathfrak{m}, \mathbf{k})$ be a complete Noetherian local ring and $E = E_R(\mathbf{k})$.*

(i) There is an order-reversing bijection $\perp$ between the set of R-submodules of E and the set of ideals of R given by: if M is a submodule of E then $(E/M)^\vee \cong M^\perp = (0 :_R M)$, and $(R/I)^\vee \cong I^\perp = (0 :_E I)$ for an ideal $I \subset R$,
(ii) E is an Artinian R-module,
(iii) an R-module is Artinian if and only if it can be embedded in E^n for some $n \in \mathbb{N}$.

Proof (*i*) Since $M \subset M^{\perp\perp}$ we have to prove that $M^{\perp\perp} \subset M$. Consider the exact sequence

$$0 \longrightarrow M \longrightarrow E \xrightarrow{\pi} E/M \longrightarrow 0,$$

dualizing with respect E, we get an injective homomorphism, Lemma 2.2.18,

$$0 \longrightarrow (E/M)^\vee \xrightarrow{\pi^\vee} E^\vee \overset{\theta^{-1}}{\cong} R.$$

Hence every $g \in (E/M)^\vee$ is mapped to an $r \in R$ such that $(\theta^{-1} \circ \pi^\vee)(g) = r$, or equivalently $g \circ \pi = \pi^\vee(g) = h_r = \theta(r)$ where $h_r : E \longrightarrow E$ is the homothety defined by r. Since $g \circ \pi(M) = g(0) = 0$ we get $rM = 0$, so $(E/M)^\vee \subset M^\perp$. On the other hand if $r \in M^\perp$ then we can consider the map $g : E/M \longrightarrow E$ such that $g(\overline{x}) = rx$ for all $x \in E$. It is easy to see that $(\theta^{-1} \circ \pi^\vee)(g) = r$, so $(E/M)^\vee \overset{\theta^{-1}\pi^\vee}{\cong} M^\perp$. Let $x \in E \setminus M$ then there is $g \in (E/M)^\vee$ such that $g(\overline{x}) \neq 0$, Lemma 2.2.18. From the above isomorphism we deduce that there is $r \in M^\vee$ such that $rx \neq 0$. This shows that $M^{\vee\vee} \subset M$ and then $M = M^{\vee\vee}$.

Let I be an ideal of R. As in the previous case we have $I \subset I^{\perp\perp}$. From the natural exact sequence

$$0 \longrightarrow I \longrightarrow R \xrightarrow{\pi} R/I \longrightarrow 0,$$

we get an injective homomorphism, Lemma 2.2.18,

$$0 \longrightarrow (R/I)^\vee \xrightarrow{\pi^\vee} R^\vee \overset{\theta^{-1}}{\cong} E.$$

As in the previous case $\theta^{-1} \circ \pi^\vee$ maps $(R/I)^\vee$ to $I^\perp$. Let $r \in R \setminus I$ then there is $g \in (R/I)^\vee$ such that $g(\overline{r}) \neq 0$, Lemma 2.2.18. Hence $x = g(\overline{1}) \in I^\perp$ and $rx \neq 0$, i.e. $r \notin (0 :_R x)$. Since $I^{\perp\perp} = \bigcap_{x \in I^\perp} (0 :_R x)$ we get $I^{\perp\perp} \subset I$ and then $I = I^{\perp\perp}$.
(ii) Since R is Noetherian, by (i) we get that E is Artinian.
(iii) We consider the set X of kernels of all homomorphisms $F : M \longrightarrow E^n$, for all $n \in \mathbb{N}$. This is a set of submodules of M. Since M is Artininan there is a minimal element $\ker(F)$ of X, where $F : M \longrightarrow E^n$ for some $n \in \mathbb{N}$. Assume that $\ker(F) \neq 0$ and pick $0 \neq x \in \ker(F)$. From Proposition 2.2.15 there is $\sigma : M \longrightarrow E$ such that $\sigma(x) \neq 0$. Let us consider $F^* : M \longrightarrow E^{n+1}$ defined by $F^*(y) = (F(y), \sigma(y))$. Since $\ker(F^*) \subsetneq \ker(F)$ we get a contradiction with the minimality of $\ker(F)$.

Assume that M is a submodule of E^n for some integer n. From (ii) we get that M is an Artin module.

In the next result we will prove Matlis' duality, see [28] Theorem 5.20.

Theorem 2.2.20 (Matlis Duality) *Let $(R, \mathfrak{m}, \mathbf{k})$ be a complete Noetherian local ring, $E = E_R(\mathbf{k})$ and let M be a R-module. Then:*

(i) If M is Noetherian then $M^\vee$ is Artinian.
(ii) If M is Artinian then $M^\vee$ is Noetherian.
(iii) If M is either Noetherian or Artinian then $M^{\vee\vee} \cong M$.
(iv) The functor $(-)^\vee$ is a contravariant, additive and exact functor.
(v) The functor $(-)^\vee$ is an anti-equivalence between $R\_mod.Noeth$ and $R\_mod.Artin$ (resp. between $R\_mod.Artin$ and $R\_mod.Noeth$). It holds $(-)^\vee \circ (-)^\vee$ is the identity functor of $R\_mod.Noeth$ (resp. $R\_mod.Artin$).

Proof

(i) Let's consider a presentation of M

$$R^m \longrightarrow R^n \longrightarrow M \longrightarrow 0$$

Since $(-)^\vee$ is exact, it induces an exact sequence:

$$0 \longrightarrow M^\vee \longrightarrow (R^n)^\vee \longrightarrow (R^m)^\vee$$

Thus $M^\vee$ can be seen as a submodule of $(R^n)^\vee \cong (R^\vee)^n \cong E^n$, Lemma 2.2.18. Since E is Artinian as we saw in the previous corollary, so is E^n and hence also $M^\vee$. Applying the functor $(-)^\vee$ again we get a commutative diagram:

$$\begin{array}{ccccccc} (R^m)^{\vee\vee} & \longrightarrow & (R^n)^{\vee\vee} & \longrightarrow & M^{\vee\vee} & \longrightarrow & 0 \\ \uparrow & & \uparrow & & \uparrow & & \\ R^m & \longrightarrow & R^n & \longrightarrow & M & \longrightarrow & 0 \end{array}$$

whose rows are exact. Since we proved that in this context $R \to R^{\vee\vee}$ is an isomorphism, $M \cong M^{\vee\vee}$

(ii) We proved that $M \hookrightarrow E^n$ for some $n \in \mathbb{N}$. Since E is Artinian, so is E^n/M and thus $E^n/M \hookrightarrow E^m$ for some $m \in \mathbb{N}$. In consequence, we have an exact sequence:

$$0 \longrightarrow M \longrightarrow E^n \longrightarrow E^m$$

As before, if we apply $(-)^\vee$ we have an exact sequence:

$$(E^m)^\vee \longrightarrow (E^n)^\vee \longrightarrow M^\vee \longrightarrow 0$$

and $M^\vee$ can be seen as a quotient of $(E^n)^\vee \cong (E^\vee)^n \cong R^n$, where the isomorphism is the one we proved in Lemma 2.2.18. This implies that $M^\vee$ is Noetherian.

(iii) Finally, we apply the functor $(-)^\vee$ to the last exact sequence we obtain the commutative diagram

$$\begin{array}{ccccccc} 0 & \longrightarrow & M^{\vee\vee} & \longrightarrow & (E^n)^{\vee\vee} & \longrightarrow & (E^m)^{\vee\vee} \\ & & \uparrow & & \uparrow & & \uparrow \\ 0 & \longrightarrow & M & \longrightarrow & E^n & \longrightarrow & E^m \end{array}$$

And again, since $E \to E^{\vee\vee}$ is an isomorphism, $M \cong M^{\vee\vee}$

(iv) This is a consequence of the previous statements.

2.3 Macaulay's Correspondence

Let $\mathbf{k}$ be an arbitrary field. Let $R = \mathbf{k}[[x_1, \ldots x_n]]$ be the ring of the formal series with maximal ideal $\mathfrak{m} = (x_1, \cdots, x_n)$ and let $S = \mathbf{k}[y_1, \ldots, y_n]$ be a polynomial ring, we denote by $\mu = (x_1, \ldots, x_n)$ the homogeneous maximal ideal of S.

It is well known that R is an S-module with the standard product. On the other hand, S can be considered as R-module with two linear structures: by derivation and by contraction.

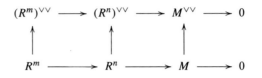

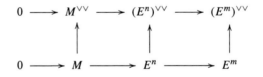

If $char(\mathbf{k}) = 0$, the R-module structure of S by derivation is defined by

$$\begin{array}{rcl} R \times S & \longrightarrow & S \\ (x^\alpha, y^\beta) & \mapsto & x^\alpha \circ y^\beta = \begin{cases} \frac{\beta!}{(\beta-\alpha)!} y^{\beta-\alpha} & \beta \geq \alpha \\ \\ 0, & \text{otherwise} \end{cases} \end{array}$$

where for all $\alpha, \beta \in \mathbb{N}^n$, $\alpha! = \prod_{i=1}^n \alpha_i!$

If $char(\mathbf{k}) \geq 0$, the R-module structure of S by contraction is defined by:

$$\begin{array}{rcl} R \times S & \longrightarrow & S \\ (x^\alpha, y^\beta) & \mapsto & x^\alpha \circ y^\beta = \begin{cases} y^{\beta-\alpha} & \beta \geq \alpha \\ \\ 0, & otherwise \end{cases} \end{array}$$

$\alpha, \beta \in \mathbb{N}^n$

Proposition 2.3.1 *For any field* $\mathbf{k}$ *there is a* R*-module homomorphism*

$$\begin{array}{rcl} \sigma : (S, der) & \longrightarrow & (S, cont) \\ y^\alpha & \mapsto & \alpha!\, y^\alpha \end{array}$$

If $char(\mathbf{k}) = 0$ *then* σ *is an isomorphism of* R*-modules.*

Proof For proving the first statement it is enough to show that

$$\sigma(x^\alpha \circ y^\beta) = x^\alpha \sigma(y^\beta).$$

This is easy:

$$\sigma(x^\alpha \circ y^\beta) = \sigma\left(\frac{\beta!}{(\beta-\alpha)!} y^{\beta-\alpha}\right) = \frac{\beta!}{(\beta-\alpha)!}((\beta-\alpha)! y^{\beta-\alpha})$$

$$= \beta!\, y^{\beta-\alpha} = x^\alpha \circ \sigma(y^\beta)$$

If $char(\mathbf{k}) = 0$ then the inverse of σ is $y^\alpha \longrightarrow (1/\alpha!) y^\alpha$

Given a family of polynomials F_j, $j \in J$, we denote by $\langle F_j, j \in J\rangle$ the submodule of S generated by F_j, $j \in J$, i.e. the $\mathbf{k}$-vector subspace of S generated by $x^\alpha \circ F_j$, $j \in J$, and $\alpha \in \mathbb{N}^n$. We denote by $\langle F_j, j \in J\rangle_{\mathbf{k}}$ the $\mathbf{k}$-vector space generated by F_j, $j \in J$.

In the next result we compute the injective hull of the residue field of a power series ring, [18, 25].

Theorem 2.3.2 *Let $R = \mathbf{k}[\![x_1, \dots x_n]\!]$ be the n-dimensional power series ring over a field* $\mathbf{k}$. *If* $\mathbf{k}$ *is of characteristic zero then*

$$E_R(\mathbf{k}) \cong (S, der) \cong (S, cont).$$

If $\mathbf{k}$ *is of positive characteristic then*

$$E_R(\mathbf{k}) \cong (S, cont).$$

Proof We write $E = E_R(\mathbf{k})$. From Corollary 2.2.17 we get

$$E = \bigcup_{i \geq 0} (0 :_E \mathfrak{m}_R^i) = \bigcup_{i \geq 0} E_{R/\mathfrak{m}_R^i}(\mathbf{k})$$

Hence the problem is reduced to the computation of $E_{R/\mathfrak{m}_R^i}(\mathbf{k}) \subset E$.

Notice that $S_{\leq i-1} := \{f \in S \mid \deg(f) \leq i-1\} \subset S$ is an sub-R-module of S, with respect to the derivation or contraction structure of S, and that $S_{\leq i-1}$ is annihilated by $\mathfrak{m}_R^i$. Hence $S_{\leq i-1}$ is an $R/\mathfrak{m}_R^i$-module. For any characteristic of the ground field $\mathbf{k}$ the extension $\mathbf{k} \subset S_{\leq i-1} := \{f \in S \mid \deg(f) \leq i-1\}$ is essential. In fact, let $0 \neq M \subset S_{\leq i-1}$ be a sub-$R/\mathfrak{m}_R^i$-module then it holds $1 \in M$.

From Theorem 2.2.13 there exists $L \cong E_{R/\mathfrak{m}_R^i}(\mathbf{k})$ such that

$$\mathbf{k} \subset S_{\leq i-1} \subset L \cong E_{R/\mathfrak{m}_R^i}(\mathbf{k}).$$

Since, Proposition 2.2.15,

$$\begin{aligned} \text{Length}_{R/\mathfrak{m}_R^i}(E_{R/\mathfrak{m}_R^i}(\mathbf{k})) &= \text{Length}_{R/\mathfrak{m}_R^i}(R/\mathfrak{m}_R^i) \\ &= \text{Length}_{R/\mathfrak{m}_R^i}(S_{\leq i-1}) \end{aligned}$$

from the last inclusions we get $S_{\leq i-1} \cong E_{R/\mathfrak{m}_R^i}(\mathbf{k})$. Hence

$$E_R(\mathbf{k}) \cong \bigcup_{i \geq 0} S_{\leq i-1} = S.$$

From the previous results we can recover the classical result of Macaulay, [23], for the power series ring, see [16, 21].

If $I \subset R$ is an ideal, then $(R/I)^\vee$ is the sub-R-module of S that we already denote by $I^\perp$, see Proposition 2.2.19,

$$I^\perp = \{g \in S \mid I \circ g = 0\},$$

this is the Macaulay's inverse system of I. Given a sub-R-module M of S then dual $M^\vee$ is an ideal of R that we already denote by $(S/M)^\perp$, see Proposition 2.2.19,

$$M^\perp = \{f \in R \mid f \circ g = 0 \text{ for all } g \in M\}.$$

We will write sometimes this module as $M^\perp = \operatorname{Ann}_R(M)$.

Proposition 2.3.3 (Macaulay's Duality) *Let $R = \mathbf{k}[[x_1, \dots x_n]]$ be the n-dimensional power series ring over a field $\mathbf{k}$. There is a order-reversing bijection $\perp$ between the set of finitely generated sub-R-submodules of $S = \mathbf{k}[[y_1, \dots y_n]]$ and the set of $\mathfrak{m}$-primary ideals of R given by: if M is a submodule of S then $M^\perp = (0 :_R M)$, and $I^\perp = (0 :_S I)$ for an ideal $I \subset R$.*

Proof The one-to-one correspondence is a particular case of Proposition 2.2.19. Theorem 2.2.20 gives the one-to-one correspondence between finitely generated sub-R-submodules of S and $\mathfrak{m}$-primary ideals of R.

Remark Macaulay proved more as we will see later on. Trough this correspondence Macaulay proved that Artin Gorenstein $\mathbf{k}$-algebras $A = R/I$ of socle degree s correspond to R-submodules of S generated by a polynomial F of degree s, see Proposition 2.4.4.

Let $A = R/I$ be an Artin quotient of R, we denote by $\mathfrak{n} = \mathfrak{m}/I$ the maximal ideal of A. The socle of A is the colon ideal $\operatorname{Soc}(A) = 0 :_A \mathfrak{n}$, notice that $\operatorname{Soc}(A)$ is a $\mathbf{k}$-vector space subspace of A. We denote by $s(A)$ the *socle degree* of A, that is the maximum integer j such that $\mathfrak{n}^j \neq 0$. The (Cohen-Macaulay) *type* of A is $t(A) := \dim_{\mathbf{k}} \operatorname{Soc}(A)$.

The Hilbert function of $A = R/I$ is by definition

$$\operatorname{HF}_A(i) = \dim_{\mathbf{k}} \left(\frac{\mathfrak{n}^i}{\mathfrak{n}^{i+1}} \right),$$

the multiplicity of A is the integer $e(A) := \dim_{\mathbf{k}}(A) = \dim_{\mathbf{k}} I^\perp$, Propositions 2.3.3 and 2.2.19. Notice that $s(A)$ is the last integer such that $\operatorname{HF}_A(i) \neq 0$ and that $e(A) = \sum_{i=0}^{s} \operatorname{HF}_A(i)$. The embedding dimension of A is $\operatorname{HF}_A(1)$.

Example 2.3.4 Let $F = y^3 + xy + x^2 \in R = \mathbf{k}[[x, y]]$ be a polynomial. We consider the R-module structure of $S = \mathbf{k}[x, y]$ defined by the contraction $\circ$. Then $\langle F \rangle = \langle F, y^2{+}x, y{+}x, x, 1 \rangle_{\mathbf{k}}$ and $\dim_{\mathbf{k}}(\langle F \rangle) = 5$. We have that $I = \operatorname{Ann}_R(\langle F \rangle) = (xy - y^3, x^2 - xy)$, i.e. I is a complete intersection ideal of R. The Hilbert function of A is $\operatorname{HF}_A = \{1, 2, 1, 1\}$, so $e(A) = 5$ and $s(A) = 3$

The associated graded ring to A is the graded $\mathbf{k}$-algebra ring $gr_{\mathfrak{n}}(A) = \oplus_{i \geq 0} \mathfrak{n}^i / \mathfrak{n}^{i+1}$. Notice that the Hilbert function of A and its associated graded ring $gr_{\mathfrak{n}}(A)$ agrees. We denote by I^* the homogeneous ideal of S generated by the initial forms of the elements I. It is well known that $gr_{\mathfrak{n}}(A) \cong S/I^*$ as graded $\mathbf{k}$-algebras, in particular $gr_{\mathfrak{n}}(A)_i \cong (S/I^*)_i$ for all $i \geq 0$.

We denote by $S_{\leq i}$ (resp. $S_{<i}$, resp. S_i), $i \in \mathbb{N}$, the $\mathbf{k}$-vector space of polynomials of S of degree less or equal (resp. less, resp. equal to) to i, and we consider the following $\mathbf{k}$-vector space

$$(I^{\perp})_i := \frac{I^{\perp} \cap S_{\leq i} + S_{<i}}{S_{<i}}.$$

Proposition 2.3.5 *For all $i \geq 0$ it holds*

$$\mathrm{HF}_A(i) = \dim_{\mathbf{k}} (I^{\perp})_i.$$

Proof Let's consider the following natural exact sequence of R-modules

$$0 \longrightarrow \frac{\mathfrak{n}^i}{\mathfrak{n}^{i+1}} \longrightarrow \frac{A}{\mathfrak{n}^{i+1}} \longrightarrow \frac{A}{\mathfrak{n}^i} \longrightarrow 0.$$

Dualizing this sequence we get

$$0 \longrightarrow (I + \mathfrak{m}^i)^{\perp} \longrightarrow (I + \mathfrak{m}^{i+1})^{\perp} \longrightarrow \left(\frac{\mathfrak{n}^i}{\mathfrak{n}^{i+1}}\right)^{\vee} \longrightarrow 0$$

so we get the following sequence of $\mathbf{k}$-vector spaces:

$$\left(\frac{\mathfrak{n}^i}{\mathfrak{n}^{i+1}}\right)^{\vee} \cong \frac{(I + \mathfrak{m}^{i+1})^{\perp}}{(I + \mathfrak{m}^i)^{\perp}} = \frac{I^{\perp} \cap S_{\leq i}}{I^{\perp} \cap S_{\leq i-1}} \cong \frac{I^{\perp} \cap S_{\leq i} + S_{<i}}{S_{<i}}.$$

From Proposition 2.2.15 we get the claim.

Consider the map

$$\begin{array}{rccc} \langle | \rangle : & R \times S & \longrightarrow & \mathbf{k} \\ & (F, G) & \mapsto & (F \circ G)(0) \end{array}$$

In the next result we collect some results on $\langle | \rangle$ that we will use later on.

Proposition 2.3.6

1. *$\langle | \rangle$ is a bilinear non-degenerate map of $\mathbf{k}$-vector spaces.*
2. *If I is an ideal of R then*

$$I^{\perp} = \{G \in S \mid \langle I \mid G \rangle = 0\}$$

3. *$\langle | \rangle$ induces a bilinear non-degenerate map of $\mathbf{k}$-vector spaces*

$$\overline{\langle | \rangle} : \frac{R}{I} \times I^{\perp} \longrightarrow \mathbf{k}$$

4. We have an isomorphism of **k***-vector spaces:*

$$\left(\frac{S}{I^*}\right)_i \cong (I^\perp)_i$$

for all $i \geq 0$.

We will denote by $*$ the duality defined by exact pairing $\overline{\langle | \rangle}$, notice that $(R/I)^* \cong I^\perp$.

If $\underline{i} = (i_1, \cdots, i_n) \in \mathbb{N}^n$ is a integer n-pla we denote by $\partial_{\underline{i}}(G)$, $G \in S$, the derivative of G with respect to $y_1^{i_1} \cdots y_n^{i_n}$, i.e. $\partial_{\underline{i}}(G) = (x_1^{i_1} \cdots x_n^{i_n}) \circ G$.

Let $\Omega = \{\omega_i\}$ be the canonical basis of $R/\mathfrak{m}^{s+1}$ as a **k**-vector space consisting of the standard monomials x^α ordered by the deg-lex order with $x_1 > \cdots > x_n$ and, then the dual basis with respect to $*$ is the basis $\Omega^* = \{\omega_i^*\}$ of $S_{\leq j}$ where

$$(x^\alpha)^* = \frac{1}{\alpha!} y^\alpha,$$

in fact $\omega_j \circ \omega_i^* = \overline{\langle \omega_j \mid \omega_i^* \rangle} = \delta_{ij}$, where $\delta_{ij} = 0$ if $i \neq j$ and $\delta_{ii} = 1$.

2.4 Gorenstein, Level and Compressed Algebras

Definition 2.4.1 An Artin ring A is Gorenstein if $t(A) = 1$; A is an Artin level algebra if $\mathrm{Soc}(A) = \mathfrak{m}^s$, where s is the socle degree of A.

Proposition 2.4.2 *Let* $A = R/I$ *be an Artin ring, the following conditions are equivalent:*

(i) *A is Gorenstein,*
(ii) $A \cong E_A(\mathbf{k})$ *as R-modules,*
(iii) *A is injective as A-module.*

Proof Assume (i). Since the extension $\mathbf{k} = \mathrm{Soc}(A) \subset A$ is essential we have the A-module extensions, Proposition 2.2.11 (ii),

$$\mathbf{k} = \mathrm{Soc}(A) \subset A \subset E_A(\mathbf{k}).$$

so $A = E_A(\mathbf{k})$, Proposition 2.2.15. Since $S \cong E_{\mathbf{k}}(\mathbf{k})$ is an injective R-module, (ii) implies (iii).

Assume that A is injective as A-module. From Proposition 2.2.13 (ii) we get the A-module extensions

$$\mathbf{k} \subset E_A(\mathbf{k}) \subset A,$$

from Proposition 2.2.15 we get (i).

Given an R-module M we denote by $\mu(M)$ the minimal number of generators of M.

Proposition 2.4.3 *Let $A = R/I$ be an Artinian local ring. Then*

$$\operatorname{Soc}(A)^\vee = \frac{I^\perp}{\mathfrak{m} \circ I^\perp}.$$

In particular the Cohen-Macaulay type of A is

$$t(A) = \dim_{\mathbf{k}}(I^\perp/\mathfrak{m} \circ I^\perp) = \mu_R(I^\perp).$$

Proof Let's consider exact sequence of R-modules

$$0 \longrightarrow \operatorname{Soc}(A) = (0 :_A \mathfrak{n}) \longrightarrow A \stackrel{(x_1,\cdots,x_n)}{\longrightarrow} A^n,$$

dualizing this sequence we get

$$(I^\perp)^n \stackrel{\sigma}{\longrightarrow} I^\perp \longrightarrow \operatorname{Soc}(A)^\vee \longrightarrow 0$$

where $\sigma(f_1, \cdots, f_n) = \sum_{i=1}^n x_i \circ f_i$. Hence

$$\operatorname{Soc}(A)^\vee = \frac{I^\perp}{(x_1, \ldots, x_n) \circ I^\perp} = \frac{I^\perp}{\mathfrak{m} \circ I^\perp}$$

Since $t(A) = \dim_{\mathbf{k}}(\operatorname{Soc}(A)) = \dim_{\mathbf{k}}(\operatorname{Soc}(A)^\vee) = \mu(I^\perp)$, Proposition 2.2.15.

Given a polynomial $F \in S$ of degree r we denote by $\operatorname{top}(F)$ the degree r form of F where $r = \deg(F)$.

Proposition 2.4.4 *Let I be an $\mathfrak{m}$-primary ideal of R. The quotient $A = R/I$ is an Artin level algebra of socle degree s and Cohen-Macaulay type t if and only if $I^\perp$ is generated by t polynomials $F_1, \cdots, F_t \in S$ such that* $\deg(F_i) = s$, *$i = 1, \cdots, t$, and* $\operatorname{top}(F_1), \cdots, \operatorname{top}(F_t)$ *are* $\mathbf{k}$*-linear independent forms of degree s. In particular, $A = R/I$ is Gorenstein of socle degree s if and only if $I^\perp$ is a cyclic R-module generated by a polynomial of degree s.*

Proof Assume that A is an Artin level algebra of socle degree s and Cohen-Macaulay type t. In particular $\operatorname{Soc}(A) = \mathfrak{n}^s = \mathfrak{m}^s + I/I$ so

$$\operatorname{Soc}(A)^\vee = \frac{I^\perp}{I^\perp \cap S_{\leq s-1}}.$$

From the last result we get

$$\mathfrak{m} \circ I^\perp = I^\perp \cap S_{\leq s-1}.$$

From this identity we deduce that $I^\vee$ is generated by t polynomials $F_1, \cdots, F_t$ of degree s and $\mathrm{top}(F_1), \cdots, \mathrm{top}(F_t)$ are $\mathbf{k}$-linear independent.

Assume that $I^\perp = \langle F_1, \cdots, F_t \rangle$ such that $\deg(F_i) = s$, $i = 1, \cdots, t$, and that $\mathrm{top}(F_1), \cdots, \mathrm{top}(F_t)$ are $\mathbf{k}$-linear independent forms of degree s. Hence $F_1, \cdots, F_t$ is a minimal system of generators of $I^\perp$, in particular $\mu_R(I^\perp) = t$ and from the last result we have that t is the Cohen-Macaulay type of A. Furthermore, since $\deg(F_i) = s$, $i = 1, \cdots, t$, we have

$$\mathfrak{m} \circ I^\perp = I^\perp \cap S_{\leq s-1}.$$

From the last result we deduce $\mathrm{Soc}(A) = \mathfrak{n}^s$, i.e. A is Artin level of socle degree s.

In the last section we will prove the following result, see Proposition 2.6.3,

Corollary 2.4.5 *Let $A = R/I$ be an Artin algebra of embedding dimension two. Then*

$$\mu(I) = t(R/I) + 1.$$

A is Gorenstein if and only if I is a complete intersection.

The initial degree of $A = R/I$ is the integer r such that $I \subseteq \mathfrak{m}^r$ and $I \not\subseteq \mathfrak{m}^{r+1}$. The *socle type* of A is the sequence $\sigma(A) = (0, \ldots, \sigma_{r-1}, \sigma_r, \ldots, \sigma_s, 0, 0, \ldots)$, s is the socle degree of A, with

$$\sigma_i := \dim_{\mathbf{k}} \left(\frac{(0 : \mathfrak{n}) \cap \mathfrak{n}^i}{(0 : \mathfrak{n}) \cap \mathfrak{n}^{i+1}} \right).$$

Notice that $\sigma_s > 0$ and $\sigma_j = 0$ for $j > s$,. See [20] for some conditions on a sequence of integers to be the socle type of an Artin algebra

Remark An Artin algebra of socle degree s and Cohen-Macaulay type t is level if and only if $\sigma_j = 0$ for $j \neq s$ and $\sigma_s = t$. The Artin algebra is Gorenstein if and only if $\sigma_j = 0$ for $j \neq s$ and $\sigma_s = 1$.

We say that the Hilbert function HF_A is maximal in the class of Artin level algebras of given embedding dimension and socle type, if for each integer i, $\mathrm{HF}_A(i) \geq \mathrm{HF}_B(i)$ for any other Artin algebra B in the same class. The existence of a maximal HF_A was shown for graded algebras by Iarrobino [20]. In the general case by Fröberg and Laksov [17], by Emsalem [16], by Iarrobino and the author of this notes in [10] in the local case.

Definition 2.4.6 An Artin algebra $A = R/I$ of socle type σ is compressed if and only if it has maximal length $e(A) = \dim_{\mathbf{k}} A$ among Artin quotients of R having socle type σ and embedding dimension n.

The maximality of the Hilbert function characterizes compressed algebras as follows. If A is an Artin algebra of socle type σ, it is known that for $i \geq 0$,

$$\mathrm{HF}_A(i) \leq \min\{\dim_{\mathbf{k}} S_i, \sigma_i \dim_{\mathbf{k}} S_0 + \sigma_{i+1} \dim_{\mathbf{k}} S_1 + \cdots + \sigma_s \dim_{\mathbf{k}} S_{s-i}\}.$$

Accordingly with [20], Definition 2.4. B, we can rephrase the previous definition in terms of the Hilbert function.

Definition 2.4.7 A local $\mathbf{k}$-algebra A of socle degree s, socle type σ and initial degree r is *compressed* if

$$\mathrm{HF}_A(i) = \begin{cases} \sum_{u=i}^{s} \sigma_u (\dim_{\mathbf{k}} S_{u-i}) & \text{if } i \geq r \\ \dim_{\mathbf{k}} S_i & \text{otherwise.} \end{cases}$$

In particular a level algebra A of socle degree s, type t and embedding dimension n is compressed if

$$\mathrm{HF}_A(i) = \min\left\{\binom{n+i-1}{i}, t\binom{n+s-i-1}{s-i}\right\}.$$

If $t = 1$ and the above equality holds then A is called *compressed Gorenstein algebra* or also *extremal Gorenstein algebra*.

It is clear that compressed algebras impose several restrictive numerical conditions on the socle sequence σ (see [20, Definition 2.2]). For instance if r is the initial degree of A, then

$$\sigma_{r-1} = \max\{0, \dim_{\mathbf{k}} S_{r-1} - \sum_{u \geq r} (\sigma_u \dim_{\mathbf{k}} S_{u-(r-1)})\}. \tag{2.1}$$

If $s \geq 2(r-1)$, then it is easy to see that $\sigma_{r-1} = 0$ because $\dim_{\mathbf{k}} S_{s-(r-1)} \geq \dim_{\mathbf{k}} S_{r-1}$. This is the case if A is Gorenstein.

The following result was proved in [20, Proposition 3.7 and Corollary 3.8].

Proposition 2.4.8 *A compressed local algebra $A = R/I$ whose dual module $I^{\perp}$ is generated by $F_1, \ldots, F_t$ of degrees $d_1, \ldots, d_t$ has a compressed associated graded ring $gr_{\mathfrak{n}}(A)$ whose dual module is generated by the leading forms of $F_1, \ldots, F_t$. Conversely if $gr_{\mathfrak{n}}(A)$ is compressed, then A is compressed and $\sigma(A) = \sigma(gr_{\mathfrak{n}}(A))$.*

It is well known that if $gr_{\mathfrak{n}}(A)$ is Gorenstein then A is Gorenstein. On the other hand, if A is Gorenstein then $gr_{\mathfrak{n}}(A)$ is no longer Gorenstein. In order to study the associated graded ring to A Iarrobino considered the following construction. For

$a = 0, \cdots, s-1$, $s = s(A)$, consider the homogeneous ideals of $gr_{\mathfrak{n}}(A)$

$$C(a) = \bigoplus_{i \geq 0} C(a)_i$$

$$C(a)_i = \frac{(0 :_A \mathfrak{n}^{s+1-a-i}) \cap \mathfrak{n}^i}{(0 :_A \mathfrak{n}^{s+1-a-i}) \cap \mathfrak{n}^{i+1}} \subset gr_{\mathfrak{n}}(A)_i$$

This defining a decreasing filtration of ideals of $gr_{\mathfrak{n}}(A)$

$$gr_{\mathfrak{m}_A}(A) = C(0) \supseteq C(1) \supseteq \cdots \supseteq C(s) = 0$$

Notice that if $a \geq 1$ then $C(a)_i = 0$ for all $i \geq s-a$ and $C(0)_i = 0$ for all $i \geq s+1$

Definition 2.4.9 (Iarrobino's Q-Decomposition of $gr_{\mathfrak{n}}(A)$) For all $a = 0, \cdots, s-1$ we consider the $gr_{\mathfrak{m}_A}(A)$-module

$$Q(a) = C(a)/C(a+1).$$

Since the Hilbert function of A and $gr_{\mathfrak{n}}(A)$ agree we have the Iarrobino's *Shell decomposition* of HF_A:

$$\mathrm{HF}_A = \sum_{a=0}^{s-1} \mathrm{HF}_{Q(a)}$$

Proposition 2.4.10 *If A is Artin Gorenstein then $Q(a)$ is a reflexive $gr_{\mathfrak{n}}(A)$-module:*

$$\mathrm{Hom}_{\mathbf{k}}(Q(a)_i, \mathbf{k}) \cong Q(a)_{s-a-i}$$

$i = 0, \cdots, s-a$. In particular, $\mathrm{HF}_{Q(a)}$ is a symmetric function w.r.t $\frac{s-a}{2}$.

Example 2.4.11 (Shell Decomposition) Assume that $\mathrm{HF} = \{1, m, n, 1\}$ is the Hilbert function of an Artin Gorenstein algebra $A = R/I$ The Shell decomposition of HF is, $s = 3$,

i	0	1	2	3
HF_A	1	m	n	1
$\mathrm{HF}_{Q(0)}$	1	n	n	1
$\mathrm{HF}_{Q(1)}$	0	$m-n$	0	0
$\mathrm{HF}_{Q(2)}$	0	0	0	0

so $m \geq n$. In fact, all function $\{1, m, n, 1\}$, $m \geq n$, is the Hilbert function of an Artin Gorenstein algebra Theorem 2.5.11. Notice that from Macaulay's characterization of Hilbert functions we get that $\{1, m, n, 1\}$ is the Hilbert function of an Artin algebra iff $1 \leq n \leq \binom{m+1}{2}$, [3, 30].

The following result is due to De Stefani, [6], it is a generalization of some results of Iarrobino.

Proposition 2.4.12 *Let $A = R/I$ be an Artin level algebra of socle degree s and Cohen-Macaulay type t. Then*

(i) *$Q(0) = gr_{\mathfrak{n}}(A)/C(1)$ is the unique (up to iso) graded level quotient of $gr_{\mathfrak{n}}(A)$ with socle degree s and Cohen-Macaulay type t.*
(ii) *Let $F_1, \cdots, F_t \in S$ be generators of $I^{\perp}$ such that such that $\deg(F_i) = s$, $i = 1, \cdots, t$, and $\text{top}(F_1), \cdots, \text{top}(F_t)$ are $\mathbf{k}$-linear independent forms of degree s, Proposition 2.4.4. Then $Q(0) \cong R/\langle \text{top}(F_1), \cdots, \text{top}(F_t)\rangle^{\perp}$.*
(iii) *The associated graded ring $gr_{\mathfrak{n}}(A)$ is an Artin level algebra of socle degree s and Cohen-Macaulay type t iff $gr_{\mathfrak{n}}(A) \cong Q(0)$.*

As corollary we get:

Proposition 2.4.13 *Let $A = R/I$ be an Artin Gorenstein algebra of socle degree s. Then the following conditions are equivalent:*

(i) *$gr_{\mathfrak{n}}(A)$ is Gorenstein,*
(ii) *$gr_{\mathfrak{n}}(A) = Q(0)$,*
(iii) *HF_A is symmetric.*

2.5 Classification of Artin Rings

It is known that there are a finite number of isomorphism classes for $e \leq 6$. J. Briançon [2] proved this result for $n = 2$, $\mathbf{k} = \mathbb{C}$; G. Mazzola [24] for $\mathbf{k} = \bar{\mathbf{k}}$ and $char(\mathbf{k}) \neq 2, 3$; finally B. Poonen [26] proved the finiteness for any $\mathbf{k} = \bar{\mathbf{k}}$. On the other hand D.A. Suprunenko [31] proved that if $\mathbf{k}$ infinite, there are infinite number of isomorphism classes for $e \geq 7$.

The problem of classification is in general very hard. For instance, before the paper [12], an open problem was the classification of Artin algebras with Hilbert function $\{1, m, n, 1\}$, even if A is Gorenstein.

Other families that has been classified are the almost stretched algebras, [14, 15]. We say that a Artin Gorenstein algebra $A = R/I$ is Almost Stretched if $\mathfrak{m}^2$ is minimally generated by two elements or equivalently, the Hilbert function of A is

$$\text{HF}_A = \{1, n, \overbrace{2, \cdots, 2}^{t-1}, \overbrace{1, \cdots, 1}^{s-t}\}$$

We assume that $3 \leq t + 1 \leq s$. We say that A is of type (s, t).

In the following result we present the possible analytic types of almost stretched algebras, [14, 15] and [9]. In fact, we proved more: we determined the pairwise analytic types of almost stretched algebras. We omit describing it here.

Theorem 2.5.1 *Let $A = R/I$ be an Almost Stretched algebra of type (s,t) with $3 \le t+1 \le s$.*
If there is not r such that $2(r+1) = s-t+1$ or $s \ge 3t-1$ then I is isomorphic to one of the following ideals:

$$I_{0,1}, I_{1,1}, \dots, I_{\min\{t-1,s-t\},1}.$$

Assume that $s \le 3t-2$ and let r be the integer such that $2(r+1) = s-t+1$, then I is isomorphic to one of the following ideals:

$$I_{0,1}, \dots, I_{r-1,1}, I_{r+1,1}, \dots, I_{\min\{t-1,s-t\},1}$$

$$\{I_{r,a}\}_{a\in \mathbf{k}^*}, \{I_{r,a+x_1}\}_{a\in \mathbf{k}^*}, \dots, \{I_{r,a+x_1^{t-r-2}}\}_{a\in \mathbf{k}^*}$$

Where $I_{p,z}$ is the ideal generated by

$$\{x_i x_j\}_{1\le i<j\le n,(i,j)\neq(1,2)}, \{x_j - x_1^s\}_{3\le j\le n}, x_2^2 - x_1^{p+1}x_2 - z x_1^{s-t+1}, x_1^t x_2$$

Example 2.5.2 ([15]) Let A be an Artin Gorenstein algebra with Hilbert function $\{1,2,2,2,1,1,1\}$. Then the analytic types are represented by

1. $I_1 = (y^2 - xy - x^4, x^3y)$
2. $I_2 = (y^2 - x^3y - x^4, x^3y)$
3. $I_c = (y^2 - x^2y - cx^4, x^3y)$, $c \in \mathbf{k}^*$

The moduli space, see [19] has two isolated points and a punctured affine line.

The main result of this section shows that some Artin algebras are isomorphic to their associated graded ring. J. Emsalem called these algebras "canonically graded".

Definition 2.5.3 (Emsalem) An Artin local algebra $A = R/I$ is canonically graded if A is analytically isomorphic to $gr_{\mathfrak{n}}(A) \cong R/I^*R$.

Notice that there are non-canonically graded algebras, for instance:

Example 2.5.4 ([15]) Let A be an Artin Gorenstein algebras with $\mathrm{HF}_A = \{1,2,3,2,1\}$ then A is is isomorphic to one and only one of the following quotients of $R = \mathbf{k}[[x_1,x_2]]$:

1. $I_1 = (x_1^4, x_2^2)$,
2. $I_2 = (x_1^4, x_1^2 + x_2^2)$, and
3. $I_3 = (x_1^4, x_2^2 - x_1^3)$.

Notice: $I_3^* = (x_1^4, x_2^2) = I_1$ and $I_1 \ncong I_3$, i.e. R/I_3 is not canonically graded.

From now on we assume that the ground field $\mathbf{k}$ is of characteristic zero.

If L is a submodule of S generated by a sequence $\underline{G} := G_1, \ldots, G_t$ of polynomials of S, then we will write

$$A_{\underline{G}} = R/\operatorname{Ann}(L).$$

Given a form G of degree s and an integer $q \leq s$, we denote by $\Delta^q(G)$ the $\binom{n-1+s-q}{n-1} \times \binom{n-1+q}{n-1}$ matrix whose columns are the coordinates of $\partial_{\underline{i}}(G)$, $|\underline{i}| = q$, with respect to $(x^L)^* = \frac{1}{L!}y^L$, $|L| = s - q$. We will denote by $(L, \underline{i})$ the corresponding position in the matrix $\Delta^q(G)$. In the following $L + \underline{i}$ denotes the sum in $\mathbb{N}^n$.

Proposition 2.5.5 ([11]) *Let $G \in S = \mathbf{k}[y_1, \cdots, y_n]$ be a form of degree s. Then*

$$\operatorname{HF}_{A_G}(s-i) = \operatorname{rank}(\Delta^i(G)) \leq \min\left\{\binom{n-1+s-i}{n-1}, \binom{n-1+i}{n-1}\right\}$$

for $i = 0, \cdots, s$. The equality holds if and only if A_G is compressed.

Given an integer $i \leq s$, then

$$\Delta^i(G) = {}^\tau \Delta^{s-i}(G)$$

where ${}^\tau$ denotes the transpose matrix.

Notice that from the last result and Proposition 2.3.5 it is easy to deduce an alternative proof of the fact that a graded Gorenstein algebra A_G has symmetric Hilbert function.

Let $A_{\underline{G}}$ be a graded level algebra. We can define for all integers $i \leq s$ the block matrix

$$\Delta^i(\underline{G}) = \begin{pmatrix} \Delta^i(G_1) \\ \hline \vdots \\ \hline \Delta^i(G_t) \end{pmatrix} \tag{2.2}$$

which is a $t\binom{n-1+s-i}{n-1} \times \binom{n-1+i}{n-1}$ matrix. We get the following result.

Proposition 2.5.6 *Let $A = A_{\underline{G}}$ be a compressed algebra of socle degree s and Cohen-Macaulay type type t. Then for every $i = 1, \ldots, s$*

$$\operatorname{HF}_A(i) = rank(\Delta^i(\underline{G}[s])) = \min\left\{\binom{n-1+i}{n-1}, t\binom{n-1+s-i}{n-1}\right\}.$$

Proof By Proposition 2.4.8 we know that $gr_\mu(A)$ is level compressed of socle degree s and type t. Since $gr_\mu(A)$ is level if and only if $gr_\mu(A) \simeq Q(0) = S/\langle \operatorname{top}(\underline{G})\rangle^\perp$, the result follows by Proposition 2.5.5.

Given a $\mathbf{k}$-algebra C, quotient of R, we will denote by $Aut(C)$ the group of the automorphisms of C as a $\mathbf{k}$-algebra and by $Aut_{\mathbf{k}}(C)$ as a $\mathbf{k}$-vector space. Since R is complete $\varphi \in Aut(R)$ is determined by

$$\varphi(x_i) \in \mathfrak{m}$$

$i = 1, \cdots, n$, i.e. φ acts by substitution of x_i by $\varphi(x_i)$.

For any $\varphi \in Aut_{\mathbf{k}}(R/\mathfrak{m}^{s+1})$ we may associate a matrix $M(\varphi)$ with respect to the basis Ω of size $r = dim_K(R/\mathfrak{m}^{s+1}) = \binom{n+s}{s}$ already defined at the end of Sect. 2.5. Given I and J ideals of R such that $\mathfrak{m}^{s+1} \subset I, J$, there exists an isomorphism of $\mathbf{k}$-algebras

$$\varphi : R/I \to R/J$$

if and only if φ is canonically induced by a $\mathbf{k}$-algebra automorphism of $R/\mathfrak{m}^{s+1}$ sending $I/\mathfrak{m}^{s+1}$ to $J/\mathfrak{m}^{s+1}$. In particular φ is an isomorphism of $\mathbf{k}$-vector spaces. Dualizing

$$\varphi^* : (R/J)^* \to (R/I)^*$$

is an isomorphism of the $\mathbf{k}$-vector subspaces where $(R/I)^* \simeq I^\perp$ and $(R/J)^* \simeq J^\perp$ of $S_{\leq s}$ according to the exact paring (2.3.6). Hence ${}^\tau M(\varphi)$ is the matrix associated to φ^* with respect to the basis Ω^* of $S_{\leq s}$.

We denote by $\mathscr{R}$ the subgroup of $Aut_{\mathbf{k}}(S_{\leq s})$ (automorphisms of $S_{\leq s}$ as a $\mathbf{k}$-vector space) represented by the matrices ${}^\tau M(\varphi)$ of $Gl_r(\mathbf{k})$ with $\varphi \in Aut(R/\mathfrak{m}^{s+1})$. For all $p \geq 1$, I_p denotes the identity matrix of order $\binom{n+p-1}{p}$. By Emsalem, [16, Proposition 15], the classification, up to analytic isomorphism, of the Artin local $\mathbf{k}$-algebras of multiplicity e, socle degree s and embedding dimension n is equivalent to the classification, up to the action of $\mathscr{R}$, of the $\mathbf{k}$-vector subspaces of $S_{\leq s}$ of dimension e, stable by derivations and containing the $\mathbf{k}$-vector space $S_{\leq 1}$.
Let $\underline{F} = F_1, \ldots, F_t$, respectively $\underline{G} = G_1, \ldots, G_t$, be polynomials of degree s. Let $\varphi \in Aut(R/\mathfrak{m}^{s+1})$, from the previous facts we have

$$\varphi(A_{\underline{F}}) = A_{\underline{G}} \quad \text{if and only if} \quad (\varphi^*)^{-1}(\langle \underline{F} \rangle_R) = \langle \underline{G} \rangle_R. \tag{2.3}$$

If $F_i = b_{i1}\omega_1^* + \ldots b_{ir}\omega_r^* \in S_{\leq s}$, then we will denote the *row vector* of the coefficients of the polynomial with respect to the basis Ω^* by

$$[F_i]_{\Omega^*} = (b_{i1}, \ldots, b_{ir}).$$

If there exists $\varphi \in Aut(R/\mathfrak{m}^{s+1})$ such that

$$[G_i]_{\Omega^*} M(\varphi) = [F_i]_{\Omega^*}, \ \text{for every}\ i = 1, \ldots, t, \ \text{then}\ \varphi(A_{\underline{F}}) = A_{\underline{G}} \tag{2.4}$$

Let φ_{s-p} be an automorphism of $R/\mathfrak{m}^{s+1}$ such that $\varphi_{s-p} = Id$ modulo $\mathfrak{m}^{p+1}$, with $1 \leq p \leq s$, that is

$$\varphi_{s-p}(x_j) = x_j + \sum_{|\underline{i}|=p+1} a_{\underline{i}}^j x^{\underline{i}} + \text{higher terms} \tag{2.5}$$

for $j = 1, \ldots, n$ and $a_{\underline{i}}^j \in \mathbf{k}$ for each n-uple $\underline{i}$ such that $|\underline{i}| = p+1$. In the following we will denote $\underline{a} := (a_{\underline{i}}^1, |\underline{i}| = p+1; \cdots ; a_{\underline{i}}^n, |\underline{i}| = p+1) \in \mathbf{k}^{n\binom{n+p}{n-1}}$.

The matrix associated to φ_{s-p}, say $M(\varphi_{s-p})$, is an element of $Gl_r(\mathbf{k})$, $r = \binom{n+s}{s+1}$, with respect to the basis Ω of $R/\mathfrak{m}^{s+1}$. We write $M(\varphi_{s-p}) = (B_{i,j})_{0 \leq i,j \leq s}$ where $B_{i,j}$ is a $\binom{n+i-1}{i} \times \binom{n+j-1}{j}$ matrix of the coefficients of monomials of degree i appearing in $\varphi(x^{\underline{j}})$ where $\underline{j} = (j_1, \ldots, j_n)$ such that $|\underline{j}| = j$. It is easy to verify that:

$$B_{i,j} = \begin{cases} 0, \ \ 0 \leq i < j \leq s, \ \text{or } j = 1, i = 1, \cdots, s, \\ \\ I_i, \ i = j = 0, \cdots, s, \\ \\ 0, \ \ j = s-p, \cdots, s-1, i = j+1, \cdots, s, \ \text{and } (i,j) \neq (s, s-p). \end{cases}$$

The matrix $M(\varphi_{s-p})$ has the following structure

$$M(\varphi_{s-p}) = \left(\begin{array}{c|c|c|c|c|c|c|c} 1 & 0 & \cdots & 0 & 0 & 0 & 0 & 0 \\ \hline 0 & I_1 & 0 & 0 & 0 & 0 & 0 & \vdots \\ \hline 0 & 0 & I_2 & 0 & 0 & 0 & 0 & \vdots \\ \hline \vdots & \vdots & 0 & \ddots & \vdots & \vdots & \vdots & \vdots \\ \hline 0 & B_{p+1,1} & 0 & \cdots & I_{s-p} & 0 & 0 & \vdots \\ \hline 0 & \cdots & B_{p+2,2} & 0 & 0 & I_{s-p+1} & 0 & \vdots \\ \hline 0 & \cdots & \cdots & \ddots & \vdots & 0 & \ddots & 0 \\ \hline 0 & B_{s,1} & B_{s,2} & \cdots & B_{s,s-p} & 0 & \cdots & I_s \end{array}\right)$$

The entries of $B_{p+1,1}, B_{p+2,2}, \ldots, B_{s,s-p}$ are linear forms in the variables $a_{\underline{i}}^j$, with $|\underline{i}| = p+1$, $j = 1, \cdots, n$. We are mainly interested in $B_{s,s-p}$ which is a $\binom{n+s-1}{s} \times \binom{n+s-p-1}{s-p}$ matrix whose columns correspond to x^W with $|W| = s-p$ and the rows correspond to the coefficients of x^L with $|L| = s$ in $\varphi(x^W)$.

Let F, G be polynomials of degree s of P and let φ_{s-p} be a $\mathbf{k}$-algebra isomorphism of type (2.5) sending A_F to A_G. We denote by $F[j]$ (respectively

$G[j]$) the homogeneous component of degree j of F (respectively of G), that is $F = F[s] + F[s-1] + \dots$ ($G = G[s] + G[s-1] + \dots$).
By (2.4) we have

$$[G]_{\Omega^*} M(\varphi_{s-p}) = [F]_{\Omega^*}, \tag{2.6}$$

in particular we deduce

$$[F[j]]_{\Omega^*} = \begin{cases} [G[s-p]]_{\Omega^*} + [G[s]]_{\Omega^*} B_{s,s-p}, & j = s-p, \\ \\ [G[j]]_{\Omega^*}, & j = s-p+1, \cdots, s. \end{cases} \tag{2.7}$$

We are going to study $[G[s]]_{\Omega^*} B_{s,s-p}$. Let $[\alpha_{\underline{i}}]$ be the vector of the coordinates of $G[s]$ w.r.t. Ω^*, i.e.

$$G[s] = \sum_{|\underline{i}|=s} \alpha_{\underline{i}} \frac{1}{\underline{i}!} y^{\underline{i}};$$

the entries of $[G[s]]_{\Omega^*} B_{s,s-p}$ are bi-homogeneous forms in the components of $[\alpha_{\underline{i}}]$ and $\underline{a} = (a^1_{\underline{i}}, \dots, a^n_{\underline{i}})$ such that $|\underline{i}| = p+1$ of bi-degree $(1,1)$. Hence there exists a matrix $M^{[s-p]}(G[s])$ of size $\binom{n-1+s-p}{n-1} \times n\binom{n+p}{n-1}$ and entries in the $\mathbf{k}[\alpha_{\underline{i}}]$ such that

$$^{\tau}([\alpha_{\underline{i}}] B_{s,s-p}) = M^{[s-p]}(G[s]) \; ^{\tau}\underline{a} \tag{2.8}$$

where $^{\tau}\underline{a}$ denotes the transpose of the row-vector $\underline{a}$. We are going to describe the entries of $M^{[s-p]}(G[s])$. We label the columns of $M^{[s-p]}(G[s])$ with the set of indexes $(j, \underline{i})$, $j = 1, \cdots, n$, $|\underline{i}| = p+1$, corresponding to the entries of $\underline{a} = (a^1_{\underline{i}}, |\underline{i}| = p+1; \cdots; a^n_{\underline{i}}, |\underline{i}| = p+1) \in \mathbf{k}^{n\binom{n+p}{n-1}}$.

For every $i = 1, \cdots, n$, we denote S^i_p the set of monomials x^α of degree p such that $x^\alpha \in x_i(x_i, \cdots, x_n)^{p-1}$, hence $\#(S^i_p) = \binom{p-1+n-i}{p-1}$.

Lemma 2.5.7 *The matrix $M^{[s-p]}(G[s])$ has the following upper-diagonal structure*

$$M^{[s-p]}(G[s]) = \left(\begin{array}{c|c|c|c|c} M_1 & * & \cdots & * & * \\ \hline 0 & M_2 & \cdots & * & * \\ \hline \vdots & \vdots & \vdots & \vdots & \vdots \\ \hline 0 & 0 & 0 & M_{n-1} & * \\ \hline 0 & 0 & 0 & 0 & M_n \end{array}\right)$$

where M_j is a matrix of size $\binom{s-p-1+n-j}{s-p-1} \times \binom{n+p}{n-1}$, $j = 1, \cdots, n$, defined as follows: the entries of M_j are the entries of $M^{[s-p]}(G[s])$ corresponding to the rows

$W \in \log(S^j_{s-p})$ *and columns* $(j, \underline{i})$, $|\underline{i}| = p + 1$. *We label the entries of* M_j *with respect to these multi-indices. Then it holds:*

(i) for all $W = (w_1, \cdots, w_n) \in \log(S^1_{s-p})$ *and* $\underline{i}$, $|\underline{i}| = p + 1$,

$$w_1 \Delta^{p+1}(G[s])_{(W-\delta_1,\underline{i})} = M_{1(W,(1,\underline{i}))},$$

(ii) for all $j = 1, \cdots, n-1$, $W \in \log(S^{j+1}_{s-p})$,

$$M_{j+1,(W,(j+1,*))} = w_{j+1} M_{j,(L,(j,*))}$$

with $L = \delta_j + W - \delta_{j+1}$,

From the last result we get the key result of this chapter.

Corollary 2.5.8 *If* $s \leq 4$ *then rank* $(M^{[s-p]}(G[s]))$ *is maximal if and only if rank* $(\Delta^{p+1}(G[s]))$ *is maximal.*

Proof Notice that $M^{[s-p]}(G[s])$ has an upper-diagonal structure where the rows of the diagonal blocks M_j are a subset of the rows of the first block matrix M_1. Let us assume that the number of rows of M_1 is not bigger than the number of columns of M_1, as a consequence the same holds for M_j with $j > 1$. Then we can compute the rank of $M^{[s-p]}(G[s])$ by rows, so rank $(M^{[s-p]}(G[s]))$ is maximal if and only if rank $(\Delta^{p+1}(G[s]))$ is maximal. Since M_1 is a $\binom{s-p-2+n}{s-p-1} \times \binom{n+p}{n-1}$ matrix, if $\binom{n+s-p-2}{s-p-1} = \binom{n+s-p-2}{n-1} \leq \binom{n+p}{n-1}$ we get the result. This inequality is equivalent to $n+s-p-2 \leq n+p$, i.e. $s \leq 2p+2$, since $p \geq 1$ we get that $s \leq 4$.

We may generalize the previous facts to a sequence $\underline{G} = G_1, \ldots, G_t$ of polynomials of degree s of S. Let φ_{s-p} be a **k**-algebra isomorphism of type (2.5) sending $A_{\underline{F}}$ to $A_{\underline{G}}$ where $\underline{F} = F_1, \ldots, F_t$. In particular we assume that, as in (2.6),

$$[G_r]_{\Omega^*} M(\varphi_{s-p}) = [F_r]_{\Omega^*},$$

for every $r = 1, \ldots, t$. We deduce the analogues of (2.7) and we restrict our interest to

$$[\underline{G}[s]]_{\Omega^*} B^{\oplus t}_{s,s-p}$$

where

$$B^{\oplus t}_{s,s-p} := \begin{pmatrix} B_{s,s-p} \\ \hline \vdots \\ \hline B_{s,s-p} \end{pmatrix}$$

obtained by gluing t times the matrix $B_{s,s-p}$ and where $[\underline{G}[s]]_{\Omega^*}$ is the row $([G_r[s]]_{\Omega^*} : r = 1, \ldots, t)$. In accordance with (2.8), it is defined the matrix

$M^{[s-p]}(G_r[s])$ of size $\binom{n-1+s-p}{n-1} \times n\binom{n+p}{n-1}$ and entries depending on $[\underline{G}[s]]_{\Omega^*}$ such that

$$^{\tau}([G_r[s]]_{\Omega^*} B_{s,s-p}) = M^{[s-p]}(G_r[s]) \ ^{\tau}\underline{a}$$

If we define

$$M^{[s-p]}(\underline{G}[s]) := \begin{pmatrix} M^{[s-p]}(G_1[s]) \\ \hline \vdots \\ \hline M^{[s-p]}(G_t[s]) \end{pmatrix} \tag{2.9}$$

which is a $t\binom{n-1+s-p}{n-1} \times n\binom{n+p}{n-1}$ matrix, we get

$$^{\tau}([\underline{G}[s]]_{\Omega^*} B^{\oplus t}_{s,s-p}) = M^{[s-p]}(\underline{G}[s]) \ ^{\tau}\underline{a}. \tag{2.10}$$

The matrix $M^{[s-p]}(\underline{G}[s])$ has the same shape of $M^{[s-p]}(G[s])$, already described in Lemma 2.5.7 and its blocks correspond to suitable submatrices of $(\Delta^{p+1}(\underline{G}[s]))$ (see (2.2)). Hence we have an analogue to (2.7) for the level case

$$[F_r[j]]_{\Omega^*} = \begin{cases} [G_r[s-p]]_{\Omega^*} + \underline{a} \ ^{\tau}(M^{[s-p]}(\underline{G}[s])), & j = s-p, \\ \\ [G_r[j]]_{\Omega^*}, & j = s-p+1, \cdots, s. \end{cases} \tag{2.11}$$

for all $r = 1, \ldots, t$.

In the next result we generalize the main result of [12].

Theorem 2.5.9 *Let A be an Artin compressed Gorenstein local* **k***-algebra. If $s \leq 4$ then A is canonically graded.*

Proof Let A be an Artin compressed Gorenstein local **k**-algebra of socle degree $s \geq 2$ and embedding dimension n. Then $A = A_G$ with $G \in S$ a polynomial of degree s and $gr_{\mu}(A) = S/\operatorname{Ann}(G[s])$ is a compressed Gorenstein graded algebra of socle degree $s \geq 2$ and embedding dimension n (see Proposition 2.4.8).

The main result of [12] shows that if $s \leq 3$ then A is canonically graded. Let us assume $s = 4$, then the Hilbert function is $\{1, n, \binom{n+1}{2}, n, 1\}$. Because $A_{G[4]}$ is a compressed Gorenstein algebra with the same Hilbert function of A, we may assume $G = G[4] + G[3]$. In fact $S_1, S_2 \subseteq \langle G[4]\rangle_R$ because of (2.5.6) and, as a consequence, it is easy to see that $\langle G[4] + G[3]\rangle_R = \langle G[4] + G[3] + G[2] + \ldots\rangle_R$. We have to prove that there exists an automorphism $\varphi \in Aut(R/\mathfrak{m}^5)$ such that

$$A_G \simeq A_{G[4]}.$$

We consider for every $j = 1, \dots, n$

$$\varphi_3(x_j) = x_j + \sum_{|\underline{i}|=2} a_{\underline{i}}^j x^{\underline{i}} + \text{higher terms}$$

If $A_F = \varphi_3^{-1}(A_G)$, then from (2.7) and (2.8) we get

$$\begin{aligned} [F[3]]_{\Omega^*} &= [G[3]]_{\Omega^*} + \underline{a}\ {}^{\tau}(M^{[3]}(G[4])) \\ [F[4]]_{\Omega^*} &= [G[4]]_{\Omega^*} \end{aligned} \tag{2.12}$$

where $\underline{a} = (a_{\underline{i}}^1, \dots, a_{\underline{i}}^n)$. By Proposition 2.5.6 and Corollary 2.5.8, we know that the matrix $M^{[3]}(G[4])$ has maximal rank and it coincides with the number of rows, so there exists a solution $\underline{a} \in \mathbf{k}^n$ of (2.12) such that $F[3] = 0$ and $F[4] = G[4]$.

The aim is now to list classes of local compressed algebras of embedding dimension n, socle degree s and socle type $\sigma = (0, \dots, \sigma_{r-1}, \sigma_r, \dots, \sigma_s, 0, 0, \dots)$ which are canonically graded. Examples will prove that the following result cannot be extended to higher socle degrees. This result extends the main result of [12] and [6].

Theorem 2.5.10 *Let $A = R/I$ be an Artin compressed $\mathbf{k}$-algebra of embedding dimension n, socle degree s and socle type σ. Then A is canonically graded in the following cases:*

(1) $s \le 3$,
(2) $s = 4$ and $e_4 = 1$,
(3) $s = 4$ and $n = 2$.

Proof Since a local ring with Hilbert function $\{1, n, t\}$ is always graded, we may assume $s \ge 3$.

If $s = 3$ and A is a compressed level algebra, then A is canonically graded by De Stefani [6]. If A is not necessarily level, but compressed, then by (2.1) the socle type is $\{0, 0, \sigma_2, \sigma_3\}$ and the Hilbert function is $\{1, n, h_2, \sigma_3\}$ where $h_2 = \min\{\dim_{\mathbf{k}} R_2, \sigma_2 + \sigma_3 n\}$. Then we may assume that in any system of coordinates $I^{\perp}$ is generated by e_2 quadratic forms and e_3 polynomials $G_1, \dots, G_{\sigma_3}$ of degree 3. Then the result follows because $R/\operatorname{Ann}_R(G_1, \dots, G_{\sigma_3})$ is a 3-level compressed algebra of type σ_3 and hence canonically graded.

Let us assume $s = 4$ and $\sigma_4 = 1$. We recall that if A is Gorenstein, then the result follows by Theorem 2.5.9. Since A is compressed, then by (2.1) the socle type is $(0, 0, 0, \sigma_3, 1)$. This means that $I^{\perp}$ is generated by e_3 polynomial of degree 3 and one polynomial of degree 4. Similarly to the above part, because $S_{\le 2} \subseteq (I^*)^{\perp}$, $I^{\perp}$ can be generated by σ_3 forms of degree 3 and one polynomial of degree 4. As before the problem is reduced to the Gorenstein case with $s = 4$ and the result follows.

Assume $s = 4$ and $n = 2$. If $\sigma_4 = 1$, then we are in case (2). If $\sigma_4 > 1$, because A is compressed, the possible socle types are: $\sigma_i = (0, 0, 0, 0, i)$ with $i = 2, \cdots, 5$

and since A is compressed, the corresponding Hilbert function is $\{1, 2, 3, 4, i\}$. In each case A is graded because the Hilbert function forces the dual module to be generated by forms of degree four.

As a corollary of the last result we get [12].

Theorem 2.5.11 *Let A be an Artinian Gorenstein $\mathbf{k}$-algebra with Hilbert function $\{1, n, m, 1\}$. Then the following conditions are equivalent:*

(i) A is canonically graded,
(ii) $m = n$,
(iii) A is compressed.

From this result we can deduce

Corollary 2.5.12 ([12]) *The classification of Artinian Gorenstein local $\mathbf{k}$-algebras with Hilbert function $\mathrm{HF}_A = \{1, n, n, 1\}$ is equivalent to the projective classification of the hypersurfaces $V(F) \subset \mathbb{P}^{n-1}_{\mathbf{k}}$ where F is a degree three non degenerate form in n variables.*

Next we will recall the classification of the Artin Gorenstein algebras for $n = 1, 2, 3$, [12].

If $n = 1$, then it is clear that $A \cong \mathbf{k}[[x]]/(x^4)$, so there is only one analytic model. If $n = 2$ we have the following result:

Proposition 2.5.13 ([12]) *Let A be an Artinian Gorenstein local K-algebra with Hilbert function $\mathrm{HF}_A = \{1, 2, 2, 1\}$. Then A is isomorphic to one and only one of the following quotients of $R = K[[x_1, x_2]]$:*

Model $A = R/I$	*Inverse system F*	*Geometry of $C = V(F) \subset \mathbb{P}^1_{\mathbf{k}}$*
(x_1^3, x_2^2)	$y_1^2 y_2$	*Double point plus a simple point*
$(x_1 x_2, x_1^3 - x_2^3)$	$y_1^3 - y_2^3$	*Three distinct points*

Finally, for $n = 3$ first we have to study with detail the classification of plane curves, in particular, the elliptic curves, see for instance [29]. Any plane elliptic cubic curve $C \subset \mathbb{P}^2_{\mathbf{k}}$ is defined, in a suitable system of coordinates, by a Weierstrass' equation, [29],

$$W_{a,b} : y_2^2 y_3 = y_1^3 + a y_1 y_3^2 + b y_3^3$$

with $a, b \in \mathbf{k}$ such that $4a^3 + 27b^2 \neq 0$. The j invariant of C is

$$j(a, b) = 1728 \frac{4a^3}{4a^3 + 27b^2}$$

It is well known that two plane elliptic cubic curves $C_i = V(W_{a_i, b_i}) \subset \mathbb{P}^2_{\mathbf{k}}$, $i = 1, 2$, are projectively isomorphic if and only if $j(a_1, b_1) = j(a_2, b_2)$.

For elliptic curves the inverse moduli problem can be done as follows. We denote by $W(j)$ the following elliptic curves with j as moduli : $W(0) = y_2^2y_1 + y_2y_3^2 - y_1^3$, $W(1728) = y_2^2y_3 - y_1y_3^2 - y_1^3$, and for $j \neq 0, 1728$

$$W(j) = (j - 1728)(y_2^2y_3 + y_1y_2y_3 - y_1^3) + 36y_1y_3^2 + y_3^3.$$

We will show by using the library `INVERSE-SYST.LIB` that:

Proposition 2.5.14 *Let A be an Artin Gorenstein local* **k***-algebra with Hilbert function* $\mathrm{HF}_A = \{1, 3, 3, 1\}$. *Then A is isomorphic to one and only one of the following quotients of* $R = \mathbf{k}[[x_1, x_2, x_3]]$ *:*

Model for $A = R/I$	*Inverse system* F	*Geometry of* $C{=}V(F){\subset}\mathbb{P}^2_{\mathbf{k}}$
(x_1^2, x_2^2, x_3^2)	$y_1y_2y_3$	*Three independent lines*
$(x_1^2, x_1x_3, x_3x_2^2, x_2^3, x_3^2 + x_1x_2)$	$y_2(y_1y_2 - y_3^2)$	*Conic and a tangent line*
$(x_1^2, x_2^2, x_3^2 + 6x_1x_2)$	$y_3(y_1y_2 - y_3^2)$	*Conic and a non-tangent line*
$(x_3^2, x_1x_2, x_1^2 + x_2^2 - 3x_1x_3)$	$y_2^2y_3 - y_1^2(y_1 + y_3)$	*Irreducible nodal cubic*
$(x_3^2, x_1x_2, x_1x_3, x_2^3, x_1^3 + 3x_2^2x_3)$	$y_2^2y_3 - y_1^3$	*Irreducible cuspidal cubic*
$(x_3^3, x_1^3 + 3x_2^2x_3, x_1x_3, x_2^2 - x_2x_3$ $+x_3^2, x_1x_2)$	$W(0) = y_2^2y_1 + y_2y_3^2 - y_1^3$	*Elliptic curve* $j = 0$
$(x_2^2 + x_1x_3, x_1x_2, x_1^2 - 3x_3^2)$	$W(1728) = y_2^2y_3 - y_1y_3^2 - y_1^3$	*Elliptic curve* $j = 1728$
$I(j) = (x_2(x_2 - 2x_1), H_j, G_j)$	$W(j),\ j \neq 0, 1728$	*Elliptic curve with* $j{\neq}0, 1728$

with:

$H_j = 6jx_1x_2 - 144(j - 1728)x_1x_3 + 72(j - 1728)x_2x_3 - (j - 1728)^2x_3^2$, *and*
$G_j = jx_1^2 - 12(j - 1728)x_1x_3 + 6(j - 1728)x_2x_3 + 144(j - 1728)x_3^2$;
$I(j_1) \cong I(j_2)$ *if and only if* $j_1 = j_2$.

Proof Let us assume that F is the product of the linear forms l_1, l_2, l_3. If l_1, l_2, l_3 are **k**-linear independent we get the first case. On the contrary, if these linear forms are **k**-linear dependent, we deduce that F is degenerate. Let us assume that F is the product of a linear form l and an irreducible quadric Q. According to the relative position of $V(l)$ and $V(Q)$ we get the second and the third case.

Let F be a degree three irreducible form. The first seven models can be obtained from the corresponding inverse system F by using the command `idealAnn` of [8]. For the last case see [8].

2.6 Computation of Betti Numbers

In this chapter we address the following problem: How can we compute the Betti numbers of I in terms of its Macaulay's inverse system $L = I^{\perp}$ without computing the ideal I? This is a longstanding problem in commutative algebra that has been considered by many authors, see for instance [22], Chap. 9, Problem L. For instance, if $A = R/I$ is an Artin Gorenstein local ring then its inverse system is a polynomial

F on the variables $x_1, \dots, x_n$ of degree the socle degree of A. In this chapter we compute the Betti numbers of A in terms of the polynomial F instead of computing I and then to compute the Betti numbers of $A = R/I$.

Let I be an $\mathfrak{m}$-primary ideal of R. Let $\mathbb{F}_\bullet$ be a minimal free resolution of the R-module R/I

$$\mathbb{F}_\bullet \qquad 0 \longrightarrow \mathbb{F}_n = R^{\beta_n} \longrightarrow \dots \longrightarrow \mathbb{F}_1 = R^{\beta_1} \longrightarrow \mathbb{F}_0 = R \longrightarrow R/I \longrightarrow 0,$$

the p-th Betti number of R/I is $\beta_p(R/I) = \operatorname{rank}_R(\mathbb{F}_p)$, $1 \le p \le n$. Tensoring $\mathbb{F}_\bullet$ by the R-module $\mathbf{k}$ we get the complex

$$\mathbb{F}_\bullet \otimes_R \mathbf{k} \qquad 0 \longrightarrow \mathbf{k}^{\beta_n} \longrightarrow \dots \longrightarrow \mathbf{k}^{\beta_1} \longrightarrow \mathbf{k} = \mathbf{k} \longrightarrow 0.$$

Since $\mathbb{F}_\bullet$ is a minimal resolution we get that the morphisms of $\mathbb{F}_\bullet \otimes_R \mathbf{k}$ are zero, so

$$\beta_p(R/I) = \dim_{\mathbf{k}}(\operatorname{Tor}_p^R(R/I, \mathbf{k})),$$

$p = 1, \dots, n$. Let us now consider Koszul's resolution of R defined by the regular sequence $x_1, \dots, x_n$

$$\mathbb{K}_\bullet \qquad 0 \xrightarrow{d_{n+1}} \mathbb{K}_n = \bigwedge^n R^n \xrightarrow{d_n} \dots \longrightarrow \mathbb{K}_1 = \bigwedge^1 R^n \xrightarrow{d_1} \mathbb{K}_0 = R \longrightarrow \mathbf{k} \longrightarrow 0.$$

We consider the R-basis of R^n: $e_i = (0, \dots, 1^{(i}, \dots, 0) \in R^n$, $i = 1, \dots, n$; for all $1 \le i_1 < \dots < i_p \le n$ we set $e_{i_1,\dots,i_p} = e_{i_1} \wedge \dots e_{i_p} \in \bigwedge^p R^n$. Since the set $e_{i_1,\dots,i_p}$, $1 \le i_1 < \dots < i_p \le n$, form a R-basis of $\mathbb{K}_p$ we define the morphism

$$d_p : \mathbb{K}_p \longrightarrow \mathbb{K}_{p-1}$$

by $d_p(e_{i_1,\dots,i_p}) = \sum_{j=1}^p (-1)^{j-1} x_{i_j} e_{i_1,\dots,i_{j-1},i_{j+1},\dots,i_p} \in \mathbb{K}_{p-1}$. Notice that the basis $e_{i_1,\dots,i_p}$, $1 \le i_1 < \dots < i_p \le n$, this defines an isomorphism of R-modules $\bigwedge^p R^n \overset{\phi_p}{\cong} R^{\binom{n}{p}}$, such that $\phi_p(e_{i_1,\dots,i_p}) = v_{i_1,\dots,i_p}$, $1 \le i_1 < \dots < i_p \le n$, is the element of $R^{\binom{n}{p}}$ with all entries zero but the $(i_1, \dots, i_p)$-th that it is equal to 1. We denote by Δ_p the associated matrix to d_p with respect the above bases of $R^{\binom{n}{p}}$ and $R^{\binom{n}{p-1}}$, notice that the entries of Δ_p are zero or $\pm x_i$, $i = 1, \dots, n$.

We can compute $\operatorname{Tor}_p^R(R/I, \mathbf{k})$ by considering the complex $R/I \otimes_R \mathbb{K}_\bullet$

$$R/I \otimes_R \mathbb{K}_\bullet \qquad 0 \longrightarrow (R/I)^{\binom{n}{n}} \longrightarrow \dots \longrightarrow (R/I)^{\binom{n}{1}} \longrightarrow R/I \longrightarrow \mathbf{k} \longrightarrow 0.$$

We denote again by d_p the morphism $\operatorname{Id}_{R/I} \otimes_R d_p$ then

$$\operatorname{Tor}_p^R(R/I, \mathbf{k}) = H_p(R/I \otimes_R \mathbb{K}_\bullet) = \frac{\ker(d_p)}{\operatorname{im}(d_{p+1})}$$

for $p = 1, \dots, n$. If we consider the dual of $R/I \otimes_R \mathbb{K}_\bullet$ with respect to E we get, $L = I^\perp$,

$$(R/I \otimes_R \mathbb{K}_\bullet)^* \qquad 0 \longrightarrow \mathbf{k} \xrightarrow{d_0^*} L \xrightarrow{d_1^*} L^{\binom{n}{1}} \longrightarrow \dots L^{\binom{n}{n}} \xrightarrow{d_{n+1}^*} 0.$$

Notice that, if $\underline{\mathrm{h}} = (h_{i_1,\dots,i_p}, 1 \le i_1 < \dots < i_{p-1} \le n) \in L^{\binom{n}{p-1}}$ then

$$d_p^*(\underline{\mathrm{h}}) = {}^\tau\Delta_p(\underline{\mathrm{h}}). \tag{2.13}$$

Proposition 2.6.1 *Let $L \subset S$ be a finitely generated sub-R-module of S of dimension $e = \dim_{\mathbf{k}}(L)$. If $I = Ann_R(L) \subset R$ then*

$$\beta_p(R/I) = e\binom{n}{p} - \dim_{\mathbf{k}}(\mathrm{im}(d_p^*)) - \dim_{\mathbf{k}}(\mathrm{im}(d_{p+1}^*)).$$

for $p = 1, \dots, n$.

Proof Since L is a finitely dimensional $\mathbf{k}$-vector space and the duallzing functor * is exact and additive we get

$$\beta_p(R/I) = \dim_{\mathbf{k}}\left(\frac{\ker(d_{p+1}^*)}{\mathrm{im}(d_p^*)}\right) = \dim_{\mathbf{k}} \ker(d_{p+1}^*) - \dim_{\mathbf{k}} \mathrm{im}(d_p^*).$$

On the other hand $\dim_{\mathbf{k}}(\ker(d_{p+1}^*)) = e\binom{n}{p} - \dim_{\mathbf{k}}(\mathrm{im}(d_{p+1}^*))$, so from these identities we get the claim.

Next step is to compute the Betti numbers effectively. For all $t \ge 0$ let W_t be the set of standard monomials x^α, $\alpha \in \mathbb{N}^n$, of degree at most t ordered by the local deg-rev-lex ordering with $x_n < \dots < x_1$. For instance, for $n = 3$ and $t = 2$, $W_2 = \{x_3^2, x_3x_2, x_3x_1, x_2^2, x_2x_1, x_1^2, x_3, x_2, x_1, 1\}$. For all $p = 1, \dots, n$ we consider the following set $\mathcal{M}_{s,p}$ of linearly independent elements of $R^{\binom{n}{p}}$

$$m_{\alpha;i_1,\dots,i_p} := \begin{cases} x^\alpha \text{ in the } (i_1,\dots,i_p)\text{-th component,} \\ \\ 0 \text{ otherwise,} \end{cases}$$

$\deg(\alpha) \le s$, $1 \le i_1 < \dots < i_p \le n$. Notice that $\#(\mathcal{M}_{s,p}) = \binom{n+s}{n}\binom{n}{p}$.

Assume that $s = \deg(L)$. Given a $\mathbf{k}$-basis $w_1, \dots, w_e$ of L we consider the following $\mathbf{k}$-basis, say B_p, of $L^{\binom{n}{p}}$:

$$w_{i_1,\dots,i_p}^i := \begin{cases} w_i \text{ in the } (i_1,\dots,i_p)\text{-th component,} \\ \\ 0 \text{ otherwise,} \end{cases}$$

$i = 1, \dots, e$, $1 \le i_1 < \dots < i_p \le n$. We denote by $\Delta_p^+(L)$ the matrix such that the columns are the coordinates of $d_p^*(w^i_{i_1,\dots,i_p})$ with respect the basis $\mathscr{M}_{s,p-1}$. This is a matrix of $\binom{n+s}{n}\binom{n}{p}$ rows and $e\binom{n}{p-1}$ columns and the entries are zero or $\pm x_i$, $i = 1, \dots, n$. Then we have:

Proposition 2.6.2 *For any finitely generated R-module L and $1 \le p \le n$ we have*

$$\dim_{\mathbf{k}}(\mathrm{im}(d_p^*)) = \mathrm{rank}\,(\Delta_p^+(L)).$$

If $e = \dim_{\mathbf{k}}(L)$ and $I = Ann_R(L) \subset R$ then

$$\beta_p(R/I) = e\binom{n}{p} - \mathrm{rank}\,(\Delta_p^+(L)) - \mathrm{rank}\,(\Delta_{p+1}^+(L)).$$

for $p = 1, \dots, n$.

From this result we get that the determination of the Betti number $\beta_p(R/I)$ involves the computation of the rank of $\Delta_p^+(L)$, $p = 1, \dots, n$. Recall that these matrices are huge, see the comments before last result, so they are difficult to manage. Moreover, this method of computation of Betti numbers implies the computation or election of a $\mathbf{k}$-basis of L. This is not possible if we want to consider a general L or the deformations of L, see Example 2.6.7.

In the next result we compute the Cohen-Macaulay type of R/I and we partially recover the classical result of Macaulay. In the second part, case $n = 2$, we prove a well known result of Serre that can be deduced from Hilbert-Burch structure theorem, i.e. the class of codimension two complete intersection ideals coincides with the class of codimension two Gorenstein ideals.

Proposition 2.6.3 *Let L be a finitely generated R-module of S of dimension e. Then the Cohen-Macaulay type of R/I, $I = Ann_R(L)$, is*

$$t(R/I) = \dim_{\mathbf{k}}(L/\mathfrak{m} \circ L) = \mu_R(L).$$

In particular, for $n = 2$ then

$$\mu(I) = t(R/I) + 1.$$

In particular, I is a complete intersection if and only if R/I is Gorenstein.

Proof The first result is Proposition 2.4.3.

Assume that $n = 2$. Then the complex $(R/I \otimes_R \mathbb{K}_\bullet)^*$ is

$$0 \longrightarrow \mathbf{k} \xrightarrow{d_0^*} L \xrightarrow{d_1^*} L^2 \xrightarrow{d_2^*} L \xrightarrow{d_3^*} 0,$$

and $\mathrm{im}(d_2^*) = \mathfrak{m} \circ L$. From Corollary 2.6.1 with $p = 1$ we get

$$\mu(I) = 2e - (e-1) - \dim_{\mathbf{k}}(\mathfrak{m} \circ L) = \dim_{\mathbf{k}}(L/\mathfrak{m} \circ L) + 1 = t(R/I) + 1,$$

so I is a complete intersection ideal, i.e. $\mu(I) = 2$, if and only if $t(R/I) = 1$, i.e. R/I is Gorenstein.

Given a finitely generated sub-R-module L of S we denote by $L :_S \mathfrak{m}$ the sub-R-module of S formed by the polynomials $h \in S$ such that $\mathfrak{m} \circ h \subset L$. Notice that if $L \subset S_{\leq s}$ then $L :_S \mathfrak{m} \subset S_{\leq s+1}$ and, in particular, $\dim_{\mathbf{k}}(L :_S \mathfrak{m}) < \infty$. We consider the $\mathbf{k}$-vector space morphism induced by d_1^*

$$d_{1,s}^* : S_{\leq s+1} \longrightarrow S_{\leq s}^n$$

with

$$d_{1,s}^*(h) = (x_1 \circ h, \dots, x_n \circ h)$$

for all $h \in S_{\leq s+1}$. It is easy to prove that $L :_S \mathfrak{m} = (d_{1,s}^*)^{-1}(L^n)$.

Proposition 2.6.4 *Let $L \subset S$ be a finitely generated sub-R-module of S of dimension $e = \dim_{\mathbf{k}}(L)$ and degree $s = \deg(L)$. If $I = Ann_R(L) \subset R$ then*

$$\mu(I) = e(n-1) + \binom{n+s+1}{n} - \dim_{\mathbf{k}}(\mathrm{im}(d_{1,s}^*) + L^n).$$

Proof If we write $V = L^n$ then we have

$$\begin{aligned}\dim_{\mathbf{k}}(L :_S \mathfrak{m}) &= \dim_{\mathbf{k}}((d_{1,s}^*)^{-1}(V)) \\ &= \dim_{\mathbf{k}}((d_{1,s}^*)^{-1}(V \cap \mathrm{im}(d_{1,s}^*)) = \dim_{\mathbf{k}}(V \cap \mathrm{im}(d_{1,s}^*)) + 1\end{aligned}$$

because $\dim_{\mathbf{k}}(\ker(\phi_s)) = 1$, so

$$\begin{aligned}\dim_{\mathbf{k}}(L :_S \mathfrak{m}) &= \dim_{\mathbf{k}}(V) + \dim_{\mathbf{k}}(\mathrm{im}(d_{1,s}^*)) - \dim_{\mathbf{k}}(\mathrm{im}(d_{1,s}^*) + V) + 1 \\ &= n.e + dim_{\mathbf{k}}(S_{\leq s+1}) - 1 - \dim_{\mathbf{k}}(\mathrm{im}(d_{1,s}^*) + V) + 1 \\ &= n.e + \tbinom{n+s+1}{n} - \dim_{\mathbf{k}}(\mathrm{im}(d_{1,s}^*) + V).\end{aligned}$$

Claim $\mu(I) = \dim_{\mathbf{k}}(L :_S \mathfrak{m}/L)$.

Proof of the Claim Let us consider the exact sequence of R-modules

$$0 \longrightarrow \frac{I}{\mathfrak{m}I} \longrightarrow \frac{R}{\mathfrak{m}I} \longrightarrow \frac{R}{I} \longrightarrow 0$$

dualizing this sequence we get the exact sequence on S-modules

$$0 \longrightarrow L \longrightarrow (\mathfrak{m}I)^{\perp} \longrightarrow \left(\frac{I}{\mathfrak{m}I}\right)^{*} \longrightarrow 0.$$

Hence

$$\begin{aligned}\dim_{\mathbf{k}}\left(\tfrac{I}{\mathfrak{m}I}\right)^{*} &= \dim_{\mathbf{k}}((\mathfrak{m}I)^{\perp}) - \dim_{\mathbf{k}}(L)\\ &= dim_{\mathbf{k}}(R/\mathfrak{m}I) - dim_{\mathbf{k}}(R/I)\\ &= dim_{\mathbf{k}}(I/\mathfrak{m}I) = \mu(I)\end{aligned}$$

by Nakayama's lemma. In particular, we get $\mu(I) = \dim_{\mathbf{k}}\left((\mathfrak{m}I)^{\perp}/L\right)$. Last step is to prove that $(\mathfrak{m}I)^{\perp} = L :_S \mathfrak{m}$. Given a polynomial $h \in S$ then $h \in (\mathfrak{m}I)^{\perp}$ if and only if $0 = (\mathfrak{m}I) \circ h = I \circ (\mathfrak{m} \circ h)$, so $(\mathfrak{m}I)^{\perp}$ is the set of polynomial h such that $\mathfrak{m} \circ h \subset L$, i.e. $(\mathfrak{m}I)^{\perp} = L :_S \mathfrak{m}$.

From the *Claim* we get

$$\mu(I) = \dim_{\mathbf{k}}(L :_S \mathfrak{m}) - \dim_{\mathbf{k}}(L) = n.e + \binom{n+s+1}{n} - \dim_{\mathbf{k}}(\mathrm{im}(d_{1,s}^{*}) + V) - e,$$

so

$$\mu(I) = (n-1)e + \binom{n+s+1}{n} - \dim_{\mathbf{k}}(\mathrm{im}(d_{1,s}^{*}) + V).$$

Next we will compute $\dim_{\mathbf{k}}(\mathrm{im}(d_{1,s}^{*}) + L^n)$ by considering a matrix that we are going to define. We denote by $\mathbb{M}_s$ the $n\binom{n+s}{n} \times \left(\binom{n+s+1}{n} - 1\right)$-matrix such that the i-th column, $i \in [1, \binom{n+s+1}{n} - 1]$, consists in the coordinates of $x_n \circ x^{\alpha}, \dots, x_1 \circ x^{\alpha}$ with respect the base W_s, where x^{α} is the i-th monomial of W_{s+1}.

Let $L \subset S$ be a finitely generated R-module of dimension e and degree s. We pick a basis $w_1, \dots, w_e$ of L and we consider the following basis, say B, of L^n: $(0, \dots, \overset{j}{w_i}, \dots, 0) \in L^n$ for $j = 1, \dots, n$, $i = 1, \dots, e$. We denote by $\mathbb{B}(L)$ the $n\binom{n+s}{n} \times (n.e)$-matrix, such that the columns consists of the coordinates of the elements of B with respect $\mathscr{M}_{s,1}$. Finally, $\mathbb{M}(L)$ is the $n\binom{n+s}{n} \times \left(\binom{n+s+1}{n} - 1 + n.e\right)$ block matrix

$$\mathbb{M}(L) = (\mathbb{M}_s \mid \mathbb{B}(L));$$

notice that

$$\dim_{\mathbf{k}}(\operatorname{im}(d_{1,s}^{*}) + L^{n}) = \operatorname{rank}(\mathbb{M}(L)).$$

If we want to consider a general $L \subset S$, see for instance Example 2.6.7, we have to avoid considering a basis of L. Let $F_1, \ldots, F_r$ be a system of generators of L as R-module. Then consider the following system of generators of L as $\mathbf{k}$-vector space $x^{\alpha} \circ F_i$ for all $\alpha \in \mathbb{N}^n$ of degree less or equal to $s = \deg(L)$ and for all $i = 1, \ldots, r$. We consider now the following system of generators, say B^{+}, of L^{n}:

$$(0, \ldots, x^{\alpha} \overset{j}{\circ} F_i, \ldots, 0) \in L^{n}$$

for $j = 1, \ldots, n$, $\alpha \in \mathbb{N}^n$ with $\deg(\alpha) \leq s$. We denote by $\mathbb{B}^{+}(L)$ the $n\binom{n+s}{n} \times r\binom{n+s}{n}$-matrix, such that the columns are the coordinates of the system of generators B^{+} with respect $\mathscr{M}_{s,1}$. This (lazy) method generates a matrix

$$\mathbb{L}(L) = (\mathbb{M}_s \mid \mathbb{B}^{+}(L))$$

with $\binom{n+s+1}{n} - 1 + nr\binom{n+s}{n}$ columns and $n\binom{n+s}{n}$ rows. Notice that the rank of $\mathbb{M}(L)$ and $\mathbb{L}(L)$ agree. Since the rank of $\mathbb{M}_s$ is $\binom{n+s+1}{n} - 1$ there is a dimension $\binom{n+s+1}{n} - 1$ square invertible matrix G_s such that

$$G_s \mathbb{M}_s = \left(\frac{Id}{Z}\right)$$

where Id is the identity matrix of dimension $\binom{n+s+1}{n} - 1$ and Z is the $\left(n\binom{n+s}{n} - \binom{n+s+1}{n} + 1\right) \times (\binom{n+s+1}{n} - 1)$ zero matrix. We denote by $\mathbb{L}^{*}(L)$, resp. $\mathbb{M}^{*}(L)$, the sub-matrix of $G_s\mathbb{L}(L)$, resp. $G_s\mathbb{M}(L)$, consisting of the last $n\binom{n+s}{n} - \binom{n+s+1}{n} + 1$ rows and the last $nr\binom{n+s}{n}$, resp. $n.e$, columns. Hence we have

Proposition 2.6.5 *Let L be a degree s finitely generated sub-R-module of S. Then*

(i) $\operatorname{rank}(\mathbb{M}(L)) = \operatorname{rank}(\mathbb{L}(L))$ *and* $\operatorname{rank}(\mathbb{M}^{*}(L)) = \operatorname{rank}(\mathbb{L}^{*}(L))$,
(ii) $\operatorname{rank}(\mathbb{M}(L)) = \operatorname{rank}(\mathbb{M}^{*}(L)) + \binom{n+s+1}{n} - 1$.

Remark Recall that $\Delta_1^{+}(L)$ is a matrix of $n\binom{n+s}{n}$ rows and e columns. If we mimic the construction of the matrix $\mathbb{M}^{*}(L)$ in the definition of $\Delta_1^{+}(L)$, i.e. considering a system of generators of L instead a $\mathbf{k}$-basis of L, we get a matrix with $n\binom{n+s}{n}$ rows and $nr\binom{n+s}{n}$ columns. Notice that $\mathbb{M}^{*}(L)$ is a smaller matrix: has $n\binom{n+s}{n} - \binom{n+s+1}{n} + 1$ rows and $nr\binom{n+s}{n}$ columns.

In the next result we compute more efficiently the minimal number of generators of an ideal by considering the matrix $\mathbb{M}^*(L)$.

Theorem 2.6.6 *Let $L \subset S$ be a finitely generated sub-R-module of S of dimension $e = \dim_{\mathbf{k}}(L)$ and degree $s = \deg(L)$. If $I = Ann_R(L) \subset R$ then*

$$\mu(I) = e(n-1) + 1 - \operatorname{rank}(\mathbb{M}^*(L)).$$

In particular, I is a complete intersection if and only if $\operatorname{rank}(\mathbb{M}^*(L)) = (e-1)(n-1)$.

Proof The statement follows from Propositions 2.6.4 and 2.6.5.

Example 2.6.7 Let $n = 2$ and consider a general polynomial of degree two

$$F = c_6 + c_5 x_1 + c_4 x_2 + c_3 x_1^2 + c_2 x_1 x_2 + c_1 x_2^2.$$

We assume that $A = R/I$, $I = Ann(\langle F \rangle)$, is an Artinian Gorenstein ring of embedding dimension two, in particular $\mu(I) = 2$. Hence the Hilbert function of A is $\{1, 2, 1\}$. $\mathbb{L}(F)$ is the $12 \times (9 + 2.6)$-matrix:

$$\mathbb{L}(F) = \left(\begin{array}{ccccccccc||cccccc|cccccc}
1&0&0&0&0&0&0&0&0 & c_1&0&0&0&0&0 & 0&0&0&0&0&0\\
0&1&0&0&0&0&0&0&0 & c_2&0&0&0&0&0 & 0&0&0&0&0&0\\
0&0&1&0&0&0&0&0&0 & c_3&0&0&0&0&0 & 0&0&0&0&0&0\\
0&0&0&0&1&0&0&0&0 & c_4&c_2&c_1&0&0&0 & 0&0&0&0&0&0\\
0&0&0&0&0&1&0&0&0 & c_5&c_3&c_2&0&0&0 & 0&0&0&0&0&0\\
0&0&0&0&0&0&0&1&0 & c_6&c_5&c_4&c_3&c_2&c_1 & 0&0&0&0&0&0\\
\hline
0&1&0&0&0&0&0&0&0 & 0&0&0&0&0&0 & c_1&0&0&0&0&0\\
0&0&1&0&0&0&0&0&0 & 0&0&0&0&0&0 & c_2&0&0&0&0&0\\
0&0&0&1&0&0&0&0&0 & 0&0&0&0&0&0 & c_3&0&0&0&0&0\\
0&0&0&0&0&1&0&0&0 & 0&0&0&0&0&0 & c_4&c_2&c_1&0&0&0\\
0&0&0&0&0&0&1&0&0 & 0&0&0&0&0&0 & c_5&c_3&c_2&0&0&0\\
0&0&0&0&0&0&0&0&1 & 0&0&0&0&0&0 & c_6&c_5&c_4&c_3&c_2&c_1
\end{array}\right)$$

The matrix $\mathbb{L}^*(F)$ is

$$\mathbb{L}^*(F) = \begin{pmatrix}
-c_5 & c_3 & -c_2 & 0 & 0 & 0 & c_4 & c_2 & c_1 & 0 & 0 & 0\\
-c_2 & 0 & 0 & 0 & 0 & 0 & c_1 & 0 & 0 & 0 & 0 & 0\\
-c_3 & 0 & 0 & 0 & 0 & 0 & c_2 & 0 & 0 & 0 & 0 & 0
\end{pmatrix}$$

after elementary transformations, the rank of $\mathbb{L}^*(F)$ agrees with the rank of

$$\begin{pmatrix}
c_5 & c_4 & c_3 & c_2 & c_1\\
c_2 & c_1 & 0 & 0 & 0\\
c_3 & c_2 & 0 & 0 & 0
\end{pmatrix}$$

Since the embedding dimension of A is two the rank of this matrix is 3. Hence $\operatorname{rank}(\mathbb{M}^*(F)) = 3$ and by Proposition 2.6.6 $\mu(I) = 4 - 3 + 1 = 2$, as expected, Proposition 2.4.3.

2.7 Examples

In this chapter we present several explicit examples proving that some results cannot be improved. We also give some explicit computations of the matrices introduced in chapter 5 and some explicit commutations of the minimal number of generators following the results of Chapter 6.

The following example shows that Theorem 2.5.9 fails if A is Gorenstein of socle degree $s = 4$, but not compressed, i.e. the Hilbert function is not maximal.

Example 2.7.1 ([15]) Let A be an Artin Gorenstein local $\mathbf{k}$-algebra with Hilbert function $\mathrm{HF}_A = \{1, 2, 2, 2, 1\}$. The local ring is called almost stretched and a classification can be found in [15]. In this case A is isomorphic to one and only one of the following rings :

(a) $A = R/I$ with $I = (x_1^4, x_2^2) \subseteq R = \mathbf{k}[[x_1, x_2]]$, and $I^\perp = \langle y_1^3 y_2 \rangle$. In this case A is canonically graded,
(b) $A = R/I$ with $I = (x_1^4, -x_1^3 + x_2^2) \subseteq R = \mathbf{k}[[x_1, x_2]]$, and $I^\perp = \langle y_1^3 y_2 + y_2^3 \rangle$. The associated graded ring is of type (a) and it is not isomorphic to R/I. Hence A is not canonically graded.
(c) $A = R/I$ with $I = (x_1^2 + x_2^2, x_2^4) \subseteq R = \mathbf{k}[[x_1, x_2]]$, and $I^\perp = \langle y_1 y_2 (y_1^2 - y_2^2) \rangle$. In this case A is graded.

The following example shows that Theorem 2.5.9 cannot be extended to compressed Gorenstein algebras of socle degree $s = 5$.

Example 2.7.2 ([11]) Let us consider the ideal

$$I = (x_1^4, x_2^3 - 2x_1^3 x_2) \subset R = \mathbf{k}[[x_1, x_2]].$$

The quotient $A = R/I$ is a compressed Gorenstein algebra with $\mathrm{HF}_A = \{1, 2, 3, 3, 2, 1\}$, $I^* = (x_1^4, x_2^3)$ and $I^\perp = \langle y_1^3 y_2^2 + y_2^4 \rangle$. Assume that there exists an analytic isomorphism φ of R mapping I into I^*. It is easy to see that the Jacobian matrix of φ is diagonal because $(I^*)^\perp = \langle y_1^3 y_2^2 \rangle$. We perform the computations modulo $(x_1, x_2)^5$, so we only have to consider the following coefficients of φ

$$\begin{cases} \varphi(x_1) = a x_1 + \dots \\ \\ \varphi(x_2) = b x_2 + i x_1^2 + j x_1 x_2 + k x_2^2 + \dots \end{cases}$$

where a, b are units, $i, j, k \in \mathbf{k}$. After the isomorphism $x_1 \to 1/a x_1$, $x_2 \to 1/b x_2$, we may assume $a = b = 1$. Then we have

$$I^* = \varphi(I) = (x_1^4, x_2^3 - 2x_1^3 x_2 + 3i x_1^2 x_2^2 + 3j x_1 x_2^3 + 3k x_2^4) \quad \text{modulo } (x_1, x_2)^5.$$

Hence there exist $\alpha \in K$, $\beta \in R$ such that

$$x_2^3 - 2x_1^3x_2 + 3ix_1^2x_2^2 + 3jx_1x_2^3 + 3kx_2^4 = \alpha x_1^4 + \beta x_2^3 \quad \text{modulo } (x_1, x_2)^5.$$

From this equality we deduce $\alpha = 0$ and

$$2x_1^3x_2 = x_2^2(x_2 + 3ix_1^2 + 3jx_1x_2 + 3kx_2^2 - \beta x_2) \quad \text{modulo } (x_1, x_2)^5,$$

a contradiction, so I is not isomorphic to I^*.

Let φ as above sending I into I^*. If we denote by $(z_i)_{i=1,\dots,6}$ the coordinates of a homogeneous form $G[5]$ of degree 5 in y_1, y_2 with respect to Ω^*, then the matrix $M^{[4]}(G[5])$ $(s = 5, p = 1)$ has the following shape

$$\begin{pmatrix} 4z_1 & 4z_2 & 4z_3 & 0 & 0 & 0 \\ 3z_2 & 3z_3 & 3z_4 & z_1 & z_2 & z_3 \\ 2z_3 & 2z_4 & 2z_5 & 2z_2 & 2z_3 & 2z_4 \\ z_4 & z_5 & z_6 & 3z_3 & 3z_4 & 3z_5 \\ 0 & 0 & 0 & 4z_4 & 4z_5 & 4z_6 \end{pmatrix}$$

In our case $G[5] = y_1^3y_2^2$, so all z_i are zero but $z_3 = 12$, hence the above matrix has rank 4 and it has not maximum rank given by Corollary 2.5.8. Since all the rows are not zero except the last one, it is easy to see that $F[4] = y_2^4$ is not in the image of $M^{[4]}(G[5])$, as (2.7) requires.

The following example shows that Theorem 2.5.10 cannot be extended to compressed type 2 level algebras of socle degree $s = 4$.

Example 2.7.3 ([11]) Let us consider the forms $G_1[4] = y_1^2y_2y_3$, $G_2[4] = y_1y_2^2y_3 + y_2y_3^3$ in $S = \mathbf{k}[y_1, y_2, y_3]$ of degree 4 and define in $R = \mathbf{k}[[x_1, x_2, x_3]]$ the ideal

$$I = Ann(G_1[4] + y_3^3, G_2[4]).$$

Then $A = R/I$ is a compressed level algebra with socle degree 4, type 2 and Hilbert function $\mathrm{HF}_A = \{1, 3, 6, 6, 2\}$. We prove that A is not canonically graded.

We know that $I^* = Ann(G_1[4], G_2[4])$ and we prove that A and $gr_{\mathfrak{n}}(A)$ are not isomorphic as $\mathbf{k}$-algebras. Let φ an analytic isomorphism sending I to I^*, then it is easy to see that $\varphi = I_3$ modulo $(x_1, x_2, x_3)^2$. The matrix $M^{[3]}(G_1[4], G_2[4])$ is of size 20×18 and, accordingly with (2.7), we will show that y_3^3 is not in the image of $M^{[3]}(G_1[4], G_2[4])$.

Let $F_1[4], F_2[4]$ be two homogeneous forms of degree 4 of $R = \mathbf{k}[y_1, y_2, y_3]$. We denote by $(z_i^j)_{i=1,\dots,15}$ the coordinates of $F_j[4]$ with respect the basis Ω^*, $j = 1, 2$. Then the 20×18 matrix $M^{[3]}(F_1[4], F_2[4])$ has the following shape, see (2.9),

$$\left(\begin{array}{cccccc|cccccc|cccccc}
3z_1^1 & 3z_2^1 & 3z_3^1 & 3z_4^1 & 3z_5^1 & 3z_6^1 & 0 & 0 & 0 & 0 & 0 & 0 & 0 & 0 & 0 & 0 & 0 & 0 \\
2z_2^1 & 2z_4^1 & 2z_5^1 & 2z_7^1 & 2z_8^1 & 2z_9^1 & z_1^1 & z_2^1 & z_3^1 & z_4^1 & z_5^1 & z_6^1 & 0 & 0 & 0 & 0 & 0 & 0 \\
2z_3^1 & 2z_5^1 & 2z_6^1 & 2z_8^1 & 2z_9^1 & 2z_{10}^1 & 0 & 0 & 0 & 0 & 0 & 0 & z_1^1 & z_2^1 & z_3^1 & z_4^1 & z_5^1 & z_6^1 \\
z_4^1 & z_7^1 & z_8^1 & z_{11}^1 & z_{12}^1 & z_{13}^1 & 2z_2^1 & 2z_4^1 & 2z_5^1 & 2z_7^1 & 2z_8^1 & 2z_9^1 & 0 & 0 & 0 & 0 & 0 & 0 \\
z_5^1 & z_8^1 & z_9^1 & z_{12}^1 & z_{13}^1 & z_{14}^1 & z_3^1 & z_5^1 & z_6^1 & z_8^1 & z_9^1 & z_{10}^1 & z_2^1 & z_4^1 & z_5^1 & z_7^1 & z_8^1 & z_9^1 \\
z_6^1 & z_9^1 & z_{10}^1 & z_{13}^1 & z_{14}^1 & z_{15}^1 & 0 & 0 & 0 & 0 & 0 & 0 & 2z_3^1 & 2z_5^1 & 2z_6^1 & 2z_8^1 & 2z_9^1 & 2z_{10}^1 \\
\hline
0 & 0 & 0 & 0 & 0 & 0 & 3z_4^1 & 3z_7^1 & 3z_8^1 & 3z_{11}^1 & 3z_{12}^1 & 3z_{13}^1 & 0 & 0 & 0 & 0 & 0 & 0 \\
0 & 0 & 0 & 0 & 0 & 0 & 2z_5^1 & 2z_8^1 & 2z_9^1 & 2z_{12}^1 & 2z_{13}^1 & 2z_{14}^1 & z_4^1 & z_7^1 & z_8^1 & z_{11}^1 & z_{12}^1 & z_{13}^1 \\
0 & 0 & 0 & 0 & 0 & 0 & z_6^1 & z_9^1 & z_{10}^1 & z_{13}^1 & z_{14}^1 & z_{15}^1 & 2z_5^1 & 2z_8^1 & 2z_9^1 & 2z_{12}^1 & 2z_{13}^1 & 2z_{14}^1 \\
\hline
0 & 0 & 0 & 0 & 0 & 0 & 0 & 0 & 0 & 0 & 0 & 0 & 3z_6^1 & 3z_9^1 & 3z_{10}^1 & 3z_{13}^1 & 3z_{14}^1 & 3z_{15}^1 \\
\hline
3z_1^2 & 3z_2^2 & 3z_3^2 & 3z_4^2 & 3z_5^2 & 3z_6^2 & 0 & 0 & 0 & 0 & 0 & 0 & 0 & 0 & 0 & 0 & 0 & 0 \\
2z_2^2 & 2z_4^2 & 2z_5^2 & 2z_7^2 & 2z_8^2 & 2z_9^2 & z_1^2 & z_2^2 & z_3^2 & z_4^2 & z_5^2 & z_6^2 & 0 & 0 & 0 & 0 & 0 & 0 \\
2z_3^2 & 2z_5^2 & 2z_6^2 & 2z_8^2 & 2z_9^2 & 2z_{10}^2 & 0 & 0 & 0 & 0 & 0 & 0 & z_1^2 & z_2^2 & z_3^2 & z_4^2 & z_5^2 & z_6^2 \\
z_4^2 & z_7^2 & z_8^2 & z_{11}^2 & z_{12}^2 & z_{13}^2 & 2z_2^2 & 2z_4^2 & 2z_5^2 & 2z_7^2 & 2z_8^2 & 2z_9^2 & 0 & 0 & 0 & 0 & 0 & 0 \\
z_5^2 & z_8^2 & z_9^2 & z_{12}^2 & z_{13}^2 & z_{14}^2 & z_3^2 & z_5^2 & z_6^2 & z_8^2 & z_9^2 & z_{10}^2 & z_2^2 & z_4^2 & z_5^2 & z_7^2 & z_8^2 & z_9^2 \\
z_6^2 & z_9^2 & z_{10}^2 & z_{13}^2 & z_{14}^2 & z_{15}^2 & 0 & 0 & 0 & 0 & 0 & 0 & 2z_3^2 & 2z_5^2 & 2z_6^2 & 2z_8^2 & 2z_9^2 & 2z_{10}^2 \\
\hline
0 & 0 & 0 & 0 & 0 & 0 & 3z_4^2 & 3z_7^2 & 3z_8^2 & 3z_{11}^2 & 3z_{12}^2 & 3z_{13}^2 & 0 & 0 & 0 & 0 & 0 & 0 \\
0 & 0 & 0 & 0 & 0 & 0 & 2z_5^2 & 2z_8^2 & 2z_9^2 & 2z_{12}^2 & 2z_{13}^2 & 2z_{14}^2 & z_4^2 & z_7^2 & z_8^2 & z_{11}^2 & z_{12}^2 & z_{13}^2 \\
0 & 0 & 0 & 0 & 0 & 0 & z_6^2 & z_9^2 & z_{10}^2 & z_{13}^2 & z_{14}^2 & z_{15}^2 & 2z_5^2 & 2z_8^2 & 2z_9^2 & 2z_{12}^2 & 2z_{13}^2 & 2z_{14}^2 \\
\hline
0 & 0 & 0 & 0 & 0 & 0 & 0 & 0 & 0 & 0 & 0 & 0 & 3z_6^2 & 3z_9^2 & 3z_{10}^2 & 3z_{13}^2 & 3z_{14}^2 & 3z_{15}^2
\end{array}\right)$$

It is enough to specialize the matrix to our case for proving that y_3^3 is not in the image of $M^{[3]}(G_1[4], G_2[4])$.

Next we will show how to apply the main result of the chapter six, Theorem 2.6.6. We assume that the ground field $\mathbf{k}$ is infinite.

Example 2.7.4 Artin Graded Level algebras of type 2.

Let F, G be two forms of degree three of $S = \mathbf{k}[x_1, x_2, x_3]$. We write $I = \mathrm{Ann}_R(\langle F, G\rangle)$. Then $\mathbb{L}^*(\langle F, G\rangle)$ is a 26×120 matrix in the coefficients of F, $c_1, \dots, c_{10}$, and the coefficients of G, $c_{11}, \dots, c_{20}$. This matrix has rank 17 considered as matrix with entries in the field K of fractions of $c_1, \dots, c_{20}$. This means that for generic $c_1, \dots, c_{20}$ the matrix $\mathbb{L}^*(\langle F, G\rangle)$ has rank 17. Moreover, there is a 17×17 submatrix of $\mathbb{L}^*(\langle F, G\rangle)$ whose determinant is non-zero in K

$$D_1 = c_1(c_5c_7 - c_3c_8)(c_1c_{12} - c_2c_{11})G_4G_8$$

where G_4 is a form of degree 4 on $c_1, \dots, c_{10}$ and G_8 is a form of degree 8 on $c_{11}, \dots, c_{20}$. The condition $c_1c_{12} - c_2c_{11} \neq 0$ implies that F, G are linearly independent over $\mathbf{k}$, so $A = R/I$ is an Artin level algebra of socle degree three and type 2. If the determinant, say D_2, of the matrix

$$\begin{pmatrix} c_9 & c_8 & c_6 \\ c_7 & c_6 & c_3 \\ c_6 & c_5 & c_2 \end{pmatrix}$$

is non-zero, the embedding dimension of A is three.

Let now consider the $\mathbf{k}$-vector space V generated by $x_i \circ F$, $i = 1, 2, 3$; $x_i \circ G$, $i = 1, 2, 3$. The dimension of V equals $\mathrm{HF}_A(2)$ and agrees with the rank of the following matrix

$$\begin{pmatrix} c_{10} & c_9 & c_8 & c_7 & c_6 & c_5 \\ c_{20} & c_{19} & c_{18} & c_{17} & c_{16} & c_{15} \\ c_9 & c_7 & c_6 & c_4 & c_3 & c_2 \\ c_{19} & c_{17} & c_{16} & c_{14} & c_{13} & c_{12} \\ c_8 & c_6 & c_5 & c_3 & c_2 & c_1 \\ c_{18} & c_{16} & c_{15} & c_{13} & c_{12} & c_{11} \end{pmatrix}$$

If the determinant D_3 of this matrix is non-zero then the Hilbert function of A is $\{1, 3, 6, 2\}$. Hence, if $D_1 D_2 D_3 \neq 0$ then A is a compressed Artin level algebra of type 2, socle degree 3, embedding dimension 3 and Hilbert function $\{1, 3, 6, 2\}$. From Theorem 2.6.6 we get

$$\mu(I) = 2.12 - \mathrm{rank}\,(\mathbb{L}^*(\langle F, G\rangle)) + 1 = 8$$

as Böij conjecture predicts, [1, Section 3.2].

Let $\mathbb{P}^9_{\mathbf{k}} \times \mathbb{P}^9_{\mathbf{k}}$ be the space parameterizing the pairs (F, G) up to scalars in each component. Since $D_1 D_2 D_3$ is bi-homogeneous form of degree 26 on $(c_1, \dots, c_{10})$ and $(c_{11}, \dots, c_{20})$, in this example we have shown a principal non-empty subset $U = \mathbb{P}^9_{\mathbf{k}} \times \mathbb{P}^9_{\mathbf{k}} \setminus V(D_1 D_2 D_3)$ parameterizing a family of compressed Artin level algebra of type 2, socle degree 3, embedding dimension 3 and Hilbert function $\{1, 3, 6, 2\}$.

Example 2.7.5 Artin Gorenstein algebras with Hilbert function $\{1, 4, 4, 1\}$.

Let us consider a general polynomial F of degree 3 of $R = \mathbf{k}[\![x_1, x_2, x_3, x_4]\!]$. We write $I = \langle F\rangle^{\perp}$. Then $\mathbb{L}^*(F)$ is a 71×140 matrix in the coefficients of F, say $c_1, \dots, c_{35}$. This matrix has rank 25 considered as matrix with coefficients in the field K of fractions of $c_1, \dots, c_{35}$. Hence for generic values of $c_1, \dots, c_{35}$ the ring $A = R/I$ is a compressed Gorenstein algebra with Hilbert function $\{1, 4, 4, 1\}$, and the matrix $\mathbb{L}^*(F)$ has rank 25 so

$$\mu(I) = 3.10 - 25 + 1 = 6$$

as it was expected, [1].

Example 2.7.6 Artin Gorenstein algebras with Hilbert function $\{1, 3, 3, 1\}$.

In this example we assume that the ground field $\mathbf{k}$ is algebraically closed. In [12] we prove that all Artin Gorenstein algebra $A = R/I$ with Hilbert function $\{1, n, n, 1\}$ is isomorphic to its associated graded ring. Hence in the case $n = 3$ we may assume that $I^{\perp}$ is generated by a form F in x_1, x_2, x_3 of degree three. In [12, Proposition 3.7] we classify such algebras in terms of the geometry of the projective plane cubic C defined by F. Next we will compute the minimal number of generators of I in the case that C is non-singular by using the main theorem of this paper.

Let is consider the Legendre form attached to C, $\lambda \neq 0, 1$,

$$U_\lambda = x_2^2 x_3 - x_1(x_1 - x_2)(x_1 - \lambda x_3)$$

the j-invariant of C is

$$j(\lambda) = 2^8 \frac{(\lambda^2 - \lambda + 1)^3}{\lambda^2(\lambda - 1)^3}.$$

$\mathbb{L}^*(U_\lambda)$ is a 20×24-matrix, after elementary transformations we get that the rank of $\mathbb{L}^*(U_\lambda)$ is 10 plus the rank of the 4×4 square matrix

$$W = \left(\begin{array}{cc|cc} \lambda + 1 & -\lambda & 0 & 0 \\ -3 & \lambda + 1 & 0 & 0 \\ \hline 0 & 0 & \lambda + 1 & -\lambda \\ 0 & 0 & -3 & \lambda + 1 \end{array}\right)$$

The determinant of W is

$$\det(K) = (\lambda^2 - \lambda + 1)^2.$$

Hence if $j(\lambda) \neq 0$ then $\mu(I) = 2.8 - (10 + 4) + 1 = 3$, i.e. I is a complete intersection, as we get in [12, Proposition 3.7]. If $j(\lambda) = 0$, i.e. C is the elliptic Fermat curve, then one gets rank $(W) = 2$ for all roots of $\det(W) = 0$. Hence $\mu(I) = 2.8 - (10 + 2) + 1 = 5$ as we get in [12, Proposition 3.7].

Acknowledgements I am grateful to Le Tuan Hoa and to Ngo Viet Trung for giving me the opportunity to speak about one of my favorite subjects. I am also grateful to the participants for their kind hospitality and mathematical discussions that made a very interesting and productive month in the city of Hanoi. We thank to Marcela Silva and Roser Homs for their useful comments and remarks. Last but not least, I am greatly indebted to M. E. Rossi for a long time collaboration on Macaulay's inverse systems and other topics. Some of the main results of these notes are made in collaboration with M. E. Rossi.

References

1. M. Böij, Betti numbers of compressed level algebras. J. Pure Appl. Algebra **134**(2), 111–131 (1999)
2. J. Briançon, Description de Hilb$^n\mathbb{C}\{x, y\}$. Inv. Math. **41**, 45–89 (1977)
3. W. Bruns, J. Herzog, *Cohen-Macaulay Rings*, revised edn. Cambridge Studies in Advanced Mathematics, vol. 39 (Cambridge University Press, Cambridge, 1997)
4. G. Casnati, J. Elias, R. Notari, M.E. Rossi, Poincaré series and deformations of Gorenstein local algebras. Commun. Algebra **41**(3), 1049–1059 (2013)

5. G. Casnati, J. Jelisiejew, R. Notari, Irreducibility of the Gorenstein loci of Hilbert schemes via ray families. Algebra Number Theory **9**, 1525–1570 (2015)
6. A. De Stefani, Artinian level algebras of low socle degree. Commun. Algebra **42**(2), 729–754 (2014)
7. W. Decker, G.-M. Greuel, G. Pfister, H. Schönemann, Singular 4-0-1 –A computer algebra system for polynomial computations (2014). www.singular.uni-kl.de
8. J. Elias, Inverse-syst.lib–Singular library for computing Macaulay's inverse systems (2015). arXiv:1501.01786
9. J. Elias, R. Homs, On the analytic type of Artin algebras. Commun. Algebra **44**, 2277–2304 (2016)
10. J. Elias, A. Iarrobino, The Hilbert function of a Cohen-Macaulay local algebra: extremal Gorenstein algebras. J. Algebra **110**, 344–356 (1987)
11. J. Elias, M.E. Rossi, Analytic isomorphisms of compressed local algebras. Trans. Am. Math. Soc. **364**(9), 4589–4604 (2012)
12. J. Elias, M.E. Rossi, Isomorphism classes of short Gorenstein local rings via Macaulay's inverse system. Proc. Am. Math. Soc. **143**(3), 973–987 (2015)
13. J. Elias, M.E. Rossi, The structure of the inverse system on Gorenstein k-algebras. Adv. Math. **314**, 306–327 (2017)
14. J. Elias, G. Valla, Structure theorems for certain Gorenstein ideals. Mich. Math. J. **57**, 269–292 (2008). Special volume in honor of Melvin Hochster
15. J. Elias, G. Valla, Isomorphism classes of certain Gorenstein ideals. Algebr. Represent. Theory **14**, 429–448 (2011)
16. J. Emsalem, Géométrie des points épais. Bull. Soc. Math. France **106**(4), 399–416 (1978)
17. R. Froberg, D. Laksov, *Compressed Algebras*. Lecture Notes in Mathematics, vol. 1092 (Springer, Berlin, 1984), pp. 121–151
18. P. Gabriel, Objects injectifs dans les catégories abéliennes. Séminaire P. Dubriel 1958/1959 (1959), pp. 17–01, 32
19. R. Hartshorne, *Deformation Theory*. Graduate Texts in Mathematics, vol. 257 (Springer, Berlin, 2010)
20. A. Iarrobino, Compressed algebras: artin algebras having given socle degrees and maximal length. Trans. Am. Math. Soc. **285**(1), 337–378 (1984)
21. A. Iarrobino, Associated graded algebra of a Gorenstein Artin algebra, Mem. Am. Math. Soc. **107**(514), viii+115 (1994)
22. A. Iarrobino, V. Kanev, *Power Sums, Gorenstein Algebras, and Determinantal Loci*. Lecture Notes in Mathematics, vol. 1721 (Springer, Berlin, 1999). Appendix C by Iarrobino and Steven L. Kleiman
23. F.S. Macaulay, *The Algebraic Theory of Modular Systems*. Revised reprint of the original 1916 original. With an introduction of P. Robert, Cambridge Mathematical Library (Cambridge University Press, Cambridge, 1994)
24. G. Mazzola, Generic finite schemes and Hochschild cocycles. Comment. Math. Helv. **55**(2), 267–293 (1980)
25. D.G. Northcott, Injective envelopes and inverse polynomials. J. Lond. Math. Soc. **8**, 290–296 (1972)
26. B. Poonen, *Isomorphism Types of Commutative Algebras of Finite Rank Over an Algebraically Closed Field*. Computational Arithmetic Geometry, Contemp. Math., vol. 463 (Amer. Math. Soc., Providence, 2008), pp. 111–120
27. J.J. Rotman, *An Introduction to Homological Algebra* (Academic, New York, 1979)
28. D.W. Sharpe, P. Vamos, *Injective Modules* (Cambridge University Press, Cambridge, 1972)
29. J.H. Silverman, *The Arithmetic of Elliptic Curves*, 2nd edn. Graduate Texts in Mathematics, vol. 106 (Springer, Dordrecht, 2009)
30. R.P. Stanley, Hilbert functions of graded algebras. Adv. Math. **28**, 57–83 (1978)
31. D.A. Suprunenko, On maximal commutative subalgebras of the full linear algebra. Uspehi Mat. Nauk (N.S.) **11**(3(69)), 181–184 (1956)

Chapter 3
Lectures on the Representation Type of a Projective Variety

Rosa M. Miró-Roig

Abstract In these notes, we construct families of non-isomorphic Arithmetically Cohen Macaulay (ACM for short) sheaves (i.e., sheaves without intermediate cohomology) on a projective variety X. The study of such sheaves has a long and interesting history behind. Since the seminal result by Horrocks characterizing ACM sheaves on $\mathbb{P}^n$ as those that split into a sum of line bundles, an important amount of research has been devoted to the study of ACM sheaves on a given variety.

ACM sheaves also provide a criterium to determine the complexity of the underlying variety. This complexity is studied in terms of the dimension and number of families of undecomposable ACM sheaves that it supports, namely, its *representation type*. Varieties that admit only a finite number of undecomposable ACM sheaves (up to twist and isomorphism) are called of *finite representation type*. These varieties are completely classified: They are either three or less reduced points in $\mathbb{P}^2$, $\mathbb{P}^n$, a smooth hyperquadric $X \subset \mathbb{P}^n$, a cubic scroll in $\mathbb{P}^4$, the Veronese surface in $\mathbb{P}^5$ or a rational normal curve.

On the other extreme of complexity we find the varieties of *wild representation type*, namely, varieties for which there exist r-dimensional families of non-isomorphic undecomposable ACM sheaves for arbitrary large r. In the case of dimension one, it is known that curves of wild representation type are exactly those of genus larger or equal than two. In dimension greater or equal than two few examples are know and in these notes, we give a brief account of the known results.

3.1 Introduction

These notes grew out of a series of lectures given by the author at the Vietnam Institute for Advanced Study in Mathematics (VIASM), Hanoi, during the period February 8–March 7, 2014. In no case do I claim it is a survey on the representation

R. M. Miró-Roig (✉)
Department de Matemàtiques i Informàtica, Universitat de Barcelona, Barcelona, Spain
e-mail: miro@ub.edu

N. Tu CUONG et al. (eds.), *Commutative Algebra and its Interactions to Algebraic Geometry*, Lecture Notes in Mathematics 2210,
https://doi.org/10.1007/978-3-319-75565-6_3

type of a projective variety. Many people have made important contributions without even being mentioned here and I apologize to those whose work I made have failed to cite properly. The author gave three lectures of length 120 min each. She attempted to cover the basic facts on the representation type of a projective variety. Given the extensiveness of the subject, it was not possible to go into great detail in every proof. Still, it was hoped that the material that she chose will be beneficial and illuminating for the participants, and for the reader.

The projective space $\mathbb{P}^n$ holds a very remarkable property: the only undecomposable vector bundle $\mathcal{E}$ without intermediate cohomology (i.e., $\mathrm{H}^i(\mathbb{P}^n, \mathcal{E}(t)) = 0$ for $t \in \mathbb{Z}$ and $1 < i < n$), up to twist, is the structural line bundle $\mathcal{O}_{\mathbb{P}^n}$. This is the famous Horrocks' Theorem, proved in [30]. Ever since this result was stated, the study of the category of undecomposable arithmetically Cohen-Macaulay bundles (i.e., bundles without intermediate cohomology) supported on a given projective variety X has raised a lot of interest since it is a natural way to understand the complexity of the underlying variety X. Mimicking an analogous trichotomy in Representation Theory, in [17] it was proposed a classification of ACM projective varieties as *finite, tame or wild* (see Definition 3.2.10) according to the complexity of their associated category of ACM vector bundles and it was proved that this trichotomy is exhaustive for the case of ACM curves: rational curves are finite, elliptic curves are tame and curves of higher genus are wild. Unfortunately very little is known for varieties of higher dimension and in this series of lectures I will give a brief account of known results.

The result due to Horrocks (cf. [30]) which asserts that, up to twist, $\mathcal{O}_{\mathbb{P}^n}$ is the only one undecomposable ACM bundle on $\mathbb{P}^n$ and the result due to Knörrer (cf. [34]) which states that on a smooth hyperquadric X the only undecomposable ACM bundles up to twist are $\mathcal{O}_X$ and the spinor bundles S match with the general philosophy that a "simple" variety should have associated a "simple" category of ACM bundles. Following these lines, a cornerstone result was the classification of ACM varieties of *finite representation type*, i.e., varieties that support (up to twist and isomorphism) only a finite number of undecomposable ACM bundles. It turned out that they fall into a very short list: $\mathbb{P}^n$, a smooth hyperquadric $Q \subset \mathbb{P}^n$, a cubic scroll in $\mathbb{P}^4$, the Veronese surface in $\mathbb{P}^5$, a rational normal curve and three or less reduced points in $\mathbb{P}^2$ (cf. [7, Theorem C] and [18, p. 348]).

For the rest of ACM varieties, it became an interesting problem to give a criterium to split them into a finer classification, i.e. it is a challenging problem to find out the representation type of the remaining ones. So far only few examples of varieties of wild representation type are known: curves of genus $g \geq 2$ (cf. [17]), del Pezzo surfaces and Fano blow-ups of points in $\mathbb{P}^n$ (cf. [45], the cases of the cubic surface and the cubic threefold have also been handled in [10]), ACM rational surfaces on $\mathbb{P}^4$ (cf. [44]), any Segre variety unless the quadric surface in $\mathbb{P}^3$ (cf. [15, Theorem 4.6]) and non-singular rational normal scrolls $S(a_0, \cdots, a_k) \subseteq \mathbb{P}^N$, $N = \sum_{i=0}^{k}(a_i) + k$, (unless $\mathbb{P}^{k+1} = S(0, \cdots, 0, 1)$, the rational normal curve $S(a)$ in $\mathbb{P}^a$, the quadric surface $S(1, 1)$ in $\mathbb{P}^3$ and the cubic scroll $S(1, 2)$ in $\mathbb{P}^4$) (cf. [40, Theorem 3.8]).

Among ACM vector bundles $\mathcal{E}$ on a given variety X, it is interesting to spot a very important subclass for which its associated module $\oplus_t \mathrm{H}^0(X, \mathcal{E}(t))$ has the maximal number of generators, which turns out to be $\deg(X)\mathrm{rk}(\mathcal{E})$. This property was isolated by Ulrich in [51], and ever since modules with this property have been called Ulrich modules and correspondingly Ulrich bundles in the geometric case (see [21] for more details on Ulrich bundles). The search of Ulrich sheaves on a particular variety is a challenging problem. In fact, few examples of varieties supporting Ulrich sheaves are known and, in [21], Eisenbud and Schreier asked the following question: Is any projective variety the support of an Ulrich sheaf? If so, what is the smallest possible rank for such a sheaf? Moreover, the recent interest in the existence of Ulrich sheaves relies among other things on the fact that a d-dimensional variety $X \subset \mathbb{P}^n$ supports an Ulrich sheaf (bundle) if and only if the cone of cohomology tables of coherent sheaves (resp. vector bundles) on X coincides with the cone of cohomology tables of coherent sheaves (resp. vector bundles) on $\mathbb{P}^d$ [19, Theorem 4.2]. It is therefore a meaningful question to find out if a given projective variety X is of wild representation type with respect to the much more restrictive category of its undecomposable Ulrich vector bundles. We will prove that all smooth del Pezzo surfaces as well as all Segre varieties unless $\mathbb{P}^1 \times \mathbb{P}^1$ are of wild representation type and wildness is witnessed by Ulrich bundles.

Next we outline the structure of these notes. In Sect. 3.2, we introduce the definitions and main properties that are going to be used throughout the paper; in particular, a brief account of ACM varieties, ACM vector bundles and Ulrich bundles on projective varieties is provided.

In Sect. 3.3, we determine the representation type of any smooth del Pezzo surface S. To this end, we have to construct families of undecomposable ACM bundles of arbitrary high rank and dimension. Our construction will rely on the existence of level set of points on S and the existence of level set of points on S is related to Mustaţă's conjecture for a general set of points on a projective variety. Roughly speaking, Mustaţă's conjecture predicts the graded Betti number of a set Z of general points on a fixed projective variety X. In Sect. 3.3.1, we will address this latter conjecture and we will prove that it holds for a general set of points Z on a smooth del Pezzo surface provided the cardinality of Z falls in certain strips explicitly described. In Sect. 3.3.30, we perform the construction of large families of simple Ulrich vector bundles on del Pezzo surfaces obtained blowing up $s \leq 8$ points in $\mathbb{P}^2$. These families are constructed as the pullback of the kernel of certain surjective morphisms

$$\mathcal{O}_{\mathbb{P}^2}(1)^b \longrightarrow \mathcal{O}_{\mathbb{P}^2}(2)^a$$

with chosen properties. It is worthwhile to point out that in the case of del Pezzo surfaces with very ample anticanonical divisor, we can show that these families of vector bundles could also be obtained through Serre's correspondence from a suitable general set of level points on the del Pezzo surface.

In Sect. 3.4, we are going to focus our attention on the case of *Segre varieties* $\Sigma_{n_1,\ldots,n_s} \subseteq \mathbb{P}^N$, $N = \prod_{i=1}^{s}(n_i + 1) - 1$ for $1 \leq n_1, \ldots, n_s$. It is a classical result that the quadric surface $\mathbb{P}^1 \times \mathbb{P}^1 \subseteq \mathbb{P}^3$ only supports three undecomposable ACM vector bundles, up to shift: $\mathcal{O}_{\mathbb{P}^1\times\mathbb{P}^1}$, $\mathcal{O}_{\mathbb{P}^1\times\mathbb{P}^1}(1,0)$ and $\mathcal{O}_{\mathbb{P}^1\times\mathbb{P}^1}(0,1)$. For the rest of

Segre varieties we construct large families of simple (and, hence, undecomposable) Ulrich vector bundles on them and this will allow us to conclude that they are of wild representation type. Up to our knowledge, they will be the first family of examples of varieties of arbitrary dimension for which wild representation type is witnessed by means of Ulrich vector bundles. In this section, we first introduce the definition and main properties of Segre varieties needed later. Then, we pay attention to the case of Segre varieties $\Sigma_{n,m} \subseteq \mathbb{P}^N$, $N := nm+n+m$, for $2 \leq n, m$ and to the case of Segre varieties of the form $\Sigma_{n_1,n_2\dots,n_s} \subseteq \mathbb{P}^N$, $N = \prod_{i=1}^s (n_1+1)-1$, for $2 \leq n_1, \cdots, n_s$. We construct families of arbitrarily large dimension of simple Ulrich vector bundles on them by pulling-back certain vector bundles on each factor. This will allow us to conclude that they are of wild representation type. Finally, we move forward to the case of Segre varieties of the form $\Sigma_{n_1,n_2\dots,n_s} \subseteq \mathbb{P}^N$, $N = \prod_{i=1}^s (n_1+1)-1$, for either $n_1 = 1$ and $s \geq 3$ or $n_1 = 1$, $s = 2$ and $n_2 \geq 2$. In this case the families of undecomposable Ulrich vector bundles of arbitrarily high rank will be obtained as iterated extensions of lower rank vector bundles.

In Sect. 3.5, we could not resist to discuss some details that perhaps only the experts will care about, but hopefully will also introduce the non-expert reader to a subtle subject. We analyze how the representation type of a projective variety change when we change the polarization. Our main goal will be to prove that for any smooth ACM projective variety $X \subset \mathbb{P}^n$ there always exists a very ample line bundle $\mathcal{L}$ on X which naturally embeds X in $\mathbb{P}^{h^0(X,\mathcal{L})-1}$ as a variety of wild representation type.

Throughout the lectures I mentioned various open problems. Some of them and further related problems are collected in the last section of these notes.

Notation Throughout these notes K will be an algebraically closed field of characteristic zero, $R = K[x_0, x_1, \cdots, x_n]$, $\mathfrak{m} = (x_0, \dots, x_n)$ and $\mathbb{P}^n = \mathrm{Proj}(R)$. Given a non-singular variety X equipped with an ample line bundle $\mathcal{O}_X(1)$, the line bundle $\mathcal{O}_X(1)^{\otimes l}$ will be denoted by $\mathcal{O}_X(l)$. For any coherent sheaf $\mathcal{E}$ on X we are going to denote the twisted sheaf $\mathcal{E} \otimes \mathcal{O}_X(l)$ by $\mathcal{E}(l)$. As usual, $\mathrm{H}^i(X, \mathcal{E})$ stands for the cohomology groups, $\mathrm{h}^i(X, \mathcal{E})$ for their dimension, $\mathrm{ext}^i(\mathcal{E}, \mathcal{F})$ for the dimension of $\mathrm{Ext}^i(\mathcal{E}, \mathcal{F})$ and $\mathrm{H}^i_*(X, \mathcal{E}) = \oplus_{l \in \mathbb{Z}} \mathrm{H}^i(X, \mathcal{E}(l))$ (or simply $\mathrm{H}^i_*\mathcal{E}$).

Given closed subschemes $X \subseteq \mathbb{P}^n$, we denote by R_X the homogeneous coordinate ring of X defined as $K[x_0, \dots, x_n]/I(X)$. As usual, the Hilbert function of X (resp. the Hilbert polynomial of X) will be denoted by $H_X(t)$ (resp. $P_X(t) \in \mathbb{Q}[t]$) and the regularity of X is defined to be the regularity of $I(X)$, i.e., $\mathrm{reg}(X) \leq m$ if and only if $\mathrm{H}^i(\mathbb{P}^n, I_X(m-i)) = 0$ for $i \geq 1$. Moreover, we know that $P_X(t) = H_X(t)$ for any $t \geq \mathrm{reg}X - 1 + \delta - n$ where δ is the projective dimension of R_X. Finally, $\Delta H_X(t)$ denotes the difference function, i.e., $\Delta H_X(t) = H_X(t) - H_X(t-1)$.

3.2 Preliminaries

In this section, we set up some preliminary notions mainly concerning the definitions and basic results on ACM schemes $X \subset \mathbb{P}^n$ as well as on ACM sheaves and Ulrich sheaves $\mathcal{E}$ on X needed in the sequel.

Definition 3.2.1 A subscheme $X \subseteq \mathbb{P}^n$ is said to be arithmetically Cohen-Macaulay (briefly, ACM) if its homogeneous coordinate ring $R_X = R/I(X)$ is a Cohen-Macaulay ring, i.e. $\mathrm{depth}(R_X) = \dim(R_X)$.

Thanks to the graded version of the Auslander-Buchsbaum formula (for any finitely generated R-module M):

$$\mathrm{pd}(M) = n + 1 - \mathrm{depth}(M),$$

we deduce that a subscheme $X \subseteq \mathbb{P}^n$ is ACM if and only if $\mathrm{pd}(R_X) = \operatorname{codim} X$. Hence, if $X \subseteq \mathbb{P}^n$ is a codimension c ACM subscheme, a graded minimal free R-resolution of $I(X)$ is of the form:

$$0 \longrightarrow F_c \xrightarrow{\varphi_c} F_{c-1} \xrightarrow{\varphi_{c-1}} \cdots \xrightarrow{\varphi_2} F_1 \xrightarrow{\varphi_1} F_0 \longrightarrow R_X \longrightarrow 0 \tag{3.1}$$

with $F_0 = R$ and $F_i = \oplus_j R(-i-j)^{b_{ij}(X)}$, $1 \leq i \leq c$. The integers $b_{ij}(X)$ are called the *graded Betti numbers* of X and they are defined as

$$b_{ij}(X) = \dim_K \mathrm{Tor}^i(R/I(X), K)_{i+j}.$$

We construct the *Betti diagram* of X writing in the (i, j)-th position the Betti number $b_{ij}(X)$. In this setting, minimal means that $\mathrm{im}\varphi_i \subset \mathfrak{m}F_{i-1}$. Therefore, the free resolution (3.1) is minimal if, after choosing basis of F_i, the matrices representing φ_i do not have any non-zero scalar.

Remark For non ACM schemes $X \subseteq \mathbb{P}^n$ of codimension c the graded minimal free R-resolution of R_X is of the form:

$$0 \longrightarrow F_p \xrightarrow{\varphi_p} F_{p-1} \xrightarrow{\varphi_{p-1}} \cdots \xrightarrow{\varphi_2} F_1 \xrightarrow{\varphi_1} F_0 \longrightarrow R_X \longrightarrow 0$$

with $F_0 = R$, $F_i = \oplus_{j=1}^{\beta_i} R(-n_j^i)$, $1 \leq i \leq p$, and $c < p \leq n$.

Notice that any zero-dimensional variety is ACM. For varieties of higher dimension we have the following characterization that will be used in this paper:

Lemma 3.2.2 (cf. [39], p. 23) *If* $\dim X \geq 1$, *then* $X \subseteq \mathbb{P}^n$ *is ACM if and only if* $\mathrm{H}^i_*(I_X) := \oplus_{t \in \mathbb{Z}} H^i(\mathbb{P}^n, I_X(t)) = 0$ *for* $1 \leq i \leq \dim X$.

Example 3.2.3

1. Any complete intersection variety $X \subset \mathbb{P}^n$ is ACM.
2. The twisted cubic $X \subset \mathbb{P}^3$ is an ACM curve.
3. The rational quartic $C \subset \mathbb{P}^3$ is not ACM since $H^1(\mathbb{P}^3, I_C(1)) \neq 0$.
4. Segre Varieties are ACM varieties.
5. Any standard determinantal variety $X \subset \mathbb{P}^n$ defined by the maximal minors of a homogeneous matrix is ACM.

Definition 3.2.4 If $X \subseteq \mathbb{P}^n$ is an ACM subscheme then, the rank of the last free R-module in a minimal free R-resolution of $I(X)$ is called the *Cohen-Macaulay type* of X.

Definition 3.2.5 A codimension c subscheme X of $\mathbb{P}^n$ is *arithmetically Gorenstein* (briefly AG) if its homogeneous coordinate ring R_X is a Gorenstein ring or, equivalently, its saturated homogeneous ideal, $I(X)$, has a minimal free graded R-resolution of the following type:

$$0 \longrightarrow R(-t) \longrightarrow \oplus_{i=1}^{\alpha_{c-1}} R(-n_{c-1,i}) \longrightarrow \dots\dots \longrightarrow \oplus_{i=1}^{\alpha_1} R(-n_{1,i}) \longrightarrow I(X) \longrightarrow 0.$$

In other words, an AG scheme is an ACM scheme with Cohen-Macaulay type 1.

Definition 3.2.6 Let $(X, \mathcal{O}_X(1))$ be a polarized variety. A coherent sheaf $\mathcal{E}$ on X is *Arithmetically Cohen Macaulay* (ACM for short) if it is locally Cohen-Macaulay (i.e., $\mathrm{depth}\mathcal{E}_x = \dim\mathcal{O}_{X,x}$ for every point $x \in X$) and has no intermediate cohomology:

$$\mathrm{H}^i_*(X, \mathcal{E}) = 0 \ \text{ for all } \ i = 1, \dots, \dim X - 1.$$

Notice that when X is a non-singular variety, which is going to be mainly our case, any coherent ACM sheaf on X is locally free. For this reason we are going to speak often of ACM bundles (since we identify locally free sheaves with their associated vector bundle). ACM sheaves are closely related to their algebraic counterpart, the maximal Cohen-Macaulay modules:

Definition 3.2.7 A graded R_X-module E is a *Maximal Cohen-Macaulay* module (MCM for short) if $\mathrm{depth}E = \dim E = \dim R_X$.

Indeed, it holds:

Proposition 3.2.8 *Let $X \subseteq \mathbb{P}^n$ be an ACM scheme. There exists a bijection between ACM sheaves $\mathcal{E}$ on X and MCM R_X-modules E given by the functors $E \to \widetilde{E}$ and $\mathcal{E} \to \mathrm{H}^0_*(X, \mathcal{E})$.*

The study of ACM bundles has a long and interesting history behind and it is well known that ACM sheaves provide a criterium to determine the complexity of the underlying variety. Indeed, this complexity can be studied in terms of the dimension and number of families of undecomposable ACM sheaves that it supports. Let us illustrate this general philosophy with a couple of examples (the simplest examples of varieties we can deal with have associated a simple category of undecomposable vector bundles).

Example 3.2.9

1. Horrocks Theorem asserts that on $\mathbb{P}^n$ a vector bundle $\mathcal{E}$ is ACM if and only if it splits into a sum of line bundles. So, up to twist, there is only one undecomposable ACM bundle on $\mathbb{P}^n$: $\mathcal{O}_{\mathbb{P}^n}$ (cf. [30]).
2. Knörrer's theorem states that on a smooth hyperquadric $Q_n \subset \mathbb{P}^{n+1}$ any ACM vector bundle $\mathcal{E}$ splits into a sum of line bundles and spinor bundles. So, up to twist and dualizing, there are only two undecomposable ACM bundles on Q_{2n+1} ($\mathcal{O}_{Q_{2n+1}}$ and the spinor bundle Σ); and three undecomposable ACM bundles on Q_{2n} ($\mathcal{O}_{Q_{2n}}$ and the spinor bundles Σ_- and Σ_+)(cf. [34]).

Recently, inspired by an analogous classification for quivers and for K-algebras of finite type, it has been proposed the classification of any ACM variety as being of *finite, tame or wild representation type* (cf. [17] for the case of curves and [9] for the higher dimensional case). Let us recall the definitions:

Definition 3.2.10 Let $X \subseteq \mathbb{P}^N$ be an ACM scheme of dimension n.

1. We say that X is of *finite representation type* if it has, up to twist and isomorphism, only a finite number of undecomposable ACM sheaves.
2. X is of *tame representation type* if either it has, up to twist and isomorphism, an infinite discrete set of undecomposable ACM sheaves or, for each rank r, the undecomposable ACM sheaves of rank r form a finite number of families of dimension at most n.
3. X is of *wild representation type* if there exist l-dimensional families of non-isomorphic undecomposable ACM sheaves for arbitrary large l.

One of the main achievements in this field has been the classification of varieties of finite representation type (cf. [7, Theorem C], and [18, p. 348]); it turns out that they fall into a very short list: three or less reduced points on $\mathbb{P}^2$, a projective space, a non-singular quadric hypersurface $X \subseteq \mathbb{P}^n$, a cubic scroll in $\mathbb{P}^4$, the Veronese surface in $\mathbb{P}^5$ or a rational normal curve. As examples of a variety of tame representation type we have the elliptic curves, the Segre product of a line and a smooth conic naturally embedded in $\mathbb{P}^5$: $\varphi_{|\mathcal{O}(2,2)|} : \mathbb{P}^1 \times \mathbb{P}^1 \hookrightarrow \mathbb{P}^8$ (cf. [23]) and the quadric cone in $\mathbb{P}^3$ (cf. [8, Proposition 6.1]). Finally, on the other extreme of complexity lie those varieties that have very large families of ACM sheaves. So far only few examples of varieties of wild representation type are known: curves of genus $g \geq 2$ (cf. [17]), smooth del Pezzo surfaces (see Sect. 3.3 of these notes) and Fano blow-ups of points in $\mathbb{P}^n$ (cf.[45], the cases of the cubic surface and the cubic threefold have also been handled in [10]), ACM rational surfaces on $\mathbb{P}^4$ (cf. [44]), Segre varieties other than the quadric in $\mathbb{P}^3$ (see Sect. 3.4 of these notes or [15, Theorem 4.6]), rational normal scrolls other than $\mathbb{P}^n$, the rational normal curve in $\mathbb{P}^n$, the quadric in $\mathbb{P}^3$ and the cubic scroll in $\mathbb{P}^4$ [40, Theorem 3.8] and hypersurfaces $X \subset \mathbb{P}^n$ of degree ≥ 4 [50, Corollary 1].

The problem of classifying ACM varieties according to the complexity of the category of ACM sheaves that they support has recently attired much attention and, in particular, the following problem is still open (for ACM varieties of dimension ≥ 2):

Problem 3.2.11 Is the trichotomy finite representation type, tame representation type and wild representation type exhaustive?

Very often the ACM bundles that we will construct will share another stronger property, namely they have the maximal possible number of global sections; they will be the so-called Ulrich bundles. Let us end this section recalling the definition of Ulrich sheaves and summarizing the properties that they share and that will be needed in the sequel.

Definition 3.2.12 Given a polarized variety $(X, \mathcal{O}X(1))$, a coherent sheaf $\mathcal{E}$ on X is said to be *initialized* if

$$\mathrm{H}^0(X, \mathcal{E}(-1)) = 0 \text{ but } \mathrm{H}^0(X, \mathcal{E}) \neq 0.$$

Notice that when $\mathcal{E}$ is a locally Cohen-Macaulay sheaf, there always exists an integer k such that $\mathcal{E}_{init} := \mathcal{E}(k)$ is initialized.

Definition 3.2.13 Given a projective scheme $X \subseteq \mathbb{P}^n$ and a coherent sheaf $\mathcal{E}$ on X, we say that $\mathcal{E}$ is an *Ulrich sheaf* if $\mathcal{E}$ is an ACM sheaf and $\mathrm{h}^0(\mathcal{E}_{init}) = \deg(X)\mathrm{rk}(\mathcal{E})$.

The following result justifies the above definition:

Theorem 3.2.14 *Let $X \subseteq \mathbb{P}^n$ be an integral ACM subscheme and let $\mathcal{E}$ be an ACM sheaf on X. Then the minimal number of generators $m(\mathcal{E})$ of the associated MCM R_X-module $\mathrm{H}^0_*(\mathcal{E})$ is bounded by*

$$m(\mathcal{E}) \leq \deg(X)\mathrm{rk}(\mathcal{E}).$$

Therefore, since it is obvious that for an initialized sheaf $\mathcal{E}$, $\mathrm{h}^0(\mathcal{E}) \leq m(\mathcal{E})$, the minimal number of generators of Ulrich sheaves is as large as possible. MCM Modules attaining this upper bound were studied by Ulrich in [51]. A complete account is provided in [21]. In particular we have:

Theorem 3.2.15 *Let $X \subseteq \mathbb{P}^N$ be an n-dimensional ACM variety and let $\mathcal{E}$ be an initialized ACM coherent sheaf on X. The following conditions are equivalent:*

1. *$\mathcal{E}$ is Ulrich.*
2. *$\mathcal{E}$ admits a linear $\mathcal{O}_{\mathbb{P}^N}$-resolution of the form:*

$$0 \to \mathcal{O}_{\mathbb{P}^N}(-N+n)^{a_{N-n}} \to \cdots \to \mathcal{O}_{\mathbb{P}^N}(-1)^{a_1} \to \mathcal{O}_{\mathbb{P}^N}^{a_0} \to \mathcal{E} \to 0.$$

3. *$\mathrm{H}^i(\mathcal{E}(-i)) = 0$ for $i > 0$ and $\mathrm{H}^i(\mathcal{E}(-i-1)) = 0$ for $i < n$.*

4. *For some (resp. all) finite linear projections* $\pi : X \to \mathbb{P}^n$, *the sheaf* $\pi_* \mathcal{E}$ *is the trivial sheaf* $\mathcal{O}_{\mathbb{P}^n}^t$ *for some* t.

In particular, initialized Ulrich sheaves are 0*-regular and therefore they are globally generated.*

Proof See [21, Proposition 2.1]. □

The search of Ulrich sheaves on a particular variety is a challenging problem. In fact, few examples of varieties supporting Ulrich sheaves are known and, in [21, p. 543], Eisenbud, Schreyer and Weyman leave open the following problem

Problem 3.2.16

1. Is every variety (or even scheme) $X \subset \mathbb{P}^n$ the support of an Ulrich sheaf?
2. If so, what is the smallest possible rank for such a sheaf?

Recently, after the Boij-Söderberg theory has been developed, the interest on these questions have grown up due to the fact that it has been proved [19, Theorem 4.2] that the existence of an Ulrich sheaf on a smooth projective variety X of dimension n implies that the cone of cohomology tables of vector bundles on X coincide with the cone of cohomology tables of vector bundles on $\mathbb{P}^n$.

In these series of lectures we are going to focus our attention on the existence of Ulrich bundles on smooth del Pezzo surfaces and on Segre varieties, providing the first example of wild varieties of arbitrary dimension whose wildness is witnessed by means of the existence of families of simple Ulrich vector bundles of arbitrary high rank and dimension.

3.3 The Representation Type of a del Pezzo Surface

In this section, we are going to construct ACM bundles and Ulrich bundles on smooth del Pezzo surfaces, and to determine their representation type. So, let us start recalling the definition and main properties of del Pezzo surfaces.

Definition 3.3.1 A *del Pezzo* surface is defined to be a smooth surface X whose anticanonical divisor $-K_X$ is ample. Its degree is defined as K_X^2. If $-K_X$ is very ample, X will be called a *strong del Pezzo surface*.

Example 3.3.2 As examples of del Pezzo surfaces we have:

1. A smooth cubic surface $X \subseteq \mathbb{P}^3$.
2. A smooth quartic surface $X \subset \mathbb{P}^4$ complete intersection of two quadrics.
3. Let Y be the blow up of $\mathbb{P}^2$ at $0 \leq s \leq 6$ general points. Consider its embedding in $\mathbb{P}^{9-s}$ through the very ample divisor $-K_Y$ and call $X \subset \mathbb{P}^{9-s}$ its image. X is a del Pezzo surface.

The classification of del Pezzo surfaces is known and we recall it for seek of completeness.

Definition 3.3.3 A set of s different points $\{p_1, \ldots, p_s\}$ on $\mathbb{P}^2$ with $s \leq 8$ is in *general position* if no three of them lie on a line, no six of them lie on a conic and no eight of them lie on a cubic with a singularity at one of these points.

Theorem 3.3.4 *Let X be a del Pezzo surface of degree d. Then $1 \leq d \leq 9$ and*

1. *If $d = 9$, then X is isomorphic to $\mathbb{P}^2$ (and $-K_{\mathbb{P}^2} = 3H_{\mathbb{P}^2}$ gives the usual Veronese embedding in $\mathbb{P}^9$).*
2. *If $d = 8$, then X is isomorphic to either $\mathbb{P}^1 \times \mathbb{P}^1$ or to a blow-up of $\mathbb{P}^2$ at one point.*
3. *If $7 \geq d \geq 1$, then X is isomorphic to a blow-up of $\mathbb{P}^2$ at $9 - d$ closed points in general position.*

Conversely, any surface described under 1., 2., 3 *is a del Pezzo surface of the corresponding degree.*

Proof See, for instance, [37, Chapter IV, Theorems 24.3 and 24.4], and [16, Proposition 8.1.9]. □

Lemma 3.3.5 *Let X be the blow-up of $\mathbb{P}^2$ on $0 \leq s \leq 8$ points in general position. Let $e_0 \in Pic(X)$ be the pull-back of a line in $\mathbb{P}^2$, e_i the exceptional divisors, $i = 1, \ldots, s$ and K_X be the canonical divisor. Then:*

1. *If $s \leq 6$, $-K_X = 3e_0 - \sum_{i=1}^{s} e_i$ is very ample and its global sections yield a closed embedding of X in a projective space of dimension*

$$\dim \mathrm{H}^0(X, \mathcal{O}_X(-K_X)) - 1 = K_X^2 = 9 - s.$$

2. *If $s = 7$, $-K_X$ is ample and generated by its global sections.*
3. *if $s = 8$, $-K_X$ is ample and $-2K_X$ is generated by its global sections.*

Proof See, for instance, [35, Proposition 3.4]. □

The construction of ACM bundles and Ulrich bundles on smooth del Pezzo surfaces is closely related (via Serre's correspondence) to the existence of level set of points.

Definition 3.3.6 A 0-dimensional scheme Z on a surface $X \subset \mathbb{P}^n$ is said to be *level of type* ρ if the last graded free module in its minimal graded free resolution has rank ρ and is concentrated in only one degree. Dualizing, this is equivalent to say that all minimal generators of the canonical module K_Z of Z have the same degree.

Example 3.3.7 Let Z be a set of 29 general points on a smooth quadric surface $Q \subset \mathbb{P}^3$. The ideal $I(Z)$ of Z has a minimal graded free resolution of the following type:

$$0 \longrightarrow R(-8)^4 \longrightarrow R(-7)^3 \oplus R(-6)^8 \longrightarrow R(-5)^7 \oplus R(-2) \longrightarrow I(Z) \longrightarrow 0.$$

Therefore, Z is level of type 4.

The existence of level set of points on a smooth del Pezzo surface is related to Mustaţă's conjecture which we will discuss in next subsection and its proof will strongly rely on the fact that we know the minimal resolution of the coordinate ring of a del Pezzo surface $X \subset \mathbb{P}^d$. Indeed, according to [29, Theorem 1], the minimal free resolution of the coordinate ring of a del Pezzo surface $X \subseteq \mathbb{P}^d$ has the form:

$$0 \longrightarrow R(-d) \longrightarrow R(-d+2)^{\alpha_{d-3}} \longrightarrow \ldots \longrightarrow R(-2)^{\alpha_1} \longrightarrow R \longrightarrow R_X \longrightarrow 0 \tag{3.2}$$

where

$$\alpha_i = i\binom{d-1}{i+1} - \binom{d-2}{i-1} \quad \text{for } 1 \leq i \leq d-3.$$

Notice that X turns out to be AG and, in particular, $\alpha_i = \alpha_{d-2-i}$ for all $i = 1, \ldots, d-2$. The Hilbert polynomial and the regularity of a del Pezzo surface X can be easily computed using the exact sequence (3.2) and we have

$$P_X(r) = \frac{d}{2}(r^2 + r) + 1 \text{ and } \operatorname{reg}(X) = 3.$$

3.3.1 *Mustaţă's Conjecture for a Set of General Points on a del Pezzo Surface*

In [46], Mustaţă predicted the minimal free resolution of a general set of points Z in an arbitrary projective variety X; he proved that the first rows of the Betti diagram of Z coincide with the Betti diagram of X and that there are two extra nontrivial rows at the bottom. Let us recall it.

Theorem 3.3.8 *Let $X \subseteq \mathbb{P}^n$ be a projective variety with $d = \dim(X) \geq 1$, $\operatorname{reg}(X) = m$ and with Hilbert polynomial P_X. Let s be an integer with $P_X(r-1) \leq s < P_X(r)$ for some $r \geq m+1$ and let Z be a set of s general points on X. Let*

$$0 \to F_n \to F_{n-1} \to \cdots \to F_2 \to F_1 \to R \to R_X \to 0$$

be a minimal graded free R-resolution of R_X. Then R_Z has a minimal free R-resolution of the following type

$$\begin{aligned} 0 &\longrightarrow F_n \oplus R(-r-n+1)^{b_{n,r-1}} \oplus R(-r-n)^{b_{n,r}} \longrightarrow \\ \cdots &\longrightarrow F_2 \oplus R(-r-1)^{b_{2,r-1}} \oplus R(-r-2)^{b_{2,r}} \longrightarrow \\ & F_1 \oplus R(-r)^{b_{1,r-1}} \oplus R(-r-1)^{b_{1,r}} \longrightarrow R \longrightarrow R_Z \longrightarrow 0. \end{aligned}$$

Moreover, if we set $Q_{i,r}(s) = b_{i+1,r-1}(Z) - b_{i,r}(Z)$,

$$Q_{i,r}(s) = \sum_{l=0}^{d-1} (-1)^l \binom{n-l-1}{i-l} \Delta^{l+1} P_X(r+l) - \binom{n}{i}(s - P_X(r-1)).$$

Conjecture 3.3.9 The *minimal resolution conjecture* (MRC for short) says that

$$b_{i+1,r-1} \cdot b_{i,r} = 0 \text{ for } i = 1, \cdots, n-1.$$

Example 3.3.10 Let $S \subset \mathbb{P}^4$ be a smooth del Pezzo surface of degree 4. S is the complete intersection of 2 hyperquadrics in $\mathbb{P}^4$, $reg(S) = 3$ and $P_S(x) = 2x^2 + 2x + 1$. Let $Z \subset S$ be a set of 45 general points on S. We have $P_S(4) = 41 \leq 45 \leq P_S(5) = 61$.

The Betti diagram of Z looks like:

	0	1	2	3	4
0	1	–	–	–	–
1	–	2	–	–	–
2	–	–	1	–	–
3	–	–	–	–	–
4	–	16	40	28	–
5	–	–	–	–	4

The first three rows of the Betti diagram of Z coincide with the Betti diagram of S and there are two extra nontrivial rows without ghost terms.

Related to it there exist two weaker conjectures that deal only with a part of the minimal resolution of a general set of points:

1. The *Ideal Generation Conjecture* (IGC for short) which says that the minimal number of generators of the ideal of a general set of points will be as small as possible; this conjecture can be translated in terms of the Betti numbers saying that

$$b_{1,r} b_{2,r-1} = 0.$$

2. On the other extreme of the resolution the *Cohen-Macaulay type Conjecture* (CMC for short) controls the ending terms of the MFR and says that the canonical module $\mathrm{Ext}^n_R(R/I(Z), R(-n-1))$ has as few generators as possible, i.e,

$$b_{n-1,r} b_{n,r-1} = 0.$$

Remark

1. When $X = \mathbb{P}^n$ the above conjecture coincides with the MRC for points in $\mathbb{P}^n$ stated in [36] which says that this resolution has no *ghost* terms, i.e, $b_{i+1,r-1}b_{i,r} = 0$ for all i. The MRC for points in $\mathbb{P}^n$ is known to hold for $n \leq 4$ (see [4, 25] and [52]) and for large values of s for any n (see [28]) but it is false in general. Eisenbud, Popescu, Schreyer and Walter showed that it fails for any $n \geq 6$, $n \neq 9$ (see [20]).
2. Regarding Mustaţă conjecture, in [27] Giuffrida, Maggioni and Ragusa proved that it holds for any general set of points when X is a smooth quadric surface in $\mathbb{P}^3$. In [43, Proposition 3.10], the authors showed that it holds for any general set of $s \geq 19$ points on a smooth cubic surface in $\mathbb{P}^3$ and, in [38], Migliore and Patnott have been able to prove it for sets of general distinct points of any cardinality on a cubic surface $X \subseteq \mathbb{P}^3$ given that X is smooth or it has at most isolated double points.

The goal of this subsection is to prove MRC for a set Z of general points on a smooth del Pezzo surface X, when the cardinality $|Z|$ of Z falls in certain interval explicitly described later. As corollary we prove IGC and CMC for a set Z of general points on a del Pezzo surface X provided $|Z| \geq P_X(3)$.

As a main tool we use the theory of liaison. Roughly speaking, Liaison Theory is an equivalence relation among schemes of the same dimension and it involves the study of the properties shared by two schemes X_1 and X_2 whose union $X_1 \cup X_2 = X$ is either a complete intersection (CI-liaison) or an arithmetically Gorenstein scheme (G-liaison). Knowing that two sets of points are G-linked, this technique will allow us to pass from the minimal resolution of the ideal of one of them to the resolution of the other one (mapping cone process) and vice versa.

Definition 3.3.11 Two subschemes X_1 and X_2 of $\mathbb{P}^n$ are *directly Gorenstein linked* (*directly G-linked* for short) by an AG scheme $G \subseteq \mathbb{P}^n$ if $I(G) \subseteq I(X_1) \cap I(X_2)$, $[I(G) : I(X_1)] = I(X_2)$ and $[I(G) : I(X_2)] = I(X_1)$. We say that X_2 is *residual* to X_1 in G. When G is a complete intersection we talk about a CI-link.

When X_1 and X_2 do not share any component, being directly G-linked by an AG scheme G is equivalent to $G = X_1 \cup X_2$.

Definition 3.3.12 Two subschemes $X_1, X_2 \subset \mathbb{P}^n$ are in the same *CI-liaison class* (resp. *G-liaison class*) if there exists $X_1 = Z_0, Z_1, \ldots, Z_t = X_2$ closed subschemes in $\mathbb{P}^n$ such that Z_i and Z_{i+1} are directly linked by a complete intersection (arithmetically Gorenstein) $X_i \subset \mathbb{P}^n$.

See [33] for more details on G-liaison.

Usually it is not easy to find out AG schemes to work with. The following theorem gives a useful way to construct them.

Definition 3.3.13 A subscheme $X \subset \mathbb{P}^n$ satisfies the *condition* G_r if every localization of $R/I(X)$ of dimension $\leq r$ is a Gorenstein ring. G_r is sometimes referred to as "Gorenstein in codimension $\leq r$", i.e. the non locally Gorenstein locus has codimension $\geq r+1$. In particular, G_0 is generically Gorenstein.

Theorem 3.3.14 *Let $S \subseteq \mathbb{P}^n$ be an ACM scheme satisfying condition G_1. Denote by K_S the canonical divisor and by H_S a general hyperplane section of S. Then any effective divisor in the linear system $|mH_S - K_S|$ is AG.*

Proof See [33, Lemma 5.4]. □

The main feature of G-liaison that is going to be exploited in this paper is that through the mapping cone process it is possible to pass from the free resolution of a scheme X_1 to the free resolution of its residual X_2 on an AG scheme. We have

Lemma 3.3.15 *Let $V_1, V_2 \subseteq \mathbb{P}^n$ be two ACM schemes of codimension c directly G-linked by an AG scheme W. Let the minimal free resolutions of $I(V_1)$ and $I(W)$ be*

$$0 \longrightarrow F_c \xrightarrow{d_c} F_{c-1} \xrightarrow{d_{c-1}} \dots F_1 \xrightarrow{d_1} I(V_1) \longrightarrow 0$$

and

$$0 \longrightarrow R(-t) \xrightarrow{e_c} G_{c-1} \xrightarrow{e_{c-1}} \dots G_1 \xrightarrow{e_1} I(W) \longrightarrow 0,$$

respectively. Then the contravariant functor $\mathrm{Hom}(-, R(-t))$ *applied to a free resolution of $I(V_1)/I(W)$ gives a (non necessarily minimal) resolution of $I(V_2)$:*

$$\begin{aligned} 0 \longrightarrow F_1^\vee(-t) \longrightarrow F_2^\vee(-t) \oplus G_1^\vee(-t) \longrightarrow \dots \\ \longrightarrow F_c^\vee(-t) \oplus G_{c-1}^\vee(-t) \longrightarrow I(V_2) \longrightarrow 0. \end{aligned}$$

In order to achieve the main result of this subsection, we define for any del Pezzo surface $X \subset \mathbb{P}^d$ of degree d the so-called critical values:

$$m(r) := \frac{d}{2}r^2 + r\frac{2-d}{2}, \quad n(r) := \frac{d}{2}r^2 + r\frac{d-2}{2}.$$

Notice that

$$P_X(r-1) < m(r) < n(r) < P_X(r).$$

Our first aim is to find out the minimal graded free resolution and to prove MRC conjecture for these two specific cardinalities $m(r)$ and $n(r)$ of general set of points on a del Pezzo surface X. Since the structure of our proof requires that X contains at least a line L and moreover that the elements of the linear system $|L + rH|$ satisfy condition G_1 in order to apply the theory of generalized divisors, we need to exclude the following two particular cases: $X \cong \mathbb{P}^2$ and $X \cong \mathbb{P}^1 \times \mathbb{P}^1$ proved in [48, Chapter

II]. Therefore, in this subsection $X \subseteq \mathbb{P}^d$ will stand for any del Pezzo surface except the two aforementioned sporadic cases. We also set the following notation.

1. L is any line on X.
2. H denotes a general hyperplane section of X.
3. If C is a curve on X, H_C will be a general hyperplane section of C and K_C the canonical divisor on C.

The strategy of the proof is as follows: firstly, we will establish the result for $m(2) = d + 2$ points which gives the starting point for our induction process. Secondly, using G-liaison, we prove that if $m(r)$ general points on any del Pezzo surface satisfy MRC then so do $n(r)$ general points. Next we observe that if $n(r)$ general points on X have the expected minimal free resolution then $n(r)+1$ general points do as well. And, finally, we show that if $n(r) + 1$ general points on a del Pezzo surface satisfy MRC then so do $m(r + 1)$.

Since the shape of the minimal free resolution of the homogeneous ideal $I(X)$ of a del Pezzo surface of degree 3 (i.e., a cubic surface) is slightly different from that of a del Pezzo surface of degree $d \geq 4$ we need to consider apart the two cases. We only sketch the proofs in the case of degree $d \geq 4$ and we leave as exercise the case of degree 3.

Lemma 3.3.16

(a) *Let $X \subseteq \mathbb{P}^d$ be any del Pezzo surface of degree $d \geq 4$ and take $C \in |(r+\varepsilon)H|$, $r \geq 2, \varepsilon \in \{0, 1\}$. Then, any effective divisor G in the linear system $|rH_C|$ is AG and it has a minimal free resolution of the following form:*

$$\begin{aligned} 0 &\longrightarrow R(-2r-d-\varepsilon) \longrightarrow R(-2r-d+2-\varepsilon)^{\alpha_{d-3}} \\ &\oplus R(-r-d)^{2-\varepsilon} \oplus R(-r-d-1)^{\varepsilon} \\ &\longrightarrow \dots \longrightarrow M_i \longrightarrow \dots \longrightarrow R(-2r-\varepsilon) \\ &\oplus R(-r-2)^{(2-\varepsilon)\alpha_1} \oplus R(-r-3)^{\varepsilon\alpha_1} \\ &\longrightarrow M_1 := R(-r)^{2-\varepsilon} \oplus R(-r-1)^{\varepsilon} \longrightarrow I(G|X) \longrightarrow 0 \end{aligned}$$

where $M_i := R(-2r-i+1-\varepsilon)^{\alpha_{i-2}} \oplus R(-r-i)^{(2-\varepsilon)\alpha_{i-1}} \oplus R(-r-i-1)^{\varepsilon\alpha_{i-1}}$ for $i = 3, \dots, d-2$ and $\alpha_i = i\binom{d-1}{i+1} - \binom{d-2}{i-1}$ for $1 \leq i \leq d-3$.

(b) *Let $X \subseteq \mathbb{P}^3$ be a del Pezzo surface of degree 3 and take $C \in |(r+\varepsilon)H|$, $r \geq 2, \varepsilon \in \{0, 1\}$. Then, any effective divisor G in the linear system $|rH_C|$ is AG and it has a minimal free resolution of the following form:*

$$\begin{aligned} 0 &\longrightarrow R(-2r-3-\varepsilon) \longrightarrow R(-2r-\varepsilon) \oplus R(-r-3)^{2-\varepsilon} \oplus R(-r-4)^{\varepsilon} \\ &\longrightarrow R(-r)^{2-\varepsilon} \oplus R(-r-1)^{\varepsilon} \longrightarrow I(G|X) \longrightarrow 0. \end{aligned}$$

Proof A curve $C \in |(r+\varepsilon)H|$ has saturated ideal $I(C|X) = \mathrm{H}^0_*(\mathcal{O}_X(-r-\varepsilon))$. From the exact sequence (3.2) we have:

$$0 \to \mathcal{O}_{\mathbb{P}^d}(-d) \to \mathcal{O}_{\mathbb{P}^d}(-d+2)^{\alpha_{d-3}} \to \cdots \to \mathcal{O}_{\mathbb{P}^d}(-2)^{\alpha_1} \to \mathcal{O}_{\mathbb{P}^d} \to \mathcal{O}_X \to 0 \tag{3.3}$$

with $\alpha_i = i\binom{d-1}{i+1} - \binom{d-2}{i-1}$ for $1 \le i \le d-3$. Twisting (3.3) with $\mathcal{O}_{\mathbb{P}^d}(-r-\varepsilon)$ and taking global sections we get the minimal graded free resolution of $I(C|X)$:

$$\begin{aligned} 0 &\longrightarrow R(-r-d-\varepsilon) \longrightarrow \ldots \longrightarrow R(-r-(i+\varepsilon))^{\alpha_{i-1}} \longrightarrow \\ &\ldots \longrightarrow R(-r-2-\varepsilon)^{\alpha_1} \longrightarrow R(-r-\varepsilon) \longrightarrow I(C|X) \longrightarrow 0. \end{aligned}$$

Now we apply the horseshoe lemma to the exact sequence

$$0 \longrightarrow I(X) \longrightarrow I(C|\mathbb{P}^d) \longrightarrow I(C|X) \longrightarrow 0$$

to obtain the minimal free resolution of $I(C|\mathbb{P}^d)$:

$$\begin{aligned} 0 &\longrightarrow R(-r-d-\varepsilon) \longrightarrow R(-r-d+2-\varepsilon)^{\alpha_{d-3}} \oplus R(-d) \longrightarrow \ldots \\ &\longrightarrow T_i := R(-r-i-\varepsilon)^{\alpha_{i-1}} \oplus R(-(i+1))^{\alpha_i} \longrightarrow \ldots \\ &\longrightarrow R(-r-\varepsilon) \oplus R(-2)^{\alpha_1} \longrightarrow I(C|\mathbb{P}^d) \longrightarrow 0. \end{aligned}$$

This sequence shows that $C \subseteq \mathbb{P}^d$ is an AG variety with canonical module

$$K_C = R_C(r-1+\varepsilon).$$

Therefore $I(G|C) = H^0_*(\mathcal{O}_C(-r)) = K_C(-2r+1-\varepsilon)$. Now, we apply the functor $\mathrm{Hom}(-, R(-d-1))$ to the previous sequence and we get a minimal free resolution of K_C:

$$\begin{aligned} 0 &\longrightarrow R(-d-1) \longrightarrow R(r-d-1+\varepsilon) \oplus R(-d+1)^{\alpha_{d-3}} \longrightarrow \ldots \\ &\longrightarrow T'_i \longrightarrow \ldots \longrightarrow R(-1) \oplus R(r-3+\varepsilon)^{\alpha_1} \longrightarrow R(r-1+\varepsilon) \longrightarrow K_C \longrightarrow 0 \end{aligned}$$

where $T'_i := T^{\vee}_{d-i}(-d-1) = R(r-i-\varepsilon)^{\alpha_{i-1}} \oplus R(-i)^{\alpha_{i-2}}$ for $i = 3, \ldots, d-2$. If we twist the previous sequence by $-2r+1-\varepsilon$ we get the minimal resolution of $I(G|C)$:

$$\begin{aligned} 0 &\longrightarrow R(-2r-d-\varepsilon) \longrightarrow R(-r-d) \oplus R(-2r-d+2-\varepsilon)^{\alpha_{d-3}} \longrightarrow \\ &\ldots \longrightarrow T'_i(-2r+1-\varepsilon) \longrightarrow \ldots \\ &\longrightarrow R(-2r-\varepsilon) \oplus R(-r-2)^{\alpha_1} \longrightarrow R(-r) \longrightarrow I(G|C) \longrightarrow 0. \end{aligned}$$

Finally, we apply the horseshoe lemma to the short exact sequence

$$0 \longrightarrow I(C|X) \longrightarrow I(G|X) \longrightarrow I(G|C) \longrightarrow 0$$

to recover the resolution of $I(G|X)$ and we finish the proof. □

Lemma 3.3.17

(a) Let $X \subseteq \mathbb{P}^d$ be a del Pezzo surface and let $L \subseteq X$ be a line on it. Take $C \in |L + rH|$, $r \geq 2$, and let G be any effective divisor in the linear system $|2rH_C - K_C|$. Then, G is AG and the minimal free resolution of $I(G|C)$ has the following form:

$$\begin{aligned} 0 &\longrightarrow R(-2r-d-1) \longrightarrow R(-2r-d+1)^{\alpha_1} \oplus R(-r-d)^{d-1} \longrightarrow \dots \\ &\longrightarrow R(-2r-i)^{\alpha_{d-i}} \oplus R(-r-i-1)^{\binom{d-1}{d-i}+\alpha_{d-i-1}} \longrightarrow \dots \\ &\longrightarrow R(-2r-1) \oplus R(-r-3)^{\binom{d-1}{d-2}+\alpha_{d-3}} \\ &\longrightarrow R(-r-1) \oplus R(-r-2) \longrightarrow I(G|C) \longrightarrow 0 \end{aligned}$$

with $\alpha_i = i\binom{d-1}{i+1} - \binom{d-2}{i-1}$ for $1 \leq i \leq d-3$.

(b) Let $X \subseteq \mathbb{P}^3$ be an integral cubic surface and let $L \subseteq X$ be a line on it. Take $C \in |L + rH|$, $r \geq 2$, and let G be any effective divisor in $|2rH_C - K_C|$. Then, G is AG and the minimal free resolution of $I(G|C)$ has the following form:

$$\begin{aligned} 0 &\longrightarrow R(-2r-4) \longrightarrow R(-2r-1) \oplus R(-r-3)^2 \\ &\longrightarrow R(-r-1) \oplus R(-r-2) \longrightarrow I(G|C) \longrightarrow 0. \end{aligned}$$

Proof Let $L \subseteq X$ be any line. Its ideal as a subvariety of $\mathbb{P}^d$ has a resolution:

$$0 \longrightarrow R(-d+1) \longrightarrow \dots \longrightarrow R(-i)^{\binom{d-1}{i}} \longrightarrow \dots \longrightarrow R(-1)^{d-1} \longrightarrow I(L) \longrightarrow 0.$$

Applying the mapping cone process to $0 \to I(X) \to I(L) \to I(L|X) \to 0$ we get

$$\begin{aligned} 0 &\longrightarrow R(-d) \oplus R(-d+1) \longrightarrow \dots \longrightarrow R(-i)^{\binom{d-1}{i}+\alpha_{i-1}} \\ &\longrightarrow \dots \longrightarrow R(-1)^{d-1} \longrightarrow I(L|X) \longrightarrow 0 \end{aligned}$$

with $\alpha_i = i\binom{d-1}{i+1} - \binom{d-2}{i-1}$ for $1 \leq i \leq d-3$. Therefore, $C \in |L + rH|$ has the following minimal graded free resolution

$$\begin{aligned} 0 &\to R(-r-d) \oplus R(-r-d+1) \to \dots \to R(-r-i)^{\binom{d-1}{i}+\alpha_{i-1}} \to \\ &\dots \to R(-r-1)^{d-1} \to I(C|X) \to 0. \end{aligned} \quad (3.4)$$

Now the horseshoe lemma applied to $0 \to I(X|\mathbb{P}^d) \to I(C) \to I(C|X) \to 0$ gives us

$$0 \longrightarrow R(-r-d) \oplus R(-r-d+1) \longrightarrow R(-r-d+2)^{\binom{d-1}{d-2}+\alpha_{d-3}} \oplus R(-d) \longrightarrow \dots$$
$$\longrightarrow R(-r-i)^{\binom{d-1}{i}+\alpha_{i-1}} \oplus R(-(i+1))^{\alpha_i}$$
$$\longrightarrow \dots \longrightarrow R(-r-1)^{d-1} \oplus R(-2)^{\alpha_1} \longrightarrow I(C) \longrightarrow 0.$$

Since C is ACM we can apply $\mathrm{Hom}(-, R(-d-1))$ to get a resolution of K_C:

$$0 \longrightarrow R(-d-1) \longrightarrow R(-d+1)^{\alpha_1} \oplus R(r-d)^{d-1} \longrightarrow$$
$$\dots \longrightarrow R(r-i-1)^{\binom{d-1}{d-i}+\alpha_{d-i-1}} \oplus R(-i)^{\alpha_{d-i}} \longrightarrow$$
$$\dots \longrightarrow R(r-3)^{\binom{d-1}{d-2}+\alpha_{d-3}} \oplus R(-1) \longrightarrow R(r-1) \oplus R(r-2) \longrightarrow K_C \longrightarrow 0.$$

Finally, since $G \in |2rH_C - K_C|$ we have:

$$0 \longrightarrow R(-2r-d-1) \longrightarrow R(-2r-d+1)^{\alpha_1} \oplus R(-r-d)^{d-1} \longrightarrow$$
$$\dots \longrightarrow R(-2r-i)^{\alpha_{d-i}} \oplus R(-r-i-1)^{\binom{d-1}{d-i}+\alpha_{d-i-1}} \longrightarrow \dots$$
$$\longrightarrow R(-2r-1) \oplus R(-r-3)^{\binom{d-1}{d-2}+\alpha_{d-3}} \longrightarrow R(-r-1) \oplus R(-r-2)$$
$$\longrightarrow I(G|C) \longrightarrow 0.$$

□

Now we fix the starting point of the induction.

Lemma 3.3.18 *A general set Z of $m(2) = d+2$ points on any del Pezzo surface $X \subseteq \mathbb{P}^d$ has a minimal free resolution of the following type:*

$$0 \longrightarrow R(-d-2) \longrightarrow R(-d)^{\gamma_{d-1}} \longrightarrow \dots$$
$$\longrightarrow R(-3)^{\gamma_2} \longrightarrow R(-2)^{2d-1} \longrightarrow I(Z|X) \longrightarrow 0$$

with

$$\gamma_i = \sum_{l=0}^{1}(-1)^l \binom{d-l-1}{i-l} \Delta^{l+1} H_X(2+l) - \binom{d}{i}(m(2) - H_X(1)).$$

Proof It follows from the fact that a general set Z of $d+2$ points on X is in linearly general position (i.e., any subset of Z of $d+1$ points spans $\mathbb{P}^d$). □

Fix an integer $r \geq 2$ and let $Z_{m(r)}$ and $Z_{n(r)}$ be general sets of points on X of cardinality $m(r)$ and $n(r)$, respectively. We will see that they are directly G-linked by an effective divisor G in $|rH_C|$ with C a curve in the linear system $|rH_X|$. Recall that we have:

$$P_X(r-1) < m(r) < n(r) < P_X(r).$$

Let us start with a general set $Z_{m(r)}$ of $m(r)$ points. Since $\mathrm{h}^0(\mathcal{O}_X(r)) > m(r)$ there exists a curve C in the linear system $|rH_X|$ such that $Z_{m(r)}$ lies on C. On the other hand, the inequality $n(r) > p_a(C)$ allows us to apply Riemann-Roch Theorem for curves and assure that there exists an effective divisor $Z_{n(r)}$ of degree $n(r)$ such that $Z_{m(r)} + Z_{n(r)}$ is linearly equivalent to a divisor rH_C.

Since this construction can also be performed starting from a general set $Z_{n(r)}$ of $n(r)$ points we see that a general set of $m(r)$ points is G-linked to a general set of $n(r)$ points and vice versa. Therefore as a direct application of the mapping cone process we get

Proposition 3.3.19 *Fix $r \geq 2$ and assume that the ideal $I(Z_{m(r)}|X)$ of $m(r)$ general points on a del Pezzo surface $X \subseteq \mathbb{P}^d$ has the minimal free resolution*

$$0 \longrightarrow R(-r-d)^{r-1} \longrightarrow R(-r-d+2)^{\gamma_{d-1,r-1}} \longrightarrow \dots$$
$$\longrightarrow R(-r-1)^{\gamma_{2,r-1}} \longrightarrow R(-r)^{(d-1)r+1} \longrightarrow I(Z_{m(r)}|X) \longrightarrow 0$$

with $\gamma_{i,r-1} = \sum_{l=0}^{1}(-1)^l\binom{d-l-1}{i-l}\Delta^{l+1}P_X(r+l) - \binom{d}{i}(m(r) - P_X(r-1))$. Then the ideal $I(Z_{n(r)}|X)$ of $n(r)$ general points has the minimal free resolution

$$0 \longrightarrow R(-r-d)^{(d-1)r-1} \longrightarrow R(-r-d+1)^{\beta_{d-1,r}} \longrightarrow \dots$$
$$\longrightarrow R(-r-2)^{\beta_{2,r}} \longrightarrow R(-r)^{r+1} \longrightarrow I(Z_{n(r)}|X) \longrightarrow 0$$

with $\beta_{i,r} = \sum_{l=0}^{1}(-1)^{l+1}\binom{d-l-1}{i-l}\Delta^{l+1}P_X(r+l) + \binom{d}{i}(n(r) - P_X(r-1))$.

Vice versa, if $n(r)$ general points on a del Pezzo surface $X \subseteq \mathbb{P}^d$ have the expected resolution then $m(r)$ general points do as well.

Lemma 3.3.20 *Let $X \subset \mathbb{P}^d$ be any del Pezzo surface. Fix $r \geq 2$ and assume that the ideal $I(Z_{n(r)}|X)$ of a set $Z_{n(r)}$ of $n(r)$ general points on $X \subseteq \mathbb{P}^d$ has the expected minimal free graded resolution. Then a set of $n(r)+1$ general points do as well.*

Proof Since $I(Z_{n(r)}|X)$ has the expected minimal free resolution, it is generated by $r+1$ forms of degree r without linear relations. Take a general point $p \in X$ and set $Z := Z_{n(r)} \cup \{p\}$. Since $I(Z|X) \subset I(Z_{n(r)}|X)$, we can take the r generators of $I(Z|X)$ in degree r to be a subset of the generators of $I(Z_{n(r)}|X)$ in degree r; in particular, they do not have linear syzygies. We must add d generators of degree $r+1$ in order to get a minimal system of generators of $I(Z|X)$. Hence the first module in the minimal free resolution of $I(Z|X)$ is $R(-r)^r \oplus R(-r-1)^d$ which forces the remaining part of the resolution. $\square$

Proposition 3.3.21 *Let $X \subseteq \mathbb{P}^d$ be a del Pezzo surface. Fix $r \geq 2$ and assume that the ideal $I(Z_{p(r)}|X)$ of $p(r) := n(r) + 1$ general points on X has the minimal free resolution*

$$0 \longrightarrow R(-r-d)^{(d-1)r} \longrightarrow R(-r-d+1)^{\delta_{d-1,r}} \longrightarrow \dots$$
$$\longrightarrow R(-r-2)^{\delta_{2,r}} \longrightarrow R(-r)^r \oplus R(-r-1)^d \longrightarrow I(Z_{p(r)}|X) \longrightarrow 0$$

with

$$\delta_{i,r} = \sum_{l=0}^{1}(-1)^{l+1}\binom{d-l-1}{i-l}\Delta^{l+1}H_X(r+l) + \binom{d}{i}(p(r) - H_X(r-1)).$$

Then the ideal $I(Z_{m(r+1)}|X)$ of $m(r+1)$ general points has the minimal free resolution

$$0 \longrightarrow R(-r-d-1)^r \longrightarrow R(-r-d+1)^{\gamma_{d-1,r}} \longrightarrow \dots$$
$$\longrightarrow R(-r-2)^{\gamma_{2,r}} \longrightarrow R(-r-1)^{(d-1)r+d} \longrightarrow I(Z_{m(r+1)}|X) \longrightarrow 0$$

with

$$\gamma_{i,r} = \sum_{l=0}^{1}(-1)^{l}\binom{d-l-1}{i-l}\Delta^{l+1}H_X(r+1+l) - \binom{d}{i}(m(r+1) - H_X(r)).$$

Proof Let $Z_{p(r)}$ be a set of $p(r)$ general points with resolution as in the statement. Let us consider the linear system $|L+rH|$. Since, $\dim|L+rH| \geq \dim|rH| = P_X(r) - 1 > p(r)$, we can find a curve $C \in |L+rH|$ passing through these $p(r)$ points. Notice that $\deg(C) = 1 + rd$ and $p_a(C) = d\binom{r}{2} + r$. Since $p_a(C) < m(r+1)$ we can find an effective divisor $Z_{m(r+1)}$ of degree $m(r+1)$ such that $Z_{p(r)}$ and $Z_{m(r+1)}$ are G-linked by a divisor of degree $p(r) + m(r+1) = dr^2 + dr + 2 = deg(2rH_C - K_C)$. This allows us to find the resolution of $I(Z_{m(r+1)}|X)$. First we find the minimal free resolution of $I(Z_{p(r)}|C)$ using the exact sequence $0 \to I(C|X) \to I(Z_{p(r)}|X) \to I(Z_{p(r)}|C) \to 0$, the resolution of $I(C|X)$ given in (3.4) and the mapping cone process. It turns out to be:

$$0 \longrightarrow R(-r-d)^{(d-1)r+1} \longrightarrow R(-r-d+1)^{c_{d-1,r}} \longrightarrow \dots$$
$$\longrightarrow R(-r-2)^{c_{2,r}} \longrightarrow R(-r)^r \oplus R(-r-1) \longrightarrow I(Z_{p(r)}|C) \longrightarrow 0.$$

Since we know the minimal free resolution of $I(G|C)$ (see Lemma 3.3.17) we apply the mapping cone process to the sequence $0 \to I(G|C) \to I(Z_{(p(r)}|C) \to$

$I(Z_{p(r)}|G) \to 0$ to get

$$
\begin{aligned}
0 \longrightarrow R(-2r-d-1) &\longrightarrow R(-r-d)^{(d-1)r+d} \oplus R(-2r-d+1)^{\alpha_1} \longrightarrow \dots \\
&\longrightarrow R(-r-i)^{d_{i,r}} \oplus R(-2r-i+1)^{\alpha_{d-i+1}} \longrightarrow \dots \\
&\longrightarrow R(-r-2)^{d_{2,r}} \longrightarrow R(-r)^r \longrightarrow I(Z_{p(r)}|G) \longrightarrow 0.
\end{aligned}
$$

$(0 \to R(-2r-4) \to R(-r-3)^{2r+2} \oplus R(-2r-1) \to R(-r-2)^{d_{2,r}} \to R(-r)^r \to I(Z_{p(r)}|G) \to 0$ if $d = 3)$.

Finally we obtain the minimal free resolution of $I(Z_{m(r+1)})$:

$$
\begin{aligned}
0 &\longrightarrow R(-r-d-1)^r \\
&\longrightarrow R(-r-d+1)^{\gamma_{d-1,r}} \longrightarrow R(-r-d+2)^{\gamma_{d-2,r}} \oplus R(-d) \longrightarrow \\
&\dots \longrightarrow R(-r-i)^{\gamma_{i,r}} \oplus R(-i)^{\alpha_i} \longrightarrow \dots \\
&\longrightarrow R(-r-1)^{(d-1)r+d} \oplus R(-2)^{\alpha_1} \longrightarrow I(Z_{m(r+1)}) \longrightarrow 0
\end{aligned}
$$

$(0 \to R(-r-4)^r \to R(-r-2)^{\gamma_{2,r}} \to R(-r-1)^{2r+3} \oplus R(-3) \to I(Z_{m(r+1)}) \to 0$ if $d = 3)$ from which it is straightforward to recover the predicted resolution of $I(Z_{m(r+1)}|X)$. □

We are ready to prove the MRC for $n(r)$ and $m(r)$ general points on a del Pezzo surface:

Theorem 3.3.22 *Let $X \subseteq \mathbb{P}^d$ be a del Pezzo surface. We have:*

1. *Let $Z_{n(r)} \subseteq X$ be a general set of $n(r)$ points, $r \geq 2$. Then the minimal graded free resolution of $I(Z_{n(r)}|X)$ has the following form:*

$$
\begin{aligned}
0 \longrightarrow R(-r-d)^{(d-1)r-1} &\longrightarrow R(-r-d+1)^{\beta_{d-1,r}} \longrightarrow R(-r-d+2)^{\beta_{d-2,r}} \longrightarrow \\
\dots &\longrightarrow R(-r-2)^{\beta_{2,r}} \longrightarrow R(-r)^{r+1} \longrightarrow I(Z_{n(r)}|X) \longrightarrow 0.
\end{aligned}
$$

where

$$
\beta_{i,r} = \sum_{l=0}^{1} (-1)^{l+1} \binom{n-l-1}{i-l} \Delta^{l+1} H_X(r+l) + \binom{n}{i} (n(r) - H_X(r-1)).
$$

2. *Let $Z_{m(r)} \subseteq X$ be a general set of $m(r)$ points, $r \geq 2$. Then its minimal graded free resolution has the following form:*

$$
\begin{aligned}
0 \longrightarrow R(-r-d)^{r-1} &\longrightarrow R(-r-d+2)^{\gamma_{d-1,r-1}} \longrightarrow \dots \\
&\longrightarrow R(-r-1)^{\gamma_{2,r-1}} \longrightarrow R(-r)^{(d-1)r+1} \longrightarrow I(Z_{m(r)}|X) \longrightarrow 0
\end{aligned}
$$

with

$$\gamma_{i,r-1} = \sum_{l=0}^{1}(-1)^l\binom{n-l-1}{i-l}\Delta^{l+1}P_X(r+l) - \binom{n}{i}(m(r) - P_X(r-1)).$$

In particular, Mustaţă's conjecture works for $n(r)$ and $m(r)$, $r \geq 4$, general points on a del Pezzo surface $X \subseteq \mathbb{P}^d$.

Proof Lemma 3.3.18 establishes the result for a set of $m(2)$ general points, the starting point of our induction process. Therefore, the result about the resolution of $I(Z_{n(r)}|X)$ and $I(Z_{m(r)}|X)$ follows using Lemma 3.3.20, Propositions 3.3.19 and 3.3.21 and applying induction. □

Next lemma controls how the bottom lines of the Betti diagram of a set of general points on a projective variety change when we add another general point.

Lemma 3.3.23 *Let $X \subseteq \mathbb{P}^n$ be a projective variety with* $\dim(X) \geq 2$, $\operatorname{reg}(X) = m$ *and with Hilbert polynomial P_X. Let s be an integer with $P_X(r-1) \leq s < P_X(r)$ for some $r \geq m+1$, let Z be a set of s general points on X and let $P \in X \setminus Z$ be a general point. We have*

1. $b_{i,r-1}(Z) \geq b_{i,r-1}(Z \cup P)$ *for every i.*
2. $b_{i,r}(Z) \leq b_{i,r}(Z \cup P)$ *for every i.*

Proof See [46, Proposition 1.7]. □

Now, we prove the main result of this subsection, namely, the MRC holds for a general set of points Z on a smooth del Pezzo surface when the cardinality of Z falls in the strips of the form $[P_X(r-1), m(r)]$ or $[n(r), P_X(r)]$, $r \geq 4$.

Theorem 3.3.24 *Let $X \subseteq \mathbb{P}^d$ be a del Pezzo surface. Let r be such that $r \geq reg(X) + 1 = 4$. Then for a general set of points Z on X such that $P_X(r-1) \leq |Z| \leq m(r)$ or $n(r) \leq |Z| \leq P_X(r)$ the MRC is true.*

Proof See [48, Chapter II], for the cases of $X \cong \mathbb{P}^2$ and $X \cong \mathbb{P}^1 \times \mathbb{P}^1$. So let X be any other smooth del Pezzo surface. Let Z' be a general set of points of cardinality $|Z'| = n(r)$ and add general points to Z' in order to get a set of points Z of cardinality $n(r) \leq |Z| \leq P_X(r)$. By Theorem 3.3.22 we have that $b_{i,r-1}(Z') = 0$ for all $i = 2, \ldots, d$. Therefore we can apply Lemma 3.3.23 to deduce that $b_{i,r-1}(Z) = 0$ for all $i = 2, \ldots, d$. Thus, by semicontinuity, MRC holds for a general set of $|Z|$ points.

Now if $|Z| \leq m(r)$, we can add general points to Z in order to have a general set Z' including Z and such that $|Z'| = m(r)$. Again from the previous Theorem we have that $b_{i,r}(Z') = 0$ for all $i = 1, \ldots, d-1$. So we can use again Lemma 3.3.23 to deduce that $b_{i,r}(Z) = 0$ for all $i = 1, \ldots, d-1$ and therefore MRC holds for Z. □

Example 3.3.25 Let Y be the blow up of $\mathbb{P}^2$ at 4 general points. Consider its embedding in $\mathbb{P}^5$ through the very ample divisor $-K_Y$ and call $X \subset \mathbb{P}^5$ its image.

X is a del Pezzo surface of degree 5, $\mathrm{reg}(X) = 3$, $P_X(t) = \frac{d}{2}(t^2 + t) + 1$ and its homogenous ideal has a minimal free R-resolution of the following type:

$$0 \longrightarrow R(-5) \longrightarrow R(-3)^5 \longrightarrow R(-2)^5 \longrightarrow I(X) \longrightarrow 0.$$

Let $Z \subset X$ be a set of 79 general points on X. We have $P_X(5) = 76 < 79 < m(6) = 81 < n(6) = 99 < P_X(6) = 106$. By Theorem 3.3.24, the minimal free resolution of $I(Z)$ has the following shape:

$$0 \longrightarrow R(-11)^3 \oplus R(-10)^{10} \longrightarrow R(-9)^{75} \longrightarrow R(-8)^{135} \oplus R(-5)$$
$$\longrightarrow R(-7)^{100} \oplus R(-3)^5 \longrightarrow R(-6)^{27} \oplus R(-2)^5 \longrightarrow I(Z) \longrightarrow 0.$$

Therefore, the Betti diagram of Z looks like:

	0	1	2	3	4	5
0	1	–	–	–	–	–
1	–	5	5	–	–	–
2	–	–	–	1	–	–
3	–	–	–	–	–	–
4	–	–	–	–	–	–
5	–	27	100	135	75	10
5	–	–	–	–	–	3

The first three rows of the Betti diagram of Z coincide with the Betti diagram of X and there are two extra nontrivial rows without ghost terms.

As a consequence of Theorem 3.3.22 we prove that the number of generators of the ideal of a general set of points on a del Pezzo surface is as small as possible and so it is the number of generators of its canonical module as well. In fact, we have:

Theorem 3.3.26 *Let $X \subseteq \mathbb{P}^d$ be a del Pezzo surface. Then for a general set of points Z on X such that $|Z| \geq P_X(3)$ the Cohen-Macaulay type Conjecture and the Ideal Generation Conjecture are true.*

Proof Let Z be a general set of points on our del Pezzo surface X. If it is the case that $n(r) \leq |Z| \leq m(r+1)$ the result has been proved on the previous theorem. So we can assume that $m(r) < |Z| < n(r)$ for some r. We know that the MRC holds for a general set $|Z'|$ of $n(r)$ points on X, $Z \subseteq Z'$ and in particular $b_{1,r}(Z') = 0$. Applying Lemma 3.3.23 inductively we see that $b_{1,r}(Z) = 0$. Analogously, since MRC holds for a general set Z'' of $m(r)$ points, $b_{d,r-1}(Z'') = 0$ with $Z'' \subseteq Z$. Applying once again the same Lemma we see that $b_{d,r-1}(Z) = 0$. □

In the particular case of the cubic surface, since the minimal free resolution of its points has length three, we recover one of the main results of [42] (see also [43]):

Theorem 3.3.27 *Let $X \subseteq \mathbb{P}^3$ be a integral cubic surface (i.e., a del Pezzo surface of degree three). Then the Minimal Resolution Conjecture holds for a general set of points on X of cardinality $\geq P_X(3) = 19$.*

3.3.2 Ulrich Bundles on del Pezzo Surfaces

In this subsection, we will construct large families of ACM vector bundles on smooth del Pezzo surfaces with the maximal allowed number of global sections (the so-called Ulrich bundles) and conclude that all smooth del Pezzo surfaces are of wild representation type. This result generalizes a previous result of Pons-Llopis and Tonini [49] (see also [10]) which states that the cubic surface $S \subset \mathbb{P}^3$ is of wild representation type.

The proof for the degree 8 smooth del Pezzo surface $X \subset \mathbb{P}^8$ isomorphic to $\mathbb{P}^1 \times \mathbb{P}^1$ (i.e. the Segre product of two conics naturally embedded in $\mathbb{P}^8$: $\varphi_{|\mathcal{O}(2,2)|} : \mathbb{P}^1 \times \mathbb{P}^1 \hookrightarrow \mathbb{P}^8$) is slightly different and the reader can consult [48]. So, from now on when speaking of a smooth del Pezzo surface we will understand the blow up of $\mathbb{P}^2$ at $s \leq 8$ points in general position.

Following notation from [22], let us consider K-vector spaces A and B of respective dimension a and b. Set $V = \mathrm{H}^0(\mathbb{P}^m, \mathcal{O}_{\mathbb{P}^m}(1))$ and let $M = \mathrm{Hom}(B, A \otimes V)$ be the space of $(a \times b)$-matrices of linear forms. M is an affine space of dimension $ab(m+1)$. It is well-known that there exists a bijection between the elements $\phi \in M$ and the morphisms $\phi : B \otimes \mathcal{O}_{\mathbb{P}^m} \to A \otimes \mathcal{O}_{\mathbb{P}^m}(1)$. Taking the tensor with $\mathcal{O}_{\mathbb{P}^m}(1)$ and considering global sections, we have morphisms

$$\mathrm{H}^0(\phi(1)) : \mathrm{H}^0(\mathbb{P}^m, \mathcal{O}_{\mathbb{P}^m}(1)^b) \longrightarrow \mathrm{H}^0(\mathbb{P}^m, \mathcal{O}_{\mathbb{P}^m}(2)^a).$$

The following result tells us under which conditions the aforementioned morphisms ϕ and $\mathrm{H}^0(\phi(1))$ are surjective:

Proposition 3.3.28 *For $a \geq 1$, $b \geq a + m$ and $2b \geq (m+2)a$, the set of elements $\phi \in M$ such that $\phi : B \otimes \mathcal{O}_{\mathbb{P}^m} \to A \otimes \mathcal{O}_{\mathbb{P}^m}(1)$ and $\mathrm{H}^0(\phi(1)) : \mathrm{H}^0(\mathbb{P}^m, \mathcal{O}_{\mathbb{P}^m}(1)^b) \to \mathrm{H}^0(\mathbb{P}^m, \mathcal{O}_{\mathbb{P}^m}(2)^a)$ are surjective forms a non-empty open dense subset of the affine variety M that we will denote by V_m.*

Proof See [22, Proposition 4.1]. □

Fix $m = 2$ and for a given $r \geq 2$, set $a := r$, $b := 2r$. Take an element ϕ of the non-empty subset $V_2 \subseteq M$ provided by Proposition 3.3.28 and consider the exact sequence

$$0 \longrightarrow \mathcal{F} \longrightarrow \mathcal{O}_{\mathbb{P}^2}(1)^{2r} \stackrel{\phi(1)}{\longrightarrow} \mathcal{O}_{\mathbb{P}^2}(2)^r \longrightarrow 0. \tag{3.5}$$

It follows immediately that $\mathcal{F}$ is a vector bundle of rank r, being kernel of a surjective morphism of vector bundles. Let $X := Bl_Z(\mathbb{P}^2) \xrightarrow{\pi} \mathbb{P}^2$ be the low up of $\mathbb{P}^2$ at $0 \leq s \leq 8$ points in general position. Pulling-back the exact sequence (3.5) we obtain the exact sequence:

$$0 \longrightarrow \pi^*\mathcal{F} \longrightarrow \mathcal{O}_X(e_0)^b \xrightarrow{\phi(1)} \mathcal{O}_X(2e_0)^a \longrightarrow 0. \tag{3.6}$$

We can prove:

Proposition 3.3.29 *Let $X \xrightarrow{\pi} \mathbb{P}^2$ be the low up of $\mathbb{P}^2$ at $0 \leq s \leq 8$ points in general position and let $r \geq 2$. Let $\mathcal{F}$ be the vector bundle obtained as the kernel of a general surjective morphism between $\mathcal{O}_{\mathbb{P}^2}(1)^{2r}$ and $\mathcal{O}_{\mathbb{P}^2}(2)^r$:*

$$0 \longrightarrow \mathcal{F} \longrightarrow \mathcal{O}_{\mathbb{P}^n}(1)^{2r} \xrightarrow{\phi(1)} \mathcal{O}_{\mathbb{P}^n}(2)^r \longrightarrow 0. \tag{3.7}$$

Then, the vector bundles $\mathcal{E}$ obtained pulling-back $\mathcal{F}$, dualizing and twisting by $H := 3e_0 - \sum_{i=1}^s e_i$

$$0 \longrightarrow \mathcal{O}_X(-2e_0 + H)^r \xrightarrow{f} \mathcal{O}_X(-e_0 + H)^{2r} \xrightarrow{g} \mathcal{E}(H) := (\pi^*\mathcal{F})^*(H) \longrightarrow 0 \tag{3.8}$$

are simple (hence, undecomposable) vector bundles of rank r on X.

Proof See [45, Corollary 4.5]. □

The Chern classes of $\mathcal{E}(H)$ can be easily computed and we get:

$$c_1(\mathcal{E}(H)) = rH \text{ and } c_2(\mathcal{E}(H)) = \frac{H^2r^2 + (2 - H^2)r}{2}.$$

Let us check that $\mathcal{E}(H)$ is an initialized Ulrich bundle. For this, we need the following computations.

Remark (Riemann-Roch for Vector Bundles on a del Pezzo Surface) Let X be a del Pezzo surface. Since X is a rational connected surface we have $\chi(\mathcal{O}_X) = 1$. In particular, the Riemann-Roch formula for a vector bundle $\mathcal{E}$ on X of rank r has the form

$$\chi(\mathcal{E}) = \frac{c_1(\mathcal{E})(c_1(\mathcal{E}) - K_X)}{2} + r - c_2(\mathcal{E}).$$

Remark The Euler characteristic of the involved vector bundles can be computed thanks to the Riemann-Roch formula:

$$\chi(\mathcal{O}_X(-2e_0 + lH)) = \frac{9-s}{2}l^2 - \frac{3+s}{2}l, \tag{3.9}$$

$$\chi(\mathcal{O}_X(-e_0+lH)) = \frac{9-s}{2}l^2 + \frac{3-s}{2}l, \tag{3.10}$$

$$\begin{aligned}\chi(\mathcal{E}(lH)) &= 2r\chi(\mathcal{O}_X(-e_0+lH)) - r\chi(\mathcal{O}_X(-2e_0+lH)) \\ &= \tfrac{9r-sr}{2}l^2 + \tfrac{9r-sr}{2}l.\end{aligned} \tag{3.11}$$

Proposition 3.3.30 *Let X be a del Pezzo surface. The bundles $\mathcal{E}(H)$ given by the exact sequence (3.8) are initialized simple Ulrich bundles. Moreover, in the case of a blow-up of ≤ 7 points, they are globally generated.*

Proof First of all, notice that $\mathrm{H}^0(\mathcal{E}^*) = \mathrm{H}^2(\mathcal{E}(-H)) = 0$. Therefore, $\mathrm{H}^2(\mathcal{E}(tH)) = 0$, for all $t \geq -1$. On the other hand, since $\mathrm{H}^2(\mathcal{O}_X(-2e_0)) = \mathrm{H}^0(\mathcal{O}_X(2e_0 - H)) = 0$ and $\mathrm{h}^1(\mathcal{O}_X(-e_0)) = -\chi(\mathcal{O}_X(-e_0)) = 0$ we obtain from the long exact sequence of cohomology associated to (3.8) that $\mathrm{H}^1(\mathcal{E}) = 0$. Since $\chi(\mathcal{E}) = 0$, we also conclude that $\mathrm{H}^0(\mathcal{E}) = 0$ and therefore $\mathrm{H}^0(\mathcal{E}(tH)) = 0$ for all $t \leq 0$. Moreover, since we also have that $\chi(\mathcal{E}(-H)) = 0$, we obtain that $\mathrm{H}^1(\mathcal{E}(-H)) = 0$.

We easily check that $\mathrm{H}^0(\mathcal{E}(H)) \neq 0$ which together with the vanishing $\mathrm{H}^0(\mathcal{E}(tH)) = 0$ for all $t \leq 0$ implies that $\mathcal{E}(H)$ is initialized.

We tensor by $\mathcal{E}$ the exact sequence

$$0 \longrightarrow \mathcal{O}_X(-H) \longrightarrow \mathcal{O}_X \longrightarrow \mathcal{O}_H \longrightarrow 0$$

and we consider the cohomology sequence associated to it. We get

$$0 = \mathrm{H}^0(\mathcal{E}) \longrightarrow \mathrm{H}^0(\mathcal{E}_{|H}) \longrightarrow \mathrm{H}^1(\mathcal{E}(-H)) = 0.$$

This shows that $\mathrm{H}^0(\mathcal{E}_{|H}(-tH)) = 0$ for all $t \geq 0$. Then we can use this last fact together with the long exact sequence associated to

$$0 \longrightarrow \mathcal{E}(-(t+1)H) \longrightarrow \mathcal{E}(-tH) \longrightarrow \mathcal{E}_{|H}(-tH) \longrightarrow 0$$

to show inductively that $\mathrm{H}^1(\mathcal{E}(-tH)) = 0$ for all $t \geq 0$.

In order to complete the proof we need to consider two different cases:

1. *X is the blow-up of $s \leq 7$ points on $\mathbb{P}^2$ in general position*. In this case, by Lemma 3.3.5, H is ample and generated by its global sections. Since we have just seen that $\mathcal{E}(H)$ is 0-regular with respect to H we can conclude that $\mathcal{E}(H)$ is ACM and globally generated. Moreover, $\mathrm{h}^0(\mathcal{E}(H)) = \chi(\mathcal{E}(H)) = (9-s)r = H^2r$, i.e., $\mathcal{E}(H)$ is an Ulrich bundle.
2. *X is the blow-up of* 8 *points on $\mathbb{P}^2$ in general position*. In this case, the argument is slightly more involved, since H is ample but not very ample. Fortunately $2H$ is ample and globally generated. First of all, since the points are in general position, $\mathrm{H}^0(\mathcal{O}_X(-e_0 + H)) = 0$ and from the exact sequence (3.8) we get the following

exact sequence:

$$0 \longrightarrow \mathrm{H}^0(\mathscr{E}(H)) \longrightarrow \mathrm{H}^1(\mathscr{O}_X(-2e_0+H)^r)$$
$$\longrightarrow \mathrm{H}^1(\mathscr{O}_X(-e_0+H)^{2r}) \longrightarrow \mathrm{H}^1(\mathscr{E}(H)) \longrightarrow 0.$$

From this sequence and the fact that $\mathrm{h}^1(\mathscr{O}_X(-2e_0+H)) = -\chi(\mathscr{O}_X(-2e_0+H)) = 5$ and $\mathrm{h}^1(\mathscr{O}_X(-e_0+H)) = -\chi(\mathscr{O}_X(-e_0+H)) = 2$ we are forced to conclude that $\mathrm{h}^0(\mathscr{E}(H)) = r$ and $\mathrm{H}^1(\mathscr{E}(H)) = 0$. Now, from what we have gathered up to now, we can affirm that $\mathscr{E}(H)$ is 1-regular with respect to the very ample line bundle $2H$ and therefore, $\mathrm{H}^1(\mathscr{E}(H+2tH)) = 0$ for all $t \geq 0$. In order to deal with the cancelation of the remaining groups of cohomology, it will be enough to show that $\mathscr{E}(2H)$ is 1-regular with respect to $2H$, i.e., it remains to show that $\mathrm{H}^1(\mathscr{E}(2H)) = 0$. In order to do this consider the exact sequence (the cancelation of $\mathrm{H}^0(\mathscr{O}_X(-e_0+2H))$ is due to the fact that the points are in general position):

$$0 \longrightarrow \mathrm{H}^0(\mathscr{E}(2H)) \longrightarrow \mathrm{H}^1(\mathscr{O}_X(-2e_0+2H)^r)$$
$$\longrightarrow \mathrm{H}^1(\mathscr{O}_X(-e_0+2H)^{2r}) \longrightarrow \mathrm{H}^1(\mathscr{E}(2H)) \longrightarrow 0.$$

Once again, we control the dimension of these vector spaces:

$$\mathrm{h}^1(\oplus^r \mathscr{O}_X(-2e_0+2H)) = -r\chi(\mathscr{O}_X(-2e_0+2H)) = 9r$$

and

$$\mathrm{h}^1(\oplus^{2r} \mathscr{O}_X(-e_0+2H)) = -2r\chi(\mathscr{O}_X(-e_0+2H)) = 6r.$$

Therefore we are forced to have $\mathrm{h}^0(\mathscr{E}(2H)) = 3r$ and $\mathrm{H}^1(\mathscr{E}(2H)) = 0$. Notice that in this case $\mathscr{E}(3H)$ is globally generated.

□

As an immediate consequence we get:

Theorem 3.3.31 *Let $X \subset \mathbb{P}^d$ be a smooth del Pezzo surface of degree d. Then for any $r \geq 2$ there exists a family of dimension r^2+1 of simple initialized Ulrich bundles of rank r on X. In particular, del Pezzo surfaces are of wild representation type.*

Proof See [45, Theorem 4.9]. □

In the last part of this subsection we consider the case of strong del Pezzo surfaces X, i.e. smooth del Pezzo surfaces with anticanonical divisor very ample. In this case, $-K_X$ provides an embedding $X \subseteq \mathbb{P}^d$, with $d = K_X^2$. Let $R := K[x_0, \dots, x_d]$ be the graded polynomial ring associated to $\mathbb{P}^d$. Using our results on Mustaţă's conjecture explained in the previous subsection, we are going to show that the

(r^2+1)-dimensional family of rank r initialized Ulrich bundles given in Theorem 3.3.31 could also be obtained through a version of Serre correspondence from a general set of $\frac{dr^2+(2-d)r}{2}$ points on X.

More precisely, as a particular case of Theorem 3.3.24, we have the following result:

Theorem 3.3.32 *Let $X \subseteq \mathbb{P}^d$ be a strong del Pezzo surface of degree d embedded in $\mathbb{P}^d$ by its very ample anticanonical divisor. Let $Z_{m(r)} \subset X$ be a general set of*

$$m(r) = \frac{1}{2}(dr^2 + (2-d)r)$$

points, $r \geq 2$. Then the minimal graded free resolution (as a R-module) of the saturated ideal of $Z_{m(r)}$ in X has the following form:

$$\begin{aligned} 0 \longrightarrow R(-r-d)^{r-1} \longrightarrow R(-r-d+2)^{\gamma_{d-1,r-1}} \longrightarrow \dots \\ \longrightarrow R(-r-1)^{\gamma_{2,r-1}} \longrightarrow R(-r)^{(d-1)r+1} \longrightarrow I(Z_{m(r)}|X) \longrightarrow 0 \end{aligned} \tag{3.12}$$

with

$$\gamma_{i,r-1} = \sum_{l=0}^{1}(-1)^l \binom{d-l-1}{i-l} \Delta^{l+1} P_X(r+l) - \binom{d}{i}(m(r) - P_X(r-1)).$$

Theorem 3.3.33 *Let $X \subseteq \mathbb{P}^d$ be a strong del Pezzo surface of degree d.*

1. *If $\mathscr{E}(H)$ is an Ulrich bundle of rank $r \geq 2$ given by the exact sequence (3.8), then there is an exact sequence*

$$0 \longrightarrow \mathcal{O}_X^{r-1} \longrightarrow \mathscr{E}(H) \longrightarrow I(Z|X)(rH) \longrightarrow 0$$

   where Z is a zero-dimensional scheme of degree $m(r) = c_2(\mathscr{E}(H)) = \frac{1}{2}(dr^2 + (2-d)r)$ and $\mathrm{h}^0(I(Z|X)(r-1)H) = 0$.
2. *Conversely, for general sets Z of $m(r) = 1/2(dr^2+(2-d)r)$ points on X, $r \geq 2$, we recover the initialized Ulrich bundles given by the exact sequence (3.8) as an extension of $I(Z|X)(rH)$ by $\mathcal{O}_X^{r-1}$.*

Proof

1. As $\mathscr{E}(H)$ is globally generated, $r-1$ general global sections define an exact sequence of the form

$$0 \longrightarrow \mathcal{O}_X^{r-1} \longrightarrow \mathscr{E}(H) \longrightarrow I(Z|X)(D) \longrightarrow 0$$

where $D = c_1(\mathscr{E}(H)) = rH$ is a divisor on X and Z is a zero-dimensional scheme of length

$$c_2(\mathscr{E}(H)) = \frac{dr^2 + (2-d)r}{2}.$$

Moreover, since $\mathscr{E}(H)$ is initialized, $h^0(I(Z|X)(r-1)H) = 0$.

2. Let Z be a general set of points of cardinality $m(r)$ with the minimal free resolution of (3.12). Let us denote by R_X and R_Z the homogeneous coordinate ring of X and Z. It is well-known that for ACM varieties, there exists a bijection between ACM bundles on X and Maximal Cohen Macaulay (MCM from now on) graded R_X-modules sending $\mathscr{E}$ to $H^0_*(\mathscr{E})$. From the exact sequence

$$0 \longrightarrow I(Z|X) \longrightarrow R_X \longrightarrow R_Z \longrightarrow 0$$

we get $\mathrm{Ext}^1(I(Z|X), R_X(-1)) \cong \mathrm{Ext}^2(R_Z, R_X(-1)) \cong K_Z$ where K_Z denotes the canonical module of R_Z (the last isomorphism is due to the fact that $R_X(-1)$ is the canonical module of X and the codimension of Z in X is 2). Dualizing the exact sequence (3.12), we obtain a minimal resolution of K_Z:

$$\dots \longrightarrow R(r-3)^{\gamma_{d-1,r-1}} \longrightarrow R(r-1)^{r-1} \longrightarrow K_Z \longrightarrow 0.$$

This shows that K_Z is generated in degree $1-r$ by $r-1$ elements. These generators provide an extension

$$0 \longrightarrow R_X^{r-1} \longrightarrow F \longrightarrow I(Z|X)(r) \longrightarrow 0 \tag{3.13}$$

via the isomorphism $K_Z \cong \mathrm{Ext}^1(I(Z|X), R_X(-1))$. F turns out to be a MCM module because $\mathrm{Ext}^1(F, K_X) = 0$ (this last cancelation follows by applying $\mathrm{Hom}_{R_X}(-, K_X)$ to (3.13)). If we sheafiffy the exact sequence (3.13) we obtain the sequence

$$0 \longrightarrow \mathscr{O}_X^{r-1} \longrightarrow \widetilde{F} \longrightarrow I(Z|X)(r) \longrightarrow 0$$

where $\widetilde{F}$ is an ACM vector bundle on X. Using the exact sequence (3.12) we can see that $H^0(I(Z|X)(r-1)H) = 0$ and $h^0(I(Z|X)(rH)) = (d-1)r+1$. Therefore $\widetilde{F}$ is an initialized Ulrich bundle (i.e., $h^0(\widetilde{F}) = dr$). By Theorem 3.2.15, $\widetilde{F}$ will be globally generated.

It only remains to show that for a generic choice of $Z_{m(r)} \subset X$, the associated bundle $\mathscr{F} := \widetilde{F}$ just constructed belongs to the family (3.8). Since $\mathscr{F}$ is an initialized Ulrich bundle of rank r with the expected Chern classes, the problem boils down to a dimension counting. We need to show that the dimension of the family of vector bundles obtained through this construction from a general set $Z_{m(r)}$ is r^2+1. Since this dimension is given by the formula $\dim Hilb^{m(r)}(X) - \dim \mathrm{Grass}(h^0(\mathscr{F}), r-1)$, an easy computation taking into account that $\dim Hilb^{m(r)}(X) = 2m(r)$ and that $\dim \mathrm{Grass}(\mathrm{h}^0(\mathscr{F}), r-1) = (r-1)(dr-r+1)$ gives the desired result. □

As a nice application we get:

Theorem 3.3.34 *Let X be a smooth del Pezzo surface of degree d. Then for any $r \geq 2$ there exists a family of dimension r^2+1 of simple Ulrich bundles of rank r with Chern classes $c_1 = rH$ and $c_2 = \frac{dr^2+r(2-d)}{2}$.*

So, we conclude:

Theorem 3.3.35 *Smooth del Pezzo surfaces $X \subset \mathbb{P}^d$ are of wild representation type.*

3.4 The Representation Type of a Segre Variety

Fix integers $1 \leq n_1, \cdots, n_s$ and set $N := \prod_{i=1}^{s}(n_i+1)-1$. The goal of this section is to prove that all Segre varieties $\Sigma_{n_1,\dots,n_s} \subseteq \mathbb{P}^N$ unless the quadric surface in $\mathbb{P}^3$ support families of arbitrarily large dimension and rank of simple Ulrich (and hence ACM) vector bundles. Therefore, they are all unless $\mathbb{P}^1 \times \mathbb{P}^1$ of wild representation type. To this end, we will give an effective method to construct ACM sheaves (i.e. sheaves without intermediate cohomology) with the maximal permitted number of global sections, the so-called Ulrich sheaves, on all Segre varieties $\Sigma_{n_1,\cdots,n_s}$ other than $\mathbb{P}^1 \times \mathbb{P}^1$. To our knowledge, they will be the first family of examples of varieties of arbitrary dimension for which wild representation type is witnessed by means of Ulrich bundles.

Let us start this section recalling the definition of Segre variety and the basic properties on Segre varieties needed later on. Given integers $1 \leq n_1, \cdots, n_s$, we denote by

$$\sigma_{n_1,\cdots,n_s} : \mathbb{P}^{n_1} \times \cdots \times \mathbb{P}^{n_s} \longrightarrow \mathbb{P}^N, \quad N = \prod_{i=1}^{s}(n_i+1) - 1$$

the Segre embedding of $\mathbb{P}^{n_1} \times \cdots \times \mathbb{P}^{n_s}$. The image of $\sigma_{n_1,\cdots,n_s}$ is the Segre variety $\Sigma_{n_1,\cdots,n_s} := \sigma_{n_1,\cdots,n_s}(\mathbb{P}^{n_1} \times \cdots \times \mathbb{P}^{n_s}) \subseteq \mathbb{P}^N$, $N = \prod_{i=1}^{s}(n_i+1) - 1$. Notice that in terms of very ample line bundles, this embedding is defined by means of $\mathscr{O}_{\mathbb{P}^{n_1} \times \cdots \times \mathbb{P}^{n_s}}(1, \cdots, 1)$.

The equations of the Segre varieties are familiar to anyone who has studied Algebraic Geometry. Indeed, if we let T be the $(n_1+1) \times \cdots \times (n_s+1)$ tensor whose entries are the homogeneous coordinates in $\mathbb{P}^N$, then it is well known that the ideal of $\Sigma_{n_1,\cdots,n_s}$ is generated by the 2×2 minors of T. Moreover, we have

Proposition 3.4.1 *Fix integers* $1 \leq n_1, \cdots, n_s$ *and denote by* $\Sigma_{n_1,\cdots,n_s} \subseteq \mathbb{P}^N$, $N = \prod_{i=1}^{s}(n_i+1) - 1$, *the Segre variety. It holds:*

1. $\dim(\Sigma_{n_1,\cdots,n_s}) = \sum_{i=1}^{s} n_i$,
2. $\deg(\Sigma_{n_1,\cdots,n_s}) = \frac{(\sum_{i=1}^{s} n_i)!}{\prod_{i=1}^{s}(n_i)!}$,
3. $\Sigma_{n_1,\cdots,n_s}$ *is ACM, and*
4. $I(\Sigma_{n_1,\cdots,n_s})$ *is generated by* $\binom{N+2}{2} - \prod_{i=1}^{s}\binom{n_i+2}{2}$ *hyperquadrics.*

Example 3.4.2

1. We consider the Segre embedding

$$\sigma_{1,1} : \mathbb{P}^1 \times \mathbb{P}^1 \longrightarrow \mathbb{P}^3$$
$$((a,b),(c,d)) \mapsto (ac, ad, bc, bd).$$

Set $\Sigma_{1,1} := \sigma_{1,1}(\mathbb{P}^1 \times \mathbb{P}^1)$. If we fix coordinates x, y, z, t in $\mathbb{P}^3$, we have: $I(\Sigma_{1,1}) = (xt - yz)$, $\dim(\Sigma_{1,1}) = 2$, $\deg(\Sigma_{1,1}) = 2$ and $\operatorname{Pic}(\Sigma_{1,1}) = \mathbb{Z}^2$.

2. We consider the Segre embedding

$$\sigma_{2,3} : \mathbb{P}^2 \times \mathbb{P}^3 \longrightarrow \mathbb{P}^{11}$$
$$((a,b,c),(d,e,f,g)) \mapsto (ad, ae, af, ag, \cdots, cg).$$

Set $\Sigma_{2,3} := \sigma_{2,3}(\mathbb{P}^2 \times \mathbb{P}^3)$. If we fix coordinates $x_{0,0}, x_{0,1}, \cdots, x_{2,3}$ in $\mathbb{P}^{11}$, we have: $\Sigma_{2,3}$ is an ACM variety and its ideal $I(\Sigma_{2,3})$ is generated by 18 hyperquadrics. In fact, $\Sigma_{2,3}$ is a determinantal variety defined by the 2×2 minors of the matrix

$$M = \begin{bmatrix} x_{0,0} & x_{0,1} & x_{0,2} & x_{0,3} \\ x_{1,0} & x_{1,1} & x_{1,2} & x_{1,3} \\ x_{2,0} & x_{2,1} & x_{2,2} & x_{2,3} \end{bmatrix}.$$

Moreover, $\dim(\Sigma_{2,3}) = 5$, $\deg(\Sigma_{2,3}) = 10$ and $\operatorname{Pic}(\Sigma_{2,3}) = \mathbb{Z}^2$.

Let p_i denote the i-th projection of $\mathbb{P}^{n_1} \times \cdots \times \mathbb{P}^{n_s}$ onto $\mathbb{P}^{n_i}$. There is a canonical isomorphism $\mathbb{Z}^s \longrightarrow \operatorname{Pic}(\Sigma_{n_1,\cdots,n_s})$, given by

$$(a_1,\cdots,a_s) \mapsto \mathcal{O}_{\Sigma_{n_1,\cdots,n_s}}(a_1,\cdots,a_s) := p_1^*(\mathcal{O}_{\mathbb{P}^{n_1}}(a_1)) \otimes \cdots \otimes p_s^*(\mathcal{O}_{\mathbb{P}^{n_s}}(a_s)).$$

For any coherent sheaves $\mathcal{E}_i$ on $\mathbb{P}^{n_i}$, we set $\mathcal{E}_1 \boxtimes \cdots \boxtimes \mathcal{E}_s := p_1^*(\mathcal{E}_1) \otimes \cdots \otimes p_s^*(\mathcal{E}_s)$. We will denote by $\pi_i : \mathbb{P}^{n_1} \times \cdots \times \mathbb{P}^{n_s} \longrightarrow X_i := \mathbb{P}^{n_1} \times \cdots \times \widehat{\mathbb{P}^{n_i}} \times \cdots \times \mathbb{P}^{n_s}$ the natural projection and given sheaves $\mathcal{E}$ and $\mathcal{F}$ on X_i and $\mathbb{P}^{n_i}$, respectively, $\mathcal{E} \boxtimes \mathcal{F}$ stands for $\pi_i^*(\mathcal{E}) \otimes p_i^*(\mathcal{F})$. By the Künneth's formula, we have

$$\mathrm{H}^{\ell}(\Sigma_{n_1,\cdots,n_s}, \mathcal{E} \boxtimes \mathcal{F}) = \bigoplus_{p+q=\ell} \mathrm{H}^p(X_i, \mathcal{E}) \otimes \mathrm{H}^q(\mathbb{P}^{n_i}, \mathcal{F}).$$

While given a coherent sheaf $\mathscr{H}$ on $\Sigma_{n_1,\cdots,n_s}$, $\mathscr{H}(t)$ stands for $\mathscr{H} \otimes \mathscr{O}_{\Sigma_{n_1,\cdots,n_s}}(t,\cdots,t)$.

Let us start by determining the complete list of initialized Ulrich line bundles on Segre varieties $\Sigma_{n_1,\cdots,n_s} \subseteq \mathbb{P}^N$, $N = \prod_{i=1}^s (n_i+1) - 1$. First of all, notice that it follows from Horrocks' Theorem [30] that

Lemma 3.4.3 *The only initialized Ulrich bundle on $\mathbb{P}^n$ is the structural sheaf $\mathscr{O}_{\mathbb{P}^n}$.*

The list of initialized Ulrich line bundles on $\Sigma_{n_1,\cdots,n_s} \subseteq \mathbb{P}^N$, $N = \prod_{i=1}^s (n_i + 1) - 1$, is given by

Proposition 3.4.4 *Let $\Sigma_{n_1,\cdots,n_s} \subseteq \mathbb{P}^N$, $N = \prod_{i=1}^s (n_i+1) - 1$, be a Segre variety. Then there exist $s!$ initialized Ulrich line bundles on $\Sigma_{n_1,\cdots,n_s}$. They are of the form*

$$\mathscr{L}_{X_i} \boxtimes \mathscr{O}_{\mathbb{P}^{n_i}}(\sum_{k\neq i} n_k),$$

where $\mathscr{L}_{X_i}$ is a rank one initialized Ulrich bundle on the Segre variety $X_i := \Sigma_{n_1,\cdots,\widehat{n_i},\cdots,n_s} \subseteq \mathbb{P}^{N'}$, $N' = \prod_{\substack{1\le j\le s\\ j\neq i}} (n_j+1) - 1$. More explicitly, the initialized Ulrich line bundles on $\Sigma_{n_1,\cdots,n_s}$ are of the form $\mathscr{O}_{\Sigma_{n_1,\cdots,n_s}}(a_1,\ldots,a_s)$ where, if we order the coefficients $0 = a_{i_1} \le \cdots \le a_{i_k} \le \cdots \le a_{i_s}$ then $a_{i_k} = \sum_{1\le j<k} n_{i_j}$.

Proof The existence of this set of initialized Ulrich line bundles is a straightforward consequence of [21, Proposition 2.6]. In order to see that this list is exhaustive, let us consider an initialized Ulrich line bundle $\mathscr{L} := \mathscr{O}_{\Sigma_{n_1,\cdots,n_s}}(a_1,\ldots,a_s)$ with $a_{i_1} \le \cdots \le a_{i_k} \le \cdots \le a_{i_s}$. Given that $\mathscr{L}$ is initialized, it holds that $a_{i_1} = 0$. Since $\mathscr{L}$ is ACM, we have

$$\mathrm{H}^{\sum_{j=1}^k n_{i_j}}(\Sigma_{n_1,\cdots,n_s}, \mathscr{L}(-\Sigma_{j=1}^k n_{i_j} - 1)) = 0$$

for $k = 1,\ldots,s-1$. In particular, using Künneth's formula, it holds

$$\prod_{l=1}^k \mathrm{h}^{n_{i_l}}(\mathbb{P}^{n_{i_l}}, \mathscr{O}_{\mathbb{P}^{n_{i_l}}}(a_{i_l} - \Sigma_{j=1}^k n_{i_j} - 1)) \cdot \prod_{l=k+1}^s \mathrm{h}^0(\mathbb{P}^{n_{i_l}}, \mathscr{O}_{\mathbb{P}^{n_{i_l}}}(a_{i_l} - \Sigma_{j=1}^k n_{i_j} - 1)) = 0,$$

from where it follows that, by induction, $a_{i_{k+1}} \le b_{i_{k+1}} := \Sigma_{1\le j\le k} n_{i_j}$ for $k = 1,\ldots,s-1$ (and $b_{i_1} := 0$). But, on the other hand, since an easy computation shows that

$$\mathrm{h}^0(\Sigma_{n_1,\cdots,n_s}, \mathscr{O}_{\Sigma_{n_1,\cdots,n_s}}(b_1,\ldots,b_s)) = \frac{(\sum_{i=1}^s n_i)!}{\prod_{i=1}^s (n_i)!} = \deg(\Sigma_{n_1,\cdots,n_s})$$

we are forced to have $a_{i_j} = b_{i_j}$ for $j = 1,\ldots,s$. □

Corollary 3.4.5 *$\mathscr{O}_{\Sigma_{n,m}}(a,b)$ is an initialized Ulrich line bundle on $\Sigma_{n,m}$ if and only if $(a,b) = (0,n)$ or $(m,0)$.*

It is natural to ask if we could use these initialized Ulrich line bundles as a bricks to construct initialized Ulrich bundles of higher rank. The answer strongly depends on the values of n_i. Assume for a while that $i = 2$, take $n = n_1$, $m = n_2$ and assume $n \leq m$. The main difference between the case $n = 1$ and $1 < n$ comes from:

$$\mathrm{Ext}^1_{\Sigma_{n,m}}(\mathscr{O}(m,0), \mathscr{O}(0,n)) \neq 0 \Leftrightarrow n = 1 \text{ and } m \geq 2.$$

So, if $1 = n < m$, we can construct a rank 2 undecomposable Ulrich bundle $\mathscr{E}$ on $\Sigma_{n,m}$ taking a non-trivial extension $0 \neq e \in \mathrm{Ext}^1_{\Sigma_{n,m}}(\mathscr{O}(m,0), \mathscr{O}(0,n))$:

$$0 \to \mathscr{O}(0,n) \to \mathscr{E} \to \mathscr{O}(m,0) \to 0.$$

Iterating the process we will be able to construct Ulrich bundles of higher rank. If $2 \leq n \leq m$ we will need an alternative construction. So, we will distinguish to cases:

1. Case 1: $2 \leq n_1, \cdots, n_s$.
2. Case 2: $1 = n_1 \leq n_2, \cdots, n_s$.

3.4.1 Representation Type of $\Sigma_{n_1,\cdots,n_s}$, $2 \leq n_1, \cdots, n_s$

The goal of this subsection is the construction of families of arbitrarily large dimension of simple (and, hence, undecomposable) Ulrich vector bundles on Segre varieties $\Sigma_{n_1,\cdots,n_s} \subseteq \mathbb{P}^N$, $N = \prod_{i=1}^s (n_i + 1) - 1$, for $2 \leq n_1, \cdots, n_s$.

For any $2 \leq m$ and any $1 \leq a$, we denote by $\mathscr{E}_{m,a}$ any vector bundle on $\mathbb{P}^m$ given by the exact sequence

$$0 \to \mathscr{E}_{m,a} \to \mathscr{O}_{\mathbb{P}^m}(1)^{(m+2)a} \stackrel{\phi(1)}{\to} \mathscr{O}_{\mathbb{P}^m}(2)^{2a} \to 0 \tag{3.14}$$

where $\phi \in V_m$ and V_m is the non-empty open dense subset of the affine scheme $M = \mathrm{Hom}(\mathscr{O}_{\mathbb{P}^m}^{(m+2)a}, \mathscr{O}_{\mathbb{P}^m}(1)^{2a})$ provided by Proposition 3.3.28.

Note that $\mathscr{E}_{m,a}$ has rank ma and in the next Proposition we summarize the properties of these vector bundles needed later:

Proposition 3.4.6 *With the above notation we have:*

1.

$$\mathrm{h}^0(\mathbb{P}^m, \mathscr{E}_{m,a}(t)) = \begin{cases} 0 & \text{for } t \leq 0, \\ a((m+2)\binom{m+t+1}{m} - 2\binom{m+t+2}{m}) & \text{for } t > 0. \end{cases}$$

2.

$$h^1(\mathbb{P}^m, \mathcal{E}_{m,a}(t)) = \begin{cases} 0 & \text{for } t < -2 \text{or } t \geq 0, \\ am & \text{for } t = -1, \\ 2a & \text{for } t = -2. \end{cases}$$

3. $h^i(\mathbb{P}^m, \mathcal{E}_{m,a}(t)) = 0$ *for all* $t \in \mathbb{Z}$ *and* $2 \leq i \leq m-1$.
4. $h^m(\mathbb{P}^m, \mathcal{E}_{m,a}(t)) = 0$ *for* $t \geq -m-1$.
5. $\mathcal{E}_{m,a}$ *is simple.*

Proof

1.–4. Since $\phi \in V_m$, by Proposition 3.3.28, $H^0(\phi(1))$ is surjective. But, since the K-vector spaces $H^0(\mathbb{P}^m, \mathcal{O}_{\mathbb{P}^m}(1)^{(m+2)a})$ and $H^0(\mathbb{P}^m, \mathcal{O}_{\mathbb{P}^m}(2)^{2a})$ have the same dimension, $H^0(\phi(1))$ is an isomorphism and therefore $H^0(\mathcal{E}_{m,a}) = 0$. *A fortiori*, $H^0(\mathcal{E}_{m,a}(t)) = 0$ for $t \leq 0$. On the other hand, again by the surjectivity of $H^0(\phi(1))$, $H^1(\mathcal{E}_{m,a}) = 0$. Since it is obvious that $H^i(\mathcal{E}_{m,a}(1-i)) = 0$ for $i \geq 2$ it turns out that $\mathcal{E}_{m,a}$ is 1-regular and in particular, $H^1(\mathcal{E}_{m,a}(t)) = 0$ for $t \geq 0$. The rest of cohomology groups can be easily deduced from the long exact cohomology sequence associated to the exact sequence (3.14).

5. It follows from Kac's theorem (see [31, Theorem 4]) arguing as in [45, Proposition 3.4] that $\mathcal{E}_{m,a}$ is simple. □

We are now ready to construct families of simple (hence undecomposable) Ulrich bundles on the Segre variety $\Sigma_{n,m} \subseteq \mathbb{P}^{nm+n+m}$, $2 \leq n, m$, of arbitrary high rank and dimension and to conclude that Segre varieties $\Sigma_{n,m}$ are of wild representation type. The main ingredient on the construction of simple Ulrich bundles on $\Sigma_{n,m} \subseteq \mathbb{P}^{nm+n+m}$, $2 \leq n \leq m$, will be the family of simple vector bundles $\mathcal{E}_{m,a}$ on $\mathbb{P}^m$ given by the exact sequence (3.14) as well as the vector bundles of p-holomorphic forms of $\mathbb{P}^n$, $\Omega^p_{\mathbb{P}^n} := \wedge^p \Omega^1_{\mathbb{P}^n}$, where $\Omega^1_{\mathbb{P}^n}$ is the cotangent bundle. The values of $h^i(\Omega^p_{\mathbb{P}^n}(t))$ are given by the *Bott's formula* (see, for instance, [47, p. 8]).

Theorem 3.4.7 *Fix integers* $2 \leq n \leq m$ *and let* $\Sigma_{n,m} \subseteq \mathbb{P}^{nm+n+m}$ *be the Segre variety. For any integer* $a \geq 1$ *there exists a family of dimension* $a^2(m^2+2m-4)+1$ *of initialized simple Ulrich vector bundles* $\mathcal{F} := \Omega^{n-2}_{\mathbb{P}^n}(n-1) \boxtimes \mathcal{E}_{m,a}(n-1)$ *of rank* $am\binom{n}{2}$.

Proof Let $\mathcal{F}$ be the vector bundle $\Omega^{n-2}_{\mathbb{P}^n}(n-1) \boxtimes \mathcal{E}_{m,a}(n-1)$ for $\mathcal{E}_{m,a}$ a general vector bundle obtained on $\mathbb{P}^m$ from the exact sequence (3.14). The first goal is to prove that $\mathcal{F}$ is ACM, namely, we should show that $H^i(\Sigma_{n,m}, \mathcal{F} \otimes \mathcal{O}_{\Sigma_{n,m}}(t,t)) = 0$ for $1 \leq i \leq n+m-1$ and $t \in \mathbb{Z}$. By Künneth's formula

$$H^i(\Sigma_{n,m}, \mathcal{F} \otimes \mathcal{O}_{\Sigma_{n,m}}(t,t)) = \bigoplus_{p+q=i} H^p(\mathbb{P}^n, \Omega^{n-2}_{\mathbb{P}^n}(n-1+t)) \otimes H^q(\mathbb{P}^m, \mathcal{E}_{m,a}(n-1+t)). \tag{3.15}$$

According to Bott's formula the only non-zero cohomology groups of $\Omega^{n-2}_{\mathbb{P}^n}(n-1+t)$ are:

$$\begin{array}{ll} \mathrm{H}^0(\mathbb{P}^n, \Omega^{n-2}_{\mathbb{P}^n}(n-1+t)) & \text{for } t \geq 0 \text{ and } n \geq 3 \text{ or } t \geq -1 \text{ and } n = 2, \\ \mathrm{H}^{n-2}(\mathbb{P}^n, \Omega^{n-2}_{\mathbb{P}^n}(n-1+t)) & \text{for } t = -n+1, \\ \mathrm{H}^n(\mathbb{P}^n, \Omega^{n-2}_{\mathbb{P}^n}(n-1+t)) & \text{for } t \leq -n-2. \end{array}$$

On the other hand, by Lemma 3.4.6, the only non-zero cohomology groups of $\mathscr{E}_{m,a}(n-1+t)$ are:

$$\begin{array}{ll} \mathrm{H}^0(\mathbb{P}^m, \mathscr{E}_{m,a}(n-1+t)) & \text{for } t \geq -n+2, \\ \mathrm{H}^1(\mathbb{P}^m, \mathscr{E}_{m,a}(n-1+t)) & \text{for } -n-1 \leq t \leq -n, \\ \mathrm{H}^m(\mathbb{P}^m, \mathscr{E}_{m,a}(n-1+t)) & \text{for } t \leq -n-m-1. \end{array}$$

Therefore, using (3.15), we get

$$\mathrm{H}^i(\Sigma_{n,m}, \mathscr{F} \otimes \mathscr{O}_{\Sigma_{n,m}}(t,t)) = 0 \text{ for } 1 \leq i \leq n+m-1 \text{ and } t \in \mathbb{Z}.$$

Since for $n \geq 3$ $\mathrm{H}^0(\mathbb{P}^n, \Omega^{n-2}_{\mathbb{P}^n}(n-2)) = 0$ and for $n = 2$ $\mathrm{H}^0(\mathbb{P}^m, \mathscr{E}_{m,a}) = 0$ (Lemma 3.4.6), $\mathscr{F}$ is an initialized ACM vector bundle on $\Sigma_{n,m}$. Let us compute the number of global sections. Recall that, by Bott's formula, $\mathrm{h}^0(\mathbb{P}^n, \Omega^{n-2}_{\mathbb{P}^n}(n-1)) = \binom{n+1}{2}$. Hence:

$$\begin{aligned} \mathrm{h}^0(\mathscr{F}) &= \mathrm{h}^0(\Sigma_{n,m}, \Omega^{n-2}_{\mathbb{P}^n}(n-1) \boxtimes \mathscr{E}_{m,a}(n-1)) \\ &= \mathrm{h}^0(\mathbb{P}^n, \Omega^{n-2}_{\mathbb{P}^n}(n-1))\mathrm{h}^0(\mathbb{P}^m, \mathscr{E}_{m,a}(n-1)) \\ &= \tbinom{n+1}{2} a((m+2)\tbinom{m+n}{m} - 2\tbinom{m+n+1}{m}) \\ &= a(\tfrac{(m+2)(m+n)!(n+1)!}{m!n!(n-1)!2!} - \tfrac{2(m+n+1)!(n+1)!}{m!(n+1)!(n-1)!2!}) \\ &= a(\tfrac{n!(m+n)!}{2!(n-2)!m!n!} \cdot \tfrac{(n+1)(m+2)-2(m+n+1)}{n-1}) \\ &= a\tbinom{n}{2}\tbinom{m+n}{m}\tfrac{m(n-1)}{n-1} \\ &= a\tbinom{n}{2}\tbinom{m+n}{m}m \\ &= \mathrm{rk}(\mathscr{F})\deg(\Sigma_{n,m}) \end{aligned}$$

where the last equality follows from the fact that $\deg(\Sigma_{n,m}) = \binom{m+n}{m}$ and $\mathrm{rk}(\mathscr{F}) = \mathrm{rk}(\mathscr{E}_{m,a})\mathrm{rk}(\Omega^{n-2}_{\mathbb{P}^n}) = am\binom{n}{2}$. Therefore, $\mathscr{F}$ is an initialized Ulrich vector bundle on $\Sigma_{n,m}$. With respect to simplicity, we need only to observe that

$$\begin{aligned} \mathrm{Hom}(\mathscr{F}, \mathscr{F}) &\cong \mathrm{H}^0(\Sigma_{n,m}, \mathscr{F}^\vee \otimes \mathscr{F}) \\ &\cong \mathrm{H}^0(\mathbb{P}^n, \Omega^{n-2}_{\mathbb{P}^n}(n-1)^\vee \otimes \Omega^{n-2}_{\mathbb{P}^n}(n-1))) \\ &\quad \otimes \mathrm{H}^0(\mathbb{P}^m, \mathscr{E}_{m,a}(n-1)^\vee \otimes \mathscr{E}_{m,a}(n-1)) \end{aligned}$$

and use the fact that $\Omega^{n-2}_{\mathbb{P}^n}$ and $\mathscr{E}_{m,a}$ are both simple.

It only remains to compute the dimension of the family of simple Ulrich bundles $\mathscr{F} := \Omega^{n-2}_{\mathbb{P}^n}(n-1) \boxtimes \mathscr{E}_{m,a}(n-1)$ on $\Sigma_{n,m}$. Since they are completely determined by a general morphism $\phi \in M := \mathrm{Hom}_{\mathbb{P}^m}(\mathscr{O}^{(m+2)a}_{\mathbb{P}^m}, \mathscr{O}_{\mathbb{P}^m}(1)^{2a})$, this dimension turns out to be:

$$\dim M - \dim \mathrm{Aut}(\mathscr{O}^{(m+2)a}_{\mathbb{P}^m}) - \dim \mathrm{Aut}(\mathscr{O}_{\mathbb{P}^m}(1)^{2a}) + 1$$

$$= 2a^2(m+2)(m+1) - a^2(m+2)^2 - 4a^2 + 1 = a^2(m^2+2m-4) + 1$$

which proves what we want. □

Corollary 3.4.8 *For any integers $2 \leq n, m$, the Segre variety $\Sigma_{n,m} \subseteq \mathbb{P}^{nm+n+m}$ is of wild representation type.*

Notice that in Theorem 3.4.7 we were able to construct simple Ulrich vector bundles on $\Sigma_{n,m} \subseteq \mathbb{P}^N$ for some scattered ranks, namely for ranks of the form $am\binom{n}{2}$, $a \geq 1$. The next goal will be to construct simple Ulrich bundles on $\Sigma_{n,m} \subseteq \mathbb{P}^{nm+n+m}$, $2 \leq n \leq m$, of the remaining ranks $r \geq m\binom{n}{2}$.

Theorem 3.4.9 *Fix integers $2 \leq n \leq m$ and let $\Sigma_{n,m} \subseteq \mathbb{P}^{nm+n+m}$ be the Segre variety. For any integer $r \geq m\binom{n}{2}$, set $r = am\binom{n}{2}+\ell$ with $a \geq 1$ and $0 \leq \ell \leq m\binom{n}{2}-1$. Then, there exists a family of dimension $a^2(m^2+2m-4)+1+\ell(am\binom{n+1}{2}-\ell)$ of simple (hence, undecomposable) initialized Ulrich vector bundles $\mathscr{G}$ on $\Sigma_{n,m}$ of rank r.*

Proof Note that for any $r \geq m\binom{n}{2}$, there exists $a \geq 1$ and $m\binom{n}{2} - 1 \geq \ell \geq 0$, such that $r = am\binom{n}{2}+\ell$. For such a, consider the family $\mathscr{P}_a$ of initialized Ulrich bundles of rank $am\binom{n}{2}$ given by Theorem 3.4.7. Notice that

$$\dim \mathscr{P}_a = a^2(m^2+2m-4)+1.$$

Hence it is enough to consider the case $\ell > 0$. To this end, for any $\ell > 0$ we construct the family $\mathscr{P}_{a,\ell}$ of vector bundles $\mathscr{G}$ given by a non-trivial extension

$$e : 0 \to \mathscr{F} \to \mathscr{G} \to \mathscr{O}_{\Sigma_{n,m}}(0,n)^{\ell} \to 0 \qquad (3.16)$$

where $\mathscr{F} \in \mathscr{P}_a$ and $e := (e_1, \ldots, e_\ell) \in \mathrm{Ext}^1(\mathscr{O}_{\Sigma_{n,m}}(0,n)^{\ell}, \mathscr{F}) \cong \mathrm{Ext}^1(\mathscr{O}_{\Sigma_{n,m}}(0,n), \mathscr{F})^{\ell}$ with $e_1, \ldots, e_\ell$ linearly independent.

Since

$$\begin{aligned} \mathrm{ext}^1(\mathscr{O}_{\Sigma_{n,m}}(0,n), \mathscr{F}) &= \mathrm{h}^1(\Sigma_{n,m}, \Omega^{n-2}_{\mathbb{P}^n}(n-1) \boxtimes \mathscr{E}(-1)) \\ &= \mathrm{h}^0(\mathbb{P}^n, \Omega^{n-2}_{\mathbb{P}^n}(n-1)) \cdot \mathrm{h}^1(\mathbb{P}^m, \mathscr{E}(-1)) \\ &= \binom{n+1}{2} am \\ &> m\binom{n}{2} \end{aligned}$$

such extension exists.

It is obvious that $\mathcal{G}$, being an extension of initialized Ulrich vector bundles, is also an initialized Ulrich vector bundle. Let us see that $\mathcal{G}$ is simple, i.e., $\mathrm{Hom}(\mathcal{G},\mathcal{G}) \cong K$. If we apply the functor $\mathrm{Hom}(-,\mathcal{G})$ to the exact sequence (3.16) we obtain:

$$0 \to \mathrm{Hom}(\mathcal{O}_{\Sigma_{n,m}}(0,n)^{\ell},\mathcal{G}) \to \mathrm{Hom}(\mathcal{G},\mathcal{G}) \to \mathrm{Hom}(\mathcal{F},\mathcal{G}).$$

On the other hand, if we apply $\mathrm{Hom}(\mathcal{F},-)$ to the same exact sequence we have

$$0 \to K \cong \mathrm{Hom}(\mathcal{F},\mathcal{F}) \to \mathrm{Hom}(\mathcal{F},\mathcal{G}) \to \mathrm{Hom}(\mathcal{F},\mathcal{O}_{\Sigma_{n,m}}(0,n)^{\ell}). \tag{3.17}$$

But

$$\begin{aligned}\mathrm{Hom}(\mathcal{F},\mathcal{O}_{\Sigma_{n,m}}(0,n)) &\cong \mathrm{Ext}^{n+m}(\mathcal{O}_{\Sigma_{n,m}}(0,n),\mathcal{F}(-n-1,-m-1)) \\ &\cong \mathrm{H}^{n+m}(\Sigma_{n,m},\mathcal{F}(-n-1,-m-n-1)) \\ &= \mathrm{H}^{n}(\mathbb{P}^n,\Omega^{n-2}_{\mathbb{P}^n}(-2)) \otimes \mathrm{H}^{m}(\mathbb{P}^m,\mathcal{E}(-m-2)) = 0\end{aligned} \tag{3.18}$$

by Serre's duality and Bott's formula. This implies that $\mathrm{Hom}(\mathcal{F},\mathcal{G}) \cong K$.

Finally, using the fact that $\mathrm{Hom}(\mathcal{O}_{\Sigma_{n,m}}(0,n),\mathcal{F}) \cong \mathrm{H}^0(\mathcal{F}(0,-n)) = 0$ and applying the functor $\mathrm{Hom}(\mathcal{O}_{\Sigma_{n,m}}(0,n),\cdot)$ to the short exact sequence (3.16), we obtain

$$\begin{aligned}0 &\longrightarrow \mathrm{Hom}(\mathcal{O}_{\Sigma_{n,m}}(0,n),\mathcal{G}) \longrightarrow \mathrm{Hom}(\mathcal{O}_{\Sigma_{n,m}}(0,n),\mathcal{O}_{\Sigma_{n,m}}(0,n)^{\ell}) \cong K^{\ell} \\ &\xrightarrow{\phi} \mathrm{Ext}^1(\mathcal{O}_{\Sigma_{n,m}}(0,n),\mathcal{F}) \longrightarrow \mathrm{Ext}^1(\mathcal{O}_{\Sigma_{n,m}}(0,n),\mathcal{G}).\end{aligned}$$

Since, by construction, the image of ϕ is the subvector space generated by $e_1,\dots,e_l$ it turns out that ϕ is injective and in particular $\mathrm{Hom}(\mathcal{O}_{\Sigma_{n,m}}(0,n),\mathcal{G}) = 0$. Summing up, $\mathrm{Hom}(\mathcal{G},\mathcal{G}) \cong K$, i.e., $\mathcal{G}$ is simple.

It only remains to compute the dimension of $\mathcal{P}_{a,l}$. Assume that there exist vector bundles $\mathcal{F},\mathcal{F}' \in \mathcal{P}_a$ giving rise to isomorphic bundles, i.e.:

$$\begin{array}{ccccccccc} 0 \to & \mathcal{F} & \xrightarrow{j_1} & \mathcal{G} & \xrightarrow{\alpha} & \mathcal{O}_{\Sigma_{n,m}}(0,n)^{\ell} & \to 0 \\ & & & i \,\|\wr & & & \\ 0 \to & \mathcal{F}' & \xrightarrow{j_2} & \mathcal{G}' & \xrightarrow{\beta} & \mathcal{O}_{\Sigma_{n,m}}(0,n)^{\ell} & \to 0. \end{array}$$

Since by (3.18), $\mathrm{Hom}(F,\mathcal{O}_{\Sigma_{n,m}}(0,n)) = 0$, the isomorphism i between $\mathcal{G}$ and $\mathcal{G}'$ lifts to an automorphism f of $\mathcal{O}_{\Sigma_{n,m}}(0,n)^{\ell}$ such that $f\alpha = \beta i$ which allows us to conclude that the morphism $ij_1 : \mathcal{F} \longrightarrow \mathcal{G}'$ factorizes through $\mathcal{F}'$ showing up the required isomorphism from $\mathcal{F}$ to $\mathcal{F}'$.

Therefore, since $\dim \operatorname{Hom}(\mathscr{F}, \mathscr{G}) = 1$, we have

$$\begin{aligned}\dim \mathscr{P}_{a,\ell} &= \dim \mathscr{P}_a + \dim \operatorname{Grass}(\ell, \operatorname{Ext}^1(\mathscr{O}_{\Sigma_{n,m}}(0,n), \mathscr{F})) \\ &= \dim \mathscr{P}_a + \ell \dim \operatorname{Ext}^1(\mathscr{O}_{\Sigma_{n,m}}(0,n), \mathscr{F}) - \ell^2 \\ &= a^2(m^2+2m-4)+1+\ell(am\binom{n+1}{2}-\ell).\end{aligned}$$

□

As a by-product of the previous results we can extend the construction of simple Ulrich bundles on $\Sigma_{n,m}$, $n \geq 2$, to the case of Segre embeddings of more than two factors and get:

Theorem 3.4.10 *Fix integers* $2 \leq n_1 \leq \cdots \leq n_s$ *and let* $\Sigma_{n_1,\ldots,n_s} \subseteq \mathbb{P}^N$, $N = \prod_{i=1}^s (n_i+1)-1$ *be a Segre variety. For any integer* $r \geq n_2\binom{n_1}{2}$, *set* $r = an_2\binom{n_1}{2}+\ell$ *with* $a \geq 1$ *and* $0 \leq \ell \leq n_2\binom{n_1}{2} - 1$. *Then there exists a family of dimension* $a^2(n_2^2+2n_2-4)+1+\ell(an_2\binom{n_1+1}{2}-\ell)$ *of simple (hence, undecomposable) initialized Ulrich vector bundles on* $\Sigma_{n_1,\ldots,n_s} \subseteq \mathbb{P}^N$ *of rank* r.

Proof By Theorem 3.4.7 we can suppose that $s \geq 3$. Therefore, by Eisenbud et al. [21, Proposition 2.6], the vector bundle of the form $\mathscr{H} := \mathscr{G} \boxtimes \mathscr{L}(n_1+n_2)$, for $\mathscr{G}$ belonging to the family constructed in Theorem 3.4.9 and $\mathscr{L}$ an Ulrich line bundle on $\mathbb{P}^{n_3} \times \cdots \times \mathbb{P}^{n_s}$ as constructed in Proposition 3.4.4, is an initialized simple Ulrich bundle. In order to show that in this way we obtain a family of the aforementioned dimension it only remains to show that whenever $\mathscr{G} \ncong \mathscr{G}'$ then $\mathscr{H} \ncong \mathscr{H}'$, or equivalently $\mathscr{G} \boxtimes \mathscr{O}_{\mathbb{P}^{n_3}\times\cdots\times\mathbb{P}^{n_s}} \ncong \mathscr{G}' \boxtimes \mathscr{O}_{\mathbb{P}^{n_3}\times\cdots\times\mathbb{P}^{n_s}}$. But if there exists an isomorphism

$$\phi : \mathscr{G} \boxtimes \mathscr{O}_{\mathbb{P}^{n_3}\times\cdots\times\mathbb{P}^{n_s}} \xrightarrow{\cong} \mathscr{G}' \boxtimes \mathscr{O}_{\mathbb{P}^{n_3}\times\cdots\times\mathbb{P}^{n_s}}$$

$\pi_*\phi$ would also be an isomorphism between

$$\pi_*(\mathscr{G} \boxtimes \mathscr{O}_{\mathbb{P}^{n_3}\times\cdots\times\mathbb{P}^{n_s}}) \cong \mathscr{G} \text{ and } \pi_*(\mathscr{G}' \boxtimes \mathscr{O}_{\mathbb{P}^{n_3}\times\cdots\times\mathbb{P}^{n_s}}) \cong \mathscr{G}'$$

in contradiction with the hypothesis. □

Corollary 3.4.11 *For any integers* $2 \leq n_1, \cdots, n_s$, *the Segre variety* $\Sigma_{n_1,\ldots,n_s} \subseteq \mathbb{P}^N$, $N = \prod_{i=1}^s (n_i+1)-1$ *is of wild representation type.*

3.4.2 Representation Type of $\Sigma_{n_1,n_2\ldots,n_s}$, $1 = n_1 \leq n_2, \cdots, n_s$

In this subsection we are going to focus our attention on the construction of simple Ulrich bundles on Segre varieties of the form $\Sigma_{n_1,n_2\ldots,n_s} \subseteq \mathbb{P}^N$ for either $n_1 = 1$ and $s \geq 3$ or $n_1 = 1$ and $n_2 \geq 2$. We are going to show that they also are of wild representation type. Opposite to the Segre varieties that we studied in the previous subsection, the Ulrich bundles on $\Sigma_{1,n_2\dots,n_s} \subseteq \mathbb{P}^N$, $N = 2\prod_{i=2}^s(n_i+1)-1$, will not be obtained as products of vector bundles constructed on each factor, but they will be obtained directly as iterated extensions.

Theorem 3.4.12 *Let $X := \Sigma_{1,n_2\dots,n_s} \subseteq \mathbb{P}^N$ for either $s \geq 3$ or $n_2 \geq 2$. Let r be an integer, $2 \leq r \leq (\Sigma_{i=2}^s n_i - 1)\prod_{i=2}^s(n_i+1)$. Then:*

1. *There exists a family Λ_r of rank r initialized simple Ulrich vector bundles $\mathscr{E}$ on X given by nontrivial extensions*

$$0 \longrightarrow \mathscr{O}_X(0,1,1+n_2,\dots,1+\Sigma_{i=2}^{s-1}n_i) \longrightarrow \mathscr{E} \longrightarrow \mathscr{O}_X(\Sigma_{i=2}^s n_i, 0, n_2, \dots, \Sigma_{i=2}^{s-1}n_i)^{r-1} \longrightarrow 0 \qquad (3.19)$$

with first Chern class $c_1(\mathscr{E}) = ((r-1)\Sigma_{i=2}^s n_i, 1, 1+rn_2, \dots, 1+r(\Sigma_{i=2}^{s-1}n_i))$.

2. *There exists a family Γ_r of rank r initialized simple Ulrich vector bundles $\mathscr{F}$ on X given by nontrivial extensions*

$$0 \longrightarrow \mathscr{O}_X(0,1+n_3,1,1+n_2+n_3,\dots,1+\Sigma_{i=2}^{s-1}n_i) \longrightarrow \mathscr{F} \longrightarrow \mathscr{O}_X(\Sigma_{i=2}^s n_i, n_3, 0, n_2+n_3, \dots, \Sigma_{i=2}^{s-1}n_i)^{r-1} \longrightarrow 0 \qquad (3.20)$$

with first Chern class $c_1(\mathscr{F}) = ((r-1)\Sigma_{i=2}^s n_i, 1+rn_3, 1, \dots, 1+r(\Sigma_{i=2}^{s-1}n_i))$.

Proof To simplify we set

$$\begin{aligned}
\mathscr{A} &:= \mathscr{O}_X(0,1,1+n_2,\dots,1+\Sigma_{i=2}^{s-1}n_i),\\
\mathscr{B} &:= \mathscr{O}_X(\Sigma_{i=2}^s n_i, 0, n_2, \dots, \Sigma_{i=2}^{s-1}n_i),\\
\mathscr{C} &:= \mathscr{O}_X(0,1+n_3,1,1+n_2+n_3,\dots,1+\Sigma_{i=2}^{s-1}n_i), \quad \text{and}\\
\mathscr{D} &:= \mathscr{O}_X(\Sigma_{i=2}^s n_i, n_3, 0, n_2+n_3, \dots, \Sigma_{i=2}^{s-1}n_i).
\end{aligned}$$

We are going to give the details of the proof of statement 1. since statement 2. is proved analogously. Recall that by Proposition 3.4.4, $\mathscr{A}$ and $\mathscr{B}$ are initialized Ulrich line bundles on X. On the other hand, the dimension of $\operatorname{Ext}^1(\mathscr{B},\mathscr{A})$ can be computed as:

$$\begin{aligned}
\dim \operatorname{Ext}^1(\mathscr{B},\mathscr{A}) &= \mathrm{h}^1(X, \mathscr{O}_X(-\Sigma_{i=2}^s n_i, 1, \dots, 1))\\
&= \mathrm{h}^1(\mathbb{P}^1, \mathscr{O}_{\mathbb{P}^1}(-\Sigma_{i=2}^s n_i))\prod_{i=2}^s \mathrm{h}^0(\mathbb{P}^{n_i}, \mathscr{O}_{\mathbb{P}^{n_i}}(1))\\
&= (\Sigma_{i=2}^s n_i - 1)\prod_{i=2}^s(n_i+1).
\end{aligned}$$

So, exactly as in the proof of Theorem 3.4.9, if we take ℓ ($\ell = r-1$) linearly independent elements $e_1,\dots,e_\ell$ in $\operatorname{Ext}^1(\mathscr{B},\mathscr{A})$, $1 \leq \ell \leq (\Sigma_{i=2}^s n_i - 1)\prod_{i=2}^s(n_i+1)-1$, these elements provide with an element $e := (e_1,\dots,e_\ell)$ of $\operatorname{Ext}^1(\mathscr{B}^\ell,\mathscr{A}) \cong \operatorname{Ext}^1(\mathscr{B},\mathscr{A})^\ell$. Then the associated extension

$$0 \longrightarrow \mathscr{A} \longrightarrow \mathscr{E} \longrightarrow \mathscr{B}^\ell \longrightarrow 0 \qquad (3.21)$$

gives a rank $\ell+1$ initialized simple Ulrich vector bundle. □

Remark

1. With the same technique, using other initialized Ulrich line bundles, it is possible to construct initialized simple Ulrich bundles of ranks covered by Theorem 3.4.12 with different first Chern class.
2. Notice that for $s = 2$, we have constructed rank r simple Ulrich vector bundles on $\Sigma_{1,m} \subseteq \mathbb{P}^{2m+1}$, $r \leq m^2$ as extensions of the form:

$$0 \longrightarrow \mathcal{O}_{\Sigma_{1,m}}(0,1) \longrightarrow \mathcal{E} \longrightarrow \mathcal{O}_{\Sigma_{1,m}}(m,0)^{r-1} \longrightarrow 0.$$

Lemma 3.4.13 *Consider the Segre variety $\Sigma_{1,n_2\dots,n_s} \subseteq \mathbb{P}^N$ for either $s \geq 3$ or $n_2 \geq 2$ and keep the notation introduced in Theorem 3.4.12. We have:*

1. *For any two non-isomorphic rank 2 initialized Ulrich bundles $\mathcal{E}$ and $\mathcal{E}'$ from the family Λ_2 obtained from the exact sequence (1), it holds that* $\operatorname{Hom}(\mathcal{E}, \mathcal{E}') = 0$. *Moreover, the set of non-isomorphic classes of elements of Λ_2 is parameterized by*

$$\mathbb{P}(\operatorname{Ext}^1(\mathcal{B}, \mathcal{A})) \cong \mathbb{P}(\mathrm{H}^1(\Sigma_{1,n_2\dots,n_s}, \mathcal{O}_{\Sigma_{1,n_2\dots,n_s}}(-\sum_{i=2}^{s} n_i, 1, \cdots, 1)))$$

and, in particular, it has dimension $(\Sigma_{i=2}^s n_i - 1)\prod_{i=2}^s (n_i + 1) - 1$.

2. *For any pair of bundles $\mathcal{E} \in \Lambda_2$ and $\mathcal{F} \in \Gamma_3$ obtained from the exact sequences (1) and (2), it holds that* $\operatorname{Hom}(\mathcal{E}, \mathcal{F}) = 0$ *and* $\operatorname{Hom}(\mathcal{F}, \mathcal{E}) = 0$.

Proof The first statement is a direct consequence of Proposition [49, Proposition 5.1.3]. Regarding the second statement, it is a straightforward computation applying the functors $\operatorname{Hom}(\mathcal{F}, -)$ and $\operatorname{Hom}(\mathcal{E}, -)$ to the short exact sequences (1) and (2) respectively, and taking into account that there are no nontrivial morphisms among the vector bundles $\mathcal{A}, \mathcal{B}, \mathcal{C}, \mathcal{D}$. □

In the next Theorem we are going to construct families of increasing dimension of simple Ulrich bundles for arbitrary large rank on the Segre variety $\Sigma_{1,n_2\dots,n_s}$. In case $s \geq 3$ we can use the two distinct families of rank 2 and rank 3 Ulrich bundles obtained in Theorem 3.4.12 to cover all the possible ranks. However, when $s = 2$, since there exists just a unique family, we will have to restraint ourselves to construct Ulrich bundles of arbitrary even rank. In any case, it will be enough to conclude that these Segre varieties are of wild representation type.

Theorem 3.4.14 *Consider the Segre variety $\Sigma_{1,n_2\dots,n_s} \subseteq \mathbb{P}^N$ for either $s \geq 3$ or $n_2 \geq 2$.*

1. *Then for any $r = 2t$, $t \geq 2$, there exists a family of dimension*

$$(2t-1)(\Sigma_{i=2}^s n_i - 1)\prod_{i=2}^{s}(n_i + 1) - 3(t-1)$$

of initialized simple Ulrich vector bundles of rank r.

2. *Let us suppose that* $s \geq 3$ *and* $n_2 = 1$. *Then for any* $r = 2t + 1$, $t \geq 2$, *there exists a family of dimension* $\geq (t-1)((\sum_{i=2}^{s} n_i - 1)(n_3+2)\prod_{i=4}^{s}(n_i+1) - 1)$ *of initialized simple Ulrich vector bundles of rank* r.
3. *Let us suppose that* $s \geq 3$ *and* $n_2 > 1$. *For any integer* $r = an_3\binom{n_2}{2} + \ell \geq n_3\binom{n_2}{2}$ *with* $a \geq 1$ *and* $0 \leq \ell \leq n_3\binom{n_2}{2} - 1$, *there exists a family of dimension* $a^2(n_3^2 + 2n_3 - 4) + \ell + l(an_3\binom{n_2+1}{2} - \ell)$ *of simple (hence, undecomposable) initialized Ulrich vector bundles of rank* r.

Proof

1. Let $r = 2t$ be an even integer and set

$$a := \operatorname{ext}^1(\mathscr{B}, \mathscr{A}) = (\Sigma_{i=2}^{s} n_i - 1)\prod_{i=2}^{s}(n_i + 1)$$

with $\mathscr{A}$ and $\mathscr{B}$ defined as in the proof of Theorem 3.4.12. Denote by U the open subset of $\mathbb{P}^a \times \overset{t)}{\cdots} \times \mathbb{P}^a$, $\mathbb{P}^a \cong \mathbb{P}(\operatorname{Ext}^1(\mathscr{B}, \mathscr{A})) \cong \Lambda_2$, parameterizing closed points $[\mathscr{E}_1, \cdots, \mathscr{E}_t] \in \mathbb{P}^a \times \overset{t)}{\cdots} \times \mathbb{P}^a$ such that $\mathscr{E}_i \not\cong \mathscr{E}_j$ for $i \neq j$ (i.e. U is $\mathbb{P}^a \times \overset{t)}{\cdots} \times \mathbb{P}^a$ minus the small diagonals). Given $[\mathscr{E}_1, \cdots, \mathscr{E}_t] \in U$, by Lemma 3.4.13, the set of vector bundles $\mathscr{E}_1, \cdots, \mathscr{E}_t$ satisfy the hypothesis of Proposition [49, Proposition 5.1.3] and therefore, there exists a family of rank r simple Ulrich vector bundles $\mathscr{E}$ parameterized by

$$\mathbb{P}(\operatorname{Ext}^1(\mathscr{E}_t, \mathscr{E}_1)) \times \cdots \times \mathbb{P}(\operatorname{Ext}^1(\mathscr{E}_t, \mathscr{E}_{t-1}))$$

and given as extensions of the form

$$0 \longrightarrow \oplus_{i=1}^{t-1}\mathscr{E}_i \longrightarrow \mathscr{E} \longrightarrow \mathscr{E}_t \longrightarrow 0.$$

Next we observe that if we consider $[\mathscr{E}_1, \cdots, \mathscr{E}_t] \neq [\mathscr{E}'_1, \cdots, \mathscr{E}'_t] \in U$ and the corresponding extensions

$$0 \longrightarrow \oplus_{i=1}^{t-1}\mathscr{E}_i \longrightarrow \mathscr{E} \longrightarrow \mathscr{E}_t \longrightarrow 0,$$

and

$$0 \longrightarrow \oplus_{i=1}^{t-1}\mathscr{E}'_i \longrightarrow \mathscr{E}' \longrightarrow \mathscr{E}'_t \longrightarrow 0$$

then $\operatorname{Hom}(\mathscr{E}, \mathscr{E}') = 0$ and in particular $\mathscr{E} \not\cong \mathscr{E}'$. Therefore, we have a family of non-isomorphic rank r simple Ulrich vector bundles $\mathscr{E}$ on $\Sigma_{1,n_2\ldots,n_s}$ parameterized by a projective bundle $\mathbb{P}$ over U of dimension

$$\dim \mathbb{P} = (t-1)\dim(\mathbb{P}(\operatorname{Ext}^1(\mathscr{E}_t, \mathscr{E}_1))) + \dim U.$$

Applying the functor $\mathrm{Hom}(-,\mathscr{E}_1)$ to the short exact sequence (1) we obtain:

$$0 \longrightarrow \mathrm{Hom}(\mathscr{A},\mathscr{E}_1) \cong K \longrightarrow \mathrm{Ext}^1(\mathscr{B},\mathscr{E}_1) \longrightarrow \mathrm{Ext}^1(\mathscr{E}_t,\mathscr{E}_1) \longrightarrow \mathrm{Ext}^1(\mathscr{A},\mathscr{E}_1) = 0.$$

On the other hand, applying $\mathrm{Hom}(\mathscr{B},-)$ to the same exact sequence we have

$$\begin{aligned} 0 = \mathrm{Hom}(\mathscr{B},\mathscr{E}_1) &\longrightarrow \mathrm{Hom}(\mathscr{B},\mathscr{B}) \cong K \longrightarrow \mathrm{Ext}^1(\mathscr{B},\mathscr{A}) \cong K^a \\ &\longrightarrow \mathrm{Ext}^1(\mathscr{B},\mathscr{E}_1) \longrightarrow \mathrm{Ext}^1(\mathscr{B},\mathscr{B}) = 0. \end{aligned}$$

Summing up, we obtain $\mathrm{ext}^1(\mathscr{E}_t,\mathscr{E}_1) = a-2$ and so

$$\dim \mathbb{P} = (t-1)(a-3) + ta = (2t-1)a - 3(t-1).$$

2. Now, let us suppose that $s \geq 3$ and $n_2 = 1$ and take $r = 2t+1$, $t \geq 2$. Let $\mathscr{E}_1,\dots,\mathscr{E}_{t-1}$ be $t-1$ non-isomorphic rank 2 Ulrich vector bundles from the exact sequence (1) and let $\mathscr{F}$ be a rank 3 Ulrich bundle from the exact sequence (2). Again, by Lemma 3.4.13, this set of vector bundles satisfies the hypothesis of [49, Proposition 5.1.3] and therefore, there exists a family $\mathbb{G}$ of rank r simple Ulrich vector bundles $\mathscr{E}$ parameterized by

$$\mathbb{P}(\mathrm{Ext}^1(\mathscr{E}_1,\mathscr{F})) \times \cdots \times \mathbb{P}(\mathrm{Ext}^1(\mathscr{E}_{t-1},\mathscr{F}))$$

and given as extensions of the form

$$0 \longrightarrow \mathscr{F} \longrightarrow \mathscr{E} \longrightarrow \oplus_{i=1}^{t-1}\mathscr{E}_i \longrightarrow 0.$$

It only remains to compute the dimension of the family

$$\dim \mathbb{G} = (t-1)\dim(\mathbb{P}(\mathrm{Ext}^1(\mathscr{E}_1,\mathscr{F}))).$$

Let us fix the notation

$$\begin{aligned} b := \mathrm{ext}^1(\mathscr{B},\mathscr{C}) &= \mathrm{h}^1(\mathbb{P}^1,\mathscr{O}_{\mathbb{P}^1}(-\textstyle\sum_{i=2}^s n_i))\mathrm{h}^0(\mathbb{P}^1,\mathscr{O}_{\mathbb{P}^1}(1+n_3))\prod_{i=4}^s \mathrm{h}^0(\mathbb{P}^{n_i},\mathscr{O}_{\mathbb{P}^{n_i}}(1)) \\ &= (\textstyle\sum_{i=2}^s n_i - 1)(n_3+2)\prod_{i=4}^s(n_i+1). \end{aligned}$$

Applying the functor $\mathrm{Hom}(-,\mathscr{F})$ to the short exact sequence (1) we obtain:

$$0 = \mathrm{Hom}(\mathscr{A},\mathscr{F}) \longrightarrow \mathrm{Ext}^1(\mathscr{B},\mathscr{F}) \longrightarrow \mathrm{Ext}^1(\mathscr{E}_1,\mathscr{F}) \longrightarrow \mathrm{Ext}^1(\mathscr{A},\mathscr{F}).$$

On the other hand, applying $\mathrm{Hom}(\mathscr{B},-)$ to the short exact sequence (2) we have

$$0 = \mathrm{Hom}(\mathscr{B},\mathscr{D}) \longrightarrow \mathrm{Ext}^1(\mathscr{B},\mathscr{C}) \cong K^b \longrightarrow \mathrm{Ext}^1(\mathscr{B},\mathscr{F}) \longrightarrow \mathrm{Ext}^1(\mathscr{B},\mathscr{D}) = 0.$$

Summing up, we obtain $\mathrm{ext}^1(\mathscr{E}_1,\mathscr{F}) \geq b$ and therefore $\dim\mathbb{G} \geq (t-1)(b-1)$.

3. It follows from Theorem 3.4.9 and [21, Proposition 2.6]. □

Corollary 3.4.15 *The Segre variety* $\Sigma_{1,n_2\dots,n_s} \subseteq \mathbb{P}^N$, $N = 2\prod_{i=2}^{s}(n_i+1)-1$, *for* $s \geq 3$ *or* $s = 2$ *and* $n_2 \geq 2$ *is of wild representation type.*

Putting together Corollaries 3.4.8, 3.4.11 and 3.4.15, we get

Theorem 3.4.16 *All Segre varieties* $\Sigma_{n_1,n_2\dots,n_s} \subseteq \mathbb{P}^N$, $N = \prod_{i=1}^{s}(n_i+1)-1$, *are of wild representation type unless the quadric surface in* $\mathbb{P}^3$ *(which is of finite representation type).*

Slightly generalizing the arguments of this section we can extend the last Theorem and determine the representation type of any non-singular rational normal scroll. Scrolls are fascinating varieties which have been largely studied in Algebraic Geometry. Let us recall one of their possible definitions. To this end, we fix $\mathscr{E} = \oplus_{i=0}^{k}\mathscr{O}_{\mathbb{P}^1}(a_i)$ a rank $k+1$ vector bundle on $\mathbb{P}^1$, where $0 \leq a_0 \leq \dots \leq a_k$, and $a_k > 0$. Let $\mathbb{P}(\mathscr{E}) = \mathbb{P}(Sym(\mathscr{E})) \xrightarrow{\pi} \mathbb{P}^1$ be the projectivized vector bundle and let $\mathscr{O}_{\mathbb{P}(\mathscr{E})}(1)$ be its tautological line bundle. Then $\mathscr{O}_{\mathbb{P}(\mathscr{E})}(1)$ is generated by global sections and defines a birational map $\mathbb{P}(\mathscr{E}) \longrightarrow \mathbb{P}^N$, $N = \sum_{i=0}^{k} a_i + k$. We write $S(\mathscr{E})$ or $S(a_0,\dots,a_k)$ for the image of this map, which is a variety of dimension $k+1$ and degree $c := \sum_{i=0}^{k} a_i$.

Definition 3.4.17 A *rational normal scroll* is one of these varieties $S(\mathscr{E})$; i.e. it is the image of the map

$$\sigma : \mathbb{P}^1 \times \mathbb{P}^k \longrightarrow \mathbb{P}^N$$

given by

$$\sigma(x,y;t_0,t_1\cdots,t_k) := (x^{a_0}t_0, x^{a_0-1}yt_0,\cdots,y^{a_0}t_0,,\cdots,x^{a_k}t_k,x^{a_k-1}yt_k,\cdots,y^{a_k}t_k)$$

where $0 \leq a_0 \leq \dots \leq a_k$, and $a_k > 0$.

The most familiar examples of rational normal scrolls are $\mathbb{P}^d \cong S(0,\dots,0,1)$, the rational normal curve $S(a)$ of degree a in $\mathbb{P}^a$, the quadric $S(1,1) \subset \mathbb{P}^3$ and the cubic scroll $S(1,2) \subset \mathbb{P}^4$.

There is a beautiful geometric description of rational normal scrolls. In $\mathbb{P}^N$, take $k+1$ complementary linear spaces $L_i \cong \mathbb{P}^{a_i}$ with $0 \leq a_0 \leq \dots \leq a_k$, and $a_k > 0$. In each L_i choose a rational normal curve C_{a_i} and an isomorphism $\phi_i : \mathbb{P}^1 \longrightarrow C_{a_i}$ (ϕ_i is constant when $a_i = 0$). Then the variety

$$S(a_0,\dots,a_k) = \bigcup_{p\in\mathbb{P}^1} \langle \phi_0(p),\cdots,\phi_k(p)\rangle \subset \mathbb{P}^N$$

is a rational normal scroll of dimension $k+1$ and degree $c := \sum_{i=0}^{k} a_i$ in $\mathbb{P}^{c+k}$. Notice that rational normal scrolls are varieties of minimal degree.

This geometric description will allow us to describe the homogeneous ideal of $S(a_0,\ldots,a_k)$. Indeed, if $S(a_0,\ldots,a_k)\subset\mathbb{P}^N$, $N=\sum_{i=0}^k a_i+k$ is a rational normal scroll defined by rational normal curves $C_{a_i}\subset L_i\cong\mathbb{P}^{a_i}$, we choose coordinates $X_0^0,\cdots,X_{a_0}^0,\cdots,X_0^k,\cdots,X_{a_k}^k$ in $\mathbb{P}^N$ such that $X_0^i,\cdots,X_{a_i}^i$ are homogeneous coordinates in L_i. Then, we consider the $2\times c$ matrix with two rows and $k+1$ catalecticant blocks

$$M_{a_0,\cdots,a_k}:=\begin{bmatrix} X_0^0 & \cdots & X_{a_0-1}^0 & \cdots & X_0^k & \cdots & X_{a_k-1}^k \\ X_1^0 & \cdots & X_{a_0}^0 & \cdots & X_1^k & \cdots & X_{a_k}^k \end{bmatrix}.$$

It is well known that the ideal of $S(a_0,\ldots,a_k)$ is generated by the maximal minors of $M_{a_0,\cdots,a_k}$ and we have:

Proposition 3.4.18 *Let $S(a_0,\ldots,a_k)\subset\mathbb{P}^N$ with $N=\sum_{i=0}^k a_i+k$, $0\le a_0\le\ldots\le a_k$, and $a_k>0$ be a rational normal scroll. Set $c:=\sum_{i=0}^k a_i$. It holds:*

1. $\dim(S(a_0,\ldots,a_k))=k+1$ *and* $\deg(S(a_0,\ldots,a_k))=\sum_{i=0}^k a_i$.
2. *$S(a_0,\ldots,a_k)$ is ACM and $I(S(a_0,\ldots,a_k))$ is generated by $\binom{c}{2}$ hyperquadrics.*
3. *$S(a_0,\ldots,a_k)$ is non-singular if and only if $a_0>0$ (so, $a_i>0$ for all $0\le i\le k$) or $S(a_0,\ldots,a_k)=S(0,\cdots,0,1)\cong\mathbb{P}^k$.*

Since we are not interested in $\mathbb{P}^k$ (according to Horrocks Theorem there is, up to twist, only one ACM bundle in $\mathbb{P}^k$, namely, $\mathcal{O}_{\mathbb{P}^k}$) and we will only deal with non-singular rational scrolls, we will assume $0<a_i$, $0\le i\le k$. It holds

Theorem 3.4.19 *All rational normal scrolls $S(a_0,\cdots,a_k)\subseteq\mathbb{P}^N$, $N=\prod_{i=1}^s(n_i+1)-1$, are of wild representation type unless $\mathbb{P}^{k+1}=S(0,\cdots,0,1)$, the rational normal curve $S(a)$ in $\mathbb{P}^a$, the quadric surface $S(1,1)$ in $\mathbb{P}^3$ and the cubic scroll $S(1,2)$ in $\mathbb{P}^4$ which are of finite representation type.*

Proof See [40, Theorem 3.8]. □

3.5 Does the Representation Type of a Projective Variety Depends on the Polarization?

The representation type of an ACM variety $X\subset\mathbb{P}^n$ strongly depends on the chosen embedding and the goal of this section will be to prove that on an ACM projective variety $X\subset\mathbb{P}^n$ there always exists a very ample line bundle $\mathcal{L}$ on X which naturally embeds X in $\mathbb{P}^{h^0(X,\mathcal{L})-1}$ as a variety of wild representation type (cf. Theorem 3.5.4). As immediate consequence we will have many new examples of ACM varieties of wild representation type.

Let us start with a precise example to illustrate such phenomena.

Example 3.5.1

1. The Segre product of two lines naturally embedded in $\mathbb{P}^3$ is an example of ACM surface of finite representation type, i.e., $\varphi_{|\mathcal{O}(1,1)|} : \mathbb{P}^1 \times \mathbb{P}^1 \hookrightarrow \mathbb{P}^3$ is a variety of finite representation type. Indeed, according to Knörrer any hyperquadric $Q_n \subset \mathbb{P}^{n+1}$ is of finite representation type [34] and, up to twist, the only undecomposable ACM bundles on $\mathbb{P}^1 \times \mathbb{P}^1 \subset \mathbb{P}^3$ are: $\mathcal{O}_{\mathbb{P}^1\times\mathbb{P}^1}$, $\mathcal{O}_{\mathbb{P}^1\times\mathbb{P}^1}(1,0)$ and $\mathcal{O}_{\mathbb{P}^1\times\mathbb{P}^1}(0,1)$.
2. The Segre product of two smooth conics naturally embedded in $\mathbb{P}^8$ is an example of variety of wild representation type, i.e., $\varphi_{|\mathcal{O}(2,2)|} : \mathbb{P}^1 \times \mathbb{P}^1 \hookrightarrow \mathbb{P}^8$ is an example of ACM surface of wild representation type. Indeed, any smooth del Pezzo surface is of wild representation type (see Theorem 3.3.35).
3. The Segre product of a line and a smooth conic naturally embedded in $\mathbb{P}^5$ is an example of smooth ACM surface of tame representation type, i.e., $\varphi_{|\mathcal{O}(1,2)|} : \mathbb{P}^1 \times \mathbb{P}^1 \hookrightarrow \mathbb{P}^5$ is a variety of tame representation type. Indeed, all continuous families of undecomposable ACM bundles are one-dimensional (see [23, Theorem 1]).

This leads to the following problems:

Problem 3.5.2

1. Given an ACM variety $X \subset \mathbb{P}^n$, is there an integer N_X such that X can be embedded in $\mathbb{P}^{N_X}$ as a variety of wild representation type?
2. If so, what is the smallest possible integer N_X?

We will answer affirmatively Problem 3.5.2 (1) and provide an upper bound for N_X. In other words, we will prove that for any smooth ACM projective variety $X \subset \mathbb{P}^n$ there is an embedding of X into a projective space $\mathbb{P}^{N_X}$ such that the corresponding homogeneous coordinate ring has arbitrary large families of non-isomorphic undecomposable graded Maximal Cohen-Macaulay modules. Actually, it is proved that such an embedding can be obtained as the composition of the "original" embedding $X \subset \mathbb{P}^n$ and the Veronese 3-uple embedding $\nu_3 : \mathbb{P}^n \longrightarrow \mathbb{P}^{\binom{n+3}{3}-1}$. The idea will be to construct on any ACM variety $X \subset \mathbb{P}^n$ of dimension $d \geq 2$ irreducible families $\mathcal{F}$ of vector bundles $\mathcal{E}$ of arbitrarily high rank and dimension with the extra feature that any $\mathcal{E} \in \mathcal{F}$ satisfy $\mathrm{H}^i(X, \mathcal{E}(t)) = 0$ for all $t \in \mathbb{Z}$ and $2 \leq i \leq d-1$ and $\mathrm{H}^1(X, \mathcal{E}(t)) = 0$ for all $t \neq -1, -2$. Therefore, X embedded in $\mathbb{P}^{\mathrm{h}^0(\mathcal{O}_X(s))-1}$ through the very ample line bundle $\mathcal{O}_X(s)$, $s \geq 3$, is of wild representation type.

Let X be a smooth ACM variety of dimension $d \geq 2$ in $\mathbb{P}^n$ with a minimal free R-resolution of the following type:

$$0 \longrightarrow F_c \xrightarrow{\varphi_c} F_{c-1} \xrightarrow{\varphi_{c-1}} \cdots \xrightarrow{\varphi_2} F_1 \xrightarrow{\varphi_1} F_0 \longrightarrow R_X \longrightarrow 0 \qquad (3.22)$$

with $c = n - d$, $F_0 = R$ and $F_i = \oplus_{j=1}^{\beta_i} R(-n_j^i)$, $1 \leq i \leq c$.

For any $2 \le n$ and any $1 \le a$, we denote by $\mathscr{E}_{n,a}$ any vector bundle on $\mathbb{P}^n$ given by the exact sequence

$$0 \to \mathscr{E}_{n,a} \to \mathscr{O}_{\mathbb{P}^n}(1)^{(n+2)a} \stackrel{\phi(1)}{\to} \mathscr{O}_{\mathbb{P}^n}(2)^{2a} \to 0 \tag{3.23}$$

where $\phi \in V_n$ being V_n the non-empty open dense subset of the affine scheme $M = \mathrm{Hom}(\mathscr{O}_{\mathbb{P}^n}(1)^{(n+2)a}, \mathscr{O}_{\mathbb{P}^n}(2)^{2a})$ provided by Proposition 3.3.28.

From now on, for any $2 \le n$ and any $1 \le a$, we call $\mathscr{F}^X_{n,a}$ the non-empty irreducible family of *general* rank na vector bundles $\mathscr{E}$ on $X \subset \mathbb{P}^n$ sitting in an exact sequence of the following type:

$$0 \to \mathscr{E} \to \mathscr{O}_X(1)^{(n+2)a} \stackrel{f}{\to} \mathscr{O}_X(2)^{2a} \to 0. \tag{3.24}$$

Proposition 3.5.3 *Let $X \subset \mathbb{P}^n$ be a smooth ACM variety of dimension $d \ge 2$. With the above notation, we have:*

1. *A general vector bundle $\mathscr{E} \in \mathscr{F}^X_{n,a}$ satisfies*

$$\begin{aligned} &\mathrm{H}^i_*\mathscr{E} = 0 && \text{for } 2 \le i \le d-1, \\ &\mathrm{H}^1(X, \mathscr{E}(t)) = 0 && \text{for } \ t \ne -1, -2. \end{aligned}$$

2. *A general vector bundle $\mathscr{E} \in \mathscr{F}^X_{n,a}$ is simple.*
3. *$\mathscr{F}^X_{n,a}$ is a non-empty irreducible family of dimension $a^2(n^2+2n-4)+1$ of simple (hence undecomposable) rank an vector bundles on X.*

Proof

1. Since $\mathrm{H}^i(X, \mathscr{E}(t)) = 0$ for all $t \in \mathbb{Z}$ and $2 \le i \le d-1$, and $\mathrm{H}^1(X, \mathscr{E}(t)) = 0$ for $t \ne -1, -2$ are open conditions, it is enough to exhibit a vector bundle $\mathscr{E} \in \mathscr{F}^X_{n,a}$ verifying these vanishing. Tensoring the exact sequence (3.23) with $\mathscr{O}_X$, we get

$$0 \to \mathscr{E} := \mathscr{E}_{n,a} \otimes \mathscr{O}_X \to \mathscr{O}_X(1)^{(n+2)a} \to \mathscr{O}_X(2)^{2a} \to 0. \tag{3.25}$$

Taking cohomology, we immediately obtain $\mathrm{H}^i(X, \mathscr{E}(t)) = 0$ for all $t \in \mathbb{Z}$ and $2 \le i \le d-1$. On the other hand, we tensor with $\mathscr{E}_{n,a}$ the exact sequence (3.22) sheafiffied

$$\begin{aligned} 0 \longrightarrow \oplus_{j=1}^{\beta_c} \mathscr{O}_{\mathbb{P}^n}(-n_j^c) \stackrel{\varphi_c}{\longrightarrow} \oplus_{j=1}^{\beta_{c-1}} \mathscr{O}_{\mathbb{P}^n}(-n_j^{c-1}) \stackrel{\varphi_{c-1}}{\longrightarrow} \\ \cdots \stackrel{\varphi_2}{\longrightarrow} \oplus_{j=1}^{\beta_1} \mathscr{O}_{\mathbb{P}^n}(-n_j^1) \stackrel{\varphi_1}{\longrightarrow} \mathscr{O}_{\mathbb{P}^n} \stackrel{\varphi_0}{\longrightarrow} \mathscr{O}_X \longrightarrow 0 \end{aligned}$$

and we get

$$\begin{aligned} 0 \longrightarrow \oplus_{j=1}^{\beta_c} \mathscr{E}_{n,a}(-n_j^c) \stackrel{\varphi_c}{\longrightarrow} \oplus_{j=1}^{\beta_{c-1}} \mathscr{E}_{n,a}(-n_j^{c-1}) \stackrel{\varphi_{c-1}}{\longrightarrow} \cdots \stackrel{\varphi_{i+1}}{\longrightarrow} \oplus_{j=1}^{\beta_i} \mathscr{E}_{n,a}(-n_j^i) \stackrel{\varphi_i}{\longrightarrow} \\ \cdots \stackrel{\varphi_2}{\longrightarrow} \oplus_{j=1}^{\beta_1} \mathscr{E}_{n,a}(-n_j^1) \stackrel{\varphi_1}{\longrightarrow} \mathscr{E}_{n,a} \stackrel{\varphi_0}{\longrightarrow} \mathscr{E} = \mathscr{E}_{n,a} \otimes \mathscr{O}_X \longrightarrow 0. \end{aligned} \tag{3.26}$$

Set $\mathcal{H}_i := \ker(\varphi_i)$, $0 \leq i \leq c-2$. Cutting the exact sequence (3.26) into short exact sequences and taking cohomology, we obtain

$$\cdots \to \mathrm{H}^1(\mathbb{P}^n, \mathcal{E}_{n,a}(t)) \to \mathrm{H}^1(X, \mathcal{E}(t)) \to \mathrm{H}^2(\mathbb{P}^n, \mathcal{H}_0(t)) \to \cdots,$$

$$\cdots \to \mathrm{H}^2(\mathbb{P}^n, \oplus_{j=1}^{\beta_1} \mathcal{E}_{n,a}(-n_j^1 + t)) \to \mathrm{H}^2(\mathbb{P}^n, \mathcal{H}_0(t)) \to \mathrm{H}^3(\mathbb{P}^n, \mathcal{H}_1(t)) \to \cdots,$$

$$\cdots$$

$$\cdots \to \mathrm{H}^{c-1}(\mathbb{P}^n, \oplus_{j=1}^{\beta_{c-2}} \mathcal{E}_{n,a}(-n_j^{c-2} + t)) \to \mathrm{H}^{c-1}(\mathbb{P}^n, \mathcal{H}_{c-3}(t))$$
$$\to \mathrm{H}^c(\mathbb{P}^n, \mathcal{H}_{c-2}(t)) \to \cdots,$$

$$\cdots \to \mathrm{H}^c(\mathbb{P}^n, \oplus_{j=1}^{\beta_{c-1}} \mathcal{E}_{n,a}(-n_j^{c-1} + t)) \to \mathrm{H}^c(\mathbb{P}^n, \mathcal{H}_{c-2}(t))$$
$$\to \mathrm{H}^{c+1}(\mathbb{P}^n, \oplus_{j=1}^{\beta_c} \mathcal{E}_{n,a}(-n_j^c + t)) \to \cdots,$$

Using Proposition 3.4.6, we conclude that $\mathrm{H}^1(X, \mathcal{E}(t)) = 0$ for $t \neq -1, -2$.

2. A general vector bundle $\mathcal{E} \in \mathcal{F}_{n,a}^X$ sits in an exact sequence

$$0 \to \mathcal{E} \stackrel{g}{\to} \mathcal{O}_X(1)^{(n+2)a} \stackrel{f}{\to} \mathcal{O}_X(2)^{2a} \to 0$$

and to check that $\mathcal{E}$ is simple is equivalent to check that $\mathcal{E}^\vee$ is simple. Notice that the morphism $f^\vee : \mathcal{O}_X(-2)^{2a} \longrightarrow \mathcal{O}_X(-1)^{(n+2)a}$ appearing in the exact sequence

$$0 \to \mathcal{O}_X(-2)^{2a} \stackrel{f^\vee}{\to} \mathcal{O}_X(-1)^{(n+2)a} \stackrel{g^\vee}{\to} \mathcal{E}^\vee \to 0 \tag{3.27}$$

is a general element of the K-vector space

$$M := \mathrm{Hom}(\mathcal{O}_X(-2)^{2a}, \mathcal{O}_X(-1)^{(n+2)a}) \cong K^{n+1} \otimes K^{2a} \otimes K^{(n+2)a}$$

because $\mathrm{Hom}(\mathcal{O}_X(-2), \mathcal{O}_X(-1)) \cong \mathrm{H}^0(X, \mathcal{O}_X(1)) \cong \mathrm{H}^0(\mathbb{P}^n, \mathcal{O}_{\mathbb{P}^n}(1)) \cong K^{n+1}$. Therefore, $f^\vee : \mathcal{O}_X(-2)^{2a} \longrightarrow \mathcal{O}_X(-1)^{(n+2)a}$ is represented by a $(n+2)a \times 2a$ matrix A with entries in $\mathrm{H}^0(\mathbb{P}^n, \mathcal{O}_{\mathbb{P}^n}(1))$. Since $\mathrm{Aut}(\mathcal{O}_X(-1)^{(n+2)a}) \cong GL((n+2)a)$ and $\mathrm{Aut}(\mathcal{O}_X(-2)^{2a}) \cong GL(2a)$, the group $GL((n+2)a) \times GL(2a)$ acts naturally on M by

$$\begin{array}{ccc} GL((n+2)a) \times GL(2a) \times M & \longrightarrow & M \\ (g_1, g_2, A) & \mapsto & g_1^{-1} A g_2. \end{array}$$

For all $A \in M$ and $\lambda \in K^*$, $(\lambda Id_{(n+2)a}, \lambda Id_{2a})$ belongs to the stabilizer of A and, hence, $\dim_K Stab(A) \geq 1$. Since $(2a)^2 + (n+2)^2a^2 - 2a(n+1)(n+2)a < 0$, it follows from [31, Theorem 4] that $\dim_K Stab(A) = 1$. We will now check that $\mathcal{E}^\vee$ is simple. Otherwise, there exists a non-trivial morphism $\phi : \mathcal{E}^\vee \to \mathcal{E}^\vee$ and composing with $g^{\vee}$ we get a morphism

$$\overline{\phi} = \phi \circ g^{\vee} : \mathcal{O}_X(-1)^{(n+2)a} \to \mathcal{E}^{\vee}.$$

Applying $\mathrm{Hom}(\mathcal{O}_X(-1)^{(n+2)a}, -)$ to the exact sequence (3.27) and taking into account that

$$\mathrm{Hom}(\mathcal{O}_X(-1)^{(n+2)a}, \mathcal{O}_X(-2)^{2a}) = \mathrm{Ext}^1(\mathcal{O}_X(-1)^{(n+2)a}, \mathcal{O}_X(-2)^{2a}) = 0$$

we obtain $\mathrm{Hom}(\mathcal{O}_X(-1)^{(n+2)a}, \mathcal{O}_X(-1)^{(n+2)a}) \cong \mathrm{Hom}(\mathcal{O}_X(-1)^{(n+2)a}, \mathcal{E}^{\vee})$. Therefore, there is a non-trivial morphism $\widetilde{\phi} \in \mathrm{Hom}(\mathcal{O}_X(-1)^{(n+2)a}, \mathcal{O}_X(-1)^{(n+2)a})$ induced by $\overline{\phi}$ and represented by a matrix $B \neq \mu Id \in \mathrm{Mat}_{(n+2)a\times(n+2)a}(K)$ such that the following diagram commutes:

$$\begin{array}{ccccccccc}
0 & \longrightarrow & \mathcal{O}_X(-2)^{2a} & \xrightarrow{f^{\vee}} & \mathcal{O}_X(-1)^{(n+2)a} & \longrightarrow & \mathcal{E} & \longrightarrow & 0 \\
 & & \downarrow C & & \downarrow B & \searrow^{g^{\vee}\overline{\phi}} & \downarrow \phi & & \\
0 & \longrightarrow & \mathcal{O}_X(-2)^{2a} & \xrightarrow{f^{\vee}} & \mathcal{O}_X(-1)^{(n+2)a} & \xrightarrow{g^{\vee}} & \mathcal{E}^{\vee} & \longrightarrow & 0
\end{array}$$

where $C \in \mathrm{Mat}_{2a\times 2a}(K)$ is the matrix associated to $\widetilde{\phi}_{|\mathcal{O}_X(-2)^{2a}}$. Then the pair $(C, B) \neq (\mu Id, \mu Id)$ verifies $AC = BA$. Let us consider an element $\alpha \in K$ that does not belong to the set of eigenvalues of B and C. Then the pair $(B - \alpha Id, C - \alpha Id) \in GL((n+2)a) \times GL(2a)$ belongs to $Stab(f)$ and therefore $\dim_K Stab(f) > 1$ which is a contradiction. Thus, $\mathcal{E}$ is simple.

3. It only remains to compute the dimension of $\mathcal{F}^X_{n,a}$. Since the isomorphism class of a general vector bundle $\mathcal{E} \in \mathcal{F}^X_{n,a}$ associated to a morphism $\phi \in M := \mathrm{Hom}(\mathcal{O}_X^{(n+2)a}, \mathcal{O}_X(1)^{2a})$ depends only on the orbit of ϕ under the action of $GL((n+2)a) \times GL(2a)$ on M, we have:

$$\begin{aligned}
\dim \mathcal{F}^X_{n,a} &= \dim M - \dim \mathrm{Aut}(\mathcal{O}_X^{(n+2)a}) - \dim \mathrm{Aut}(\mathcal{O}_X(1)^{2a}) + 1 \\
&= 2a^2(n+2)(n+1) - a^2(n+2)^2 - 4a^2 + 1 = a^2(n^2 + 2n - 4) + 1.
\end{aligned}$$

□

As an immediate consequence of the above result we can answer affirmatively Problem 3.5.2 (1) and provide an upper bound for N_X. Indeed, we have:

Theorem 3.5.4 *Let $X \subset \mathbb{P}^n$ be a smooth ACM variety of dimension $d \geq 2$. The very ample line bundle $\mathcal{O}_X(s)$, $s \geq 3$, embeds X in $\mathbb{P}^{\mathrm{h}^0(\mathcal{O}_X(s))-1}$ as a variety of wild representation type.*

Proof See [41, Theorem 3.4]. □

Corollary 3.5.5 *The smallest possible integer N_X such that X embeds as a variety of wild representation type is bounded by $N_X \leq \binom{n+3}{3} - 1$.*

Proof See [41, Corollary 3.5]. □

3.6 Open Problems

In this section we collect the open problems that were mentioned in the lectures, and add some more.

1. Does Mustaţă's conjecture holds for a set of general points on a smooth surface S of degree d in $\mathbb{P}^3$?
   The answer is yes if $d = 2$ (see [27]) or $d = 3$ (see Theorem 3.3.27).
   More general, does Mustaţă's conjecture holds for a set of general points on a smooth hypersurface X of degree d in $\mathbb{P}^n$?
   To my knowledge these two problems are open.
2. Fix a projective variety $X \subset \mathbb{P}^n$. As we have seen in these notes ACM bundles on X provide a criterium to determine the complexity of X. Indeed, the complexity is studied in terms of the dimension and number of families of undecomposable ACM bundles that it supports. Mimicking an analogous trichotomy in representation theory, it was proposed a classification of ACM projective varieties as finite, tame or wild representation type. We would like to know:
   Is the trichotomy finite representation type, tame representation type and wild representation type exhaustive?
   The answer is yes for smooth ACM curves. In fact, an ACM curve is of finite representation type if its genus $g(C) = 0$, of tame representation type if $g(C) = 1$, and of wild representation type if $g(C) \geq 2$. For ACM varieties of dimension ≥ 2 the answer is not known.
3. In Sect. 3.5, we have seen that the representation type of an ACM projective variety strongly depends on the embedding and we have proved that given an ACM variety $X \subset \mathbb{P}^n$, there is an integer N_X such that X can be naturally embedded in $\mathbb{P}^{N_X}$ as a variety of wild representation type. So, the following question arise in a natural way:
   Given an ACM projective variety X, what is the smallest possible N_X such that X embeds in $\mathbb{P}^{N_X}$ as a variety of wild representation type?
4. In Sect. 3.4, we saw that all Segre varieties $\Sigma_{n_1,\cdots,n_s} \subset \mathbb{P}^N$, $N = \prod_{i=1}^s (n_i + 1) - 1$ are of wild representation type unless $\mathbb{P}^1 \times \mathbb{P}^1$; it follows from Sect. 3.5 that the Veronese embedding $\nu_d : \mathbb{P}^n \longrightarrow \mathbb{P}^{\binom{n+d}{d}-1}$, $d \geq 3$, embeds $\mathbb{P}^n$ into $\mathbb{P}^{\binom{n+d}{d}-1}$ as a variety of wild representation type. So we are led to pose the following question:
   Let $G(k, n)$ be the Grassmannian variety which parameterizes linear subspaces of $\mathbb{P}^n = \mathbb{P}(V)$ of dimension k. Embed $G(k, n)$ into $\mathbb{P}^{\binom{n+1}{k+1}-1}$ using Plücker embedding.
   Is $G(k, n) \subset \mathbb{P}^{\binom{n+1}{k+1}-1} = \mathbb{P}(\wedge^{k+1} V)$ a variety of wild representation type?

5. In Sect. 3.3, we have constructed Ulrich bundles on smooth del Pezzo surfaces and, in Sect. 3.4, on Segree varieties. Nevertheless few examples of varieties supporting Ulrich sheaves are known. In [21, p. 43], Eisenbud, Schreyer and Weyman leave open the following interesting problems:
   (a) Is every variety (or even scheme) $X \subset \mathbb{P}^n$ the support of an Ulrich sheaf?
   (b) If so, what is the smallest possible rank for such a sheaf?
6. In Sect. 3.3.1, we have addressed Mustaţă's conjecture for a general set of points on a del Pezzo surface. As a main tool we have used Liaison Theory and we will end these notes with a couple of open problems/questions on this fascinating Theory.
   (a) Does any zero-dimensional scheme $Z \subset \mathbb{P}^n$ belong to the G-liaison class of a complete intersection? In other words, is it glicci?
   (b) More general, is any ACM scheme $X \subset \mathbb{P}^n$ glicci?
   (c) Find new graded R-modules invariant under G-liaison.

These notes and list of open problems were written for a course held in 2014. Some of these questions have been studied and even solved. For seek of completeness we add a list of recent results on the subject where the reader could find more information.

Problem 1 has been solved for a set of general points on a smooth surface $S \subset \mathbb{P}^3$ of degree 4 and remains open for a set of general points on a smooth surface S of degree $d > 4$ in $\mathbb{P}^3$ (see [6]).

For more information on Problem 2 the reader could read [32] and [24]. For the existence of homogeneous ACM (resp. Ulrich) bundles on Grassmannians $G(k, n)$ and on flag manifolds $F(k_1, \cdots, k_r)$, as well as for the representation type of $G(k, n)$ the reader can see [13, 14] and [12].

Finally, new contributions to Problem 6 could be found, for instance, in [1–3, 5, 11] and [26].

Acknowledgements The author is grateful to Professor Le tuan Hoa and to Professor Ngo Viet Trung for giving her the opportunity to speak about one of her favorite subjects: Arithmetically Cohen-Macaulay bundles on projective varieties and their algebraic counterpart Maximal Cohen-Macaulay modules. She is also grateful to the participants for their kind hospitality and mathematical discussions that made for a very interesting and productive month in the lovely city of Hanoi.

Last but not least, I am greatly indebted to L. Costa and J. Pons-Llopis for a long time and enjoyable collaboration which led to part of the material described in these notes.

References

1. M. Aprodu, Y. Kim, Ulrich line bundles on Enriques surfaces with a polarization of degree four. Ann. Univ. Ferrara Sez. VII Sci. Mat. **63**(1), 9–23 (2017)
2. M. Aprodu, G. Farkas, A. Ortega, Minimal resolutions, Chow forms and Ulrich bundles on K3 surfaces. J. Reine Angew. Math. **730**, 225–249 (2017)

3. M. Aprodu, L. Costa, R.M. Miró-Roig, Ulrich bundles on ruled surfaces. J. Pure Appl. Algebra **222**, 131–138 (2018). http://dx.doi.org/10.1016/j.jpaa.2017.03.007
4. E. Ballico, A.V. Geramita, The minimal free resolution of the ideal of s general points in $\mathbb{P}^3$, in *Proceeding of the 1984 Vancouver Conference in Algebraic Geometry*. Canadian Mathematical Society Conference Proceedings, vol. 6 (American Mathematical Society, Providence, 1986), pp. 1–10
5. A. Beauville, Ulrich bundles on abelian surfaces. Proc. Am. Math. Soc. **144**(11), 4609–4611 (2016)
6. M. Boij, J. Migliore, R.M. Miró-Roig, U. Nagel, The Minimal Resolution Conjecture on a general Quartic Surface in $\mathbb{P}^3$ (preprint). arXiv:1707.05646
7. R. Buchweitz, G. Greuel, F.O. Schreyer, Cohen-Macaulay modules on hypersurface singularities, II. Invent. Math. **88**(1), 165–182 (1987)
8. M. Casanellas, R. Hartshorne, Gorenstein Biliaison and ACM sheaves. J. Algebra **278**, 314–341 (2004)
9. M. Casanellas, R. Hartshorne, ACM bundles on cubic surfaces. J. Eur. Math. Soc. **13**, 709–731 (2011)
10. M. Casanellas, R. Hartshorne, Stable Ulrich bundles. Int. J. Math. **23**(8), 1250083, 50 pp. (2012)
11. G. Casnati, Special Ulrich bundles on non-special surfaces with pg=q=0. Int. J. Math. **28**(8), 1750061, 18 pp. (2017)
12. I. Coskun, L. Costa, J. Huizenga, R.M. Miró-Roig, M. Woolf, Ulrich Schur bundles on flag varieties. J. Algorithm **474**, 49–96 (2017)
13. L. Costa, R.M. Miró-Roig, GL(V)-invariants Ulrich bundles on Grasmannians. Math. Ann. **361**, 443–457 (2015)
14. L. Costa, R.M. Miró-Roig, Homogeneous ACM bundles on a Grasmannian. Adv. Math. **289**, 95–113 (2016)
15. L. Costa, R.M Miró-Roig, J. Pons-Llopis, The representation type of Segre varieties. Adv. Math. **230**, 1995–2013 (2012)
16. I. Dolgachev, *Topics in Classical Algebraic Geometry. A Modern View* (Cambridge University Press, Cambridge, 2012)
17. Y. Drozd, G.M. Greuel, Tame and wild projective curves and classification of vector bundles. J. Algebra **246**, 1–54 (2001)
18. D. Eisenbud, J. Herzog, The classification of homogeneous Cohen-Macaulay rings of finite representation type. Math. Ann. **280**(2), 347–352 (1988)
19. D. Eisenbud, F.O. Schreyer, Boij-Söderberg theory, in *Combinatorial Aspects of Commutative Algebra and Algebraic Geometry*. Abel Symposia, vol. 6 (Springer, Berlin, 2009), pp. 35–48
20. D. Eisenbud, S. Popescu, F. Schreyer, C. Walter, Exterior algebra methods for the minimal resolution conjecture. Duke Math. J. **112**, 379–395 (2002)
21. D. Eisenbud, F. Schreyer, J. Weyman, Resultants and Chow forms via exterior syzygies. J. Am. Math. Soc. **16**, 537–579 (2003)
22. P. Ellia, A. Hirschowitz, Génération de certain fibrés sur les spaces projectifs et application. J. Algebraic Geom. **1**, 531–547 (1992)
23. D. Faenzi, F. Malaspina, Surfaces of minimal degree of tame representation type and mutations of Cohen-Macaulay modules. Adv. Math. **310**, 663–695 (2017)
24. D. Faenzi, J. Pons-Llopis, The CM representation type of projective varieties (preprint). arXiv:1504.03819
25. F. Gaeta, Sur la distribution des degrés des formes appartenant à la matrice de l'idéal homogène attaché à un groupe de *N points génèriques du plan*. C. R. Acad. Sci. Paris **233**, 912–913 (1951)
26. O. Genc, Stable Ulrich bundles on Fano threefolds with Picard number 2. J. Pure Appl. Algebra **222**(1), 213–240 (2018)
27. S. Giuffrida, R. Maggioni, A. Ragusa, Resolution of generic points lying on a smooth quadric. Manuscr. Math. **91**, 421–444 (1996)
28. A. Hirshowitz, C. Simpson, La résolution minimale d'un arrangement general d'un grand nombre de points dans $\mathbb{P}^n$. Invent. Math. **126**, 467–503 (1996)

29. L.T. Hoa, On minimal free resolutions of projective varieties of degree=codimension + 2. J. Pure Appl. Algebra **87**, 241–250 (1993)
30. G. Horrocks, Vector bundles on the punctual spectrum of a local ring. Proc. Lond. Math. Soc. **14**(3), 689–713 (1964)
31. V. Kac, Infinite root systems, representations of graphs and invariant theory. Invent. Math. **56**, 57–92 (1980)
32. J. Kleppe, R.M. Miró-Roig, The representation type of determinantal varieties. Algebr. Represent. Theory **20**(4), 1029–1059 (2017)
33. J. Kleppe, J. Migliore, R.M. Miró-Roig, U. Nagel, C. Peterson, *Gorenstein Liaison. Complete Intersection liaison invariants and unobstructedness*. Memoirs of the American Mathematical Society, no. 732, vol. 154 (American Mathematical Society, Providence, 2001)
34. Knörrer, Cohen-Macaulay modules on hypersurface singularities, I. Inv. Math. **88**, 153–164 (1987)
35. J. Kollar, *Rational Curves on Algebraic Varieties* (Springer, Berlin, 2004)
36. A. Lorenzini, The minimal resolution conjecture. J. Algebra **156**, 5–35 (1993)
37. Y. Manin, *Cubic Forms* (North-Holland, Amsterdam, 1986)
38. J. Migliore, M. Patnott, *Minimal free resolutions of general points lying on cubic surfaces in* $\mathbb{P}^3$. J. Pure Appl. Algebra **215**, 1737–1746 (2011)
39. R.M. Miró-Roig, *Determinantal Ideals*. Progress in Mathematics, vol. 264 (Birkhaüser, Basel, 2008)
40. R.M. Miró-Roig, The representation type of a rational normal scroll. Rend. Circ. Mat. Palermo **62**, 153–164 (2013)
41. R.M. Miró-Roig, On the representation type of a projective variety. Proc. Am. Math. Soc. **143**(1), 61–68 (2015)
42. R.M. Miró-Roig, J. Pons-Llopis, The minimal resolution conjecture for points on a del Pezzo surface. Algebra Number Theory **6**, 27–46 (2012)
43. R.M. Miró-Roig, J. Pons-Llopis, Minimal free resolution for points on surfaces. J. Algebra **357**, 304–318 (2012)
44. R.M. Miró-Roig, J. Pons-Llopis, Representation type of rational ACM surfaces $X \subseteq \mathbb{P}^4$. Algebr. Represent. Theory **16**, 1135–1157 (2013)
45. R.M. Miró-Roig, J. Pons-Llopis, N-dimensional Fano varieties of wild representation type. J. Pure Appl. Algebra **218**(10), 1867–1884 (2014)
46. M. Mustaţă, Graded Betti numbers of general finite subsets of points on projective varieties. Le Matematiche **53**, 53–81 (1998)
47. C. Okonek, M. Schneider, H. Spindler, *Vector Bundles on Complex Projective Spaces* (Birkhäuser, Basel, 1980)
48. J. Pons-Llopis, Ulrich bundles and varieties of wild representation type. PhD. thesis, University of Barcelona, 2011
49. J. Pons-Llopis, F. Tonini, ACM bundles on Del Pezzo surfaces. Le Matematiche **64**(2), 177–211 (2009)
50. O. Tovpyha, Graded Cohen-Macaulay rings of wild representation type (preprint). Available from arXiv: 1212.6557v1
51. B. Ulrich, Gorenstein rings and modules with high number of generators. Math. Z. **188**, 23–32 (1984)
52. C. Walter, The minimal free resolution of the homogeneous ideal of s general points in $\mathbb{P}^4$. Math. Z. **219**, 231–234 (1995)

Chapter 4
Simplicial Toric Varieties Which Are Set-Theoretic Complete Intersections

Marcel Morales

Abstract We say that a polynomial ideal I is set-theoretically generated by a family of elements $f_1, \ldots, f_k$ in I if the radical of I coincides with the radical of the ideal generated by $f_1, \ldots, f_k$. Over an algebraically closed field, the smallest number among all such possible k is the minimal number of equations needed to define the zero set of I. To find this number is a classical problem in both Commutative Algebra and Algebraic Geometry. This problem is even not solved for the defining ideals of toric varieties, whose zeros are given parametrically by monomials. In this lecture notes we study set-theoretically generation of the defining ideals of simplicial toric varieties, which are defined by the property that the exponents of the parametrizing monomials span a simplicial complex. We review and improve most of results on simplicial toric varieties which are set-theoretic complete intersections, previously obtained by the author in collaboration with M. Barile and A. Thoma.

4.1 Introduction

In the beginning of Algebraic Geometry, varieties were described by equations. However, such description is ambiguous. In order to be more precise, the notion of ideal (defining a variety) was introduced. But if we define a variety as the zero set of a polynomial ideal, there is still ambiguity because different ideals can have the same zero set. The famous Hilbert Nullstellensatz helps us to understand this phenomenon better.

More precisely, let $S := K[X_1, \ldots, X_n]$ be a polynomial ring over a field K. Let $\mathbb{A}^n_K$ be the affine n-dimensional space over K. Given a set $f_1, \ldots, f_k$ of polynomials, the zero set

$$Z(f_1, \ldots, f_k) = \{P \in \mathbb{A}^n_K \mid f_i(P) = 0 \; \forall i = 1, \ldots, k\}$$

M. Morales (✉)
Université Grenoble Alpes, Institut Fourier, UMR 5582, Saint-Martin D'Hères Cedex, France
e-mail: marcel.morales@univ-grenoble-alpes.fr

N. Tu CUONG et al. (eds.), *Commutative Algebra and its Interactions to Algebraic Geometry*, Lecture Notes in Mathematics 2210,
https://doi.org/10.1007/978-3-319-75565-6_4

is called an algebraic set. It is also the zero set $Z(I)$ of the ideal $I = (f_1, \ldots, f_k)$.

For any subset $Y \subset \mathbb{A}^n_K$, we define the ideal of Y by

$$I(Y) = \{f \in S \mid f(P) = 0\, \forall P \in Y\}.$$

For an algebraic set V, the ideal $I(V)$ is called the defining ideal of V. It is clear that if $I(V) = (f_1, \ldots, f_s)$, then $V = \cap_{i=1}^s Z(f_i)$, i.e. V is the intersection of the hypersurfaces $Z(f_i)$. However, there are many ways to define V as an intersection of hypersurfaces. An important problem in Algebraic Geometry is to determine the minimum number of equations needed to define an algebraic set V set-theoretically, that is the minimal number s such that $V = \cap_{i=1}^s Z(f_i)$ for a family of s polynomials $f_1, \ldots, f_s \in K[X_1, \ldots, X_n]$. An important tool in the study of this problem is:

Theorem 4.1.1 (Hilbert's Nullstellensatz) *Let K be an algebraically closed field. Then for any family of polynomials $f_1, \ldots, f_s$, we have*

$$I(Z(f_1, \ldots, f_s)) = \text{rad}\,(f_1, \ldots, f_s).$$

This result leads to the following definition.

Definition 4.1.2 The arithmetical rank of an algebraic set $V \subset \mathbb{A}^n_K$ is the number

$$ara(V) = \min\{k \mid \exists f_1, \ldots, f_k \in S : I(V) = \text{rad}\,(f_1, \ldots, f_k)\},$$

and the arithmetical rank of an ideal I is

$$ara(I) = \min\{k \mid \exists f_1, \ldots, f_k \in S : \text{rad}\, I = \text{rad}\,(f_1, \ldots, f_k)\}.$$

Let $\mathbb{P}^n_K$ be the projective n-dimensional space over K. Similarly, one can define an algebraic set in $\mathbb{P}^{n-1}_K$ as the zero set of a family of homogeneous polynomials in S. For any subset $Y \subset \mathbb{P}^{n-1}_K$, one define $I(Y)$ to be the ideal generated by the homogeneous polynomials $f \in K[X_0, \ldots, X_n]$ vanishing on Y. Then we also have the homogeneous Hilbert Nullstellensatz, and we can define the arithmetical rank of an algebraic set in $\mathbb{P}^n_K$ or of a homogeneous ideal in $K[X_0, \ldots, X_n]$.

Thus, if K is an algebraically closed field, we have $ara(Z(I)) = ara(I)$ for any ideal I (homogeneous or not). However, it is more convenient to work over any field K and on set-theoretic generation of ideals. From now on, when we consider affine or projective algebraic sets, we only take care of their defining ideals. For an arbitrary ideal I, we always have the following inequalities:

$$ht(I) \leq ara(I) \leq \mu(I).$$

Here, $ht(I)$ denotes the height and $\mu(I)$ the minimal number of generators of I. When $h(I) = ara(I)$, the ideal I as well as the algebraic set $V = Z(I)$ is called a *set-theoretic complete intersection (s.t.c.i)*. When $ht(I) = \mu(I)$, it is called a *complete intersection*. It is called an *almost set-theoretic complete intersection* if $ara(I) \leq ht(I) + 1$.

In this lecture notes we focus on toric ideals and toric varieties whose precise definition will be given in Sect. 4.2. Toric ideals and toric varieties play an important role in both Commutative Algebra and Algebraic Geometry because they serve as models for general algebraic varieties. Toric ideals are generated by binomials. Moreover, each binomial is a difference of two monomials with coefficients equal to 1. A rather systematic study of binomial ideals (i.e. generated by binomials) was done by Eisenbud and Sturmfels in [7]. There are numerous publications on minimal generation of a binomial ideal or of its radical, see, for example, [12] Chapter V and [1, 2, 4, 9, 10, 13–15, 19, 22].

The *binomial arithmetical rank* $bar(I)$ of a binomial ideal I is the smallest integer s for which there exist binomials $f_1, \ldots, f_s$ in S such that $rad(I) = rad(f_1, \ldots, f_s)$. This intermediate invariant is, on one side, easier to compute. On the other side, it gives an upper bound for the arithmetical rank of a binomial ideal I as we always have:

$$ht(I) \leq ara(I) \leq bar(I) \leq \mu(I).$$

Using binomial arithmetic rank, one has obtained many results on set-theoretic complete intersections. In this lecture notes we review, and sometimes improve, some of these results.

The main results are (see Sects. 4.2, 4.3 for the used notations):

1. *In characteristic $p > 0$, every simplicial toric affine or projective variety with almost full parametrization is a set-theoretic complete intersection.* This extends previous results by Hartshorne [10], Moh [13], and Barile et al. [2].
2. *In any characteristic, every simplicial toric affine or projective variety with full parametrization is an almost set-theoretic complete intersection.* We give a more transparent proof of this result, which is due to Barile et al. [2].
3. *Let $\overline{V}(p, q, r)$ be the projective toric curve in $\mathbb{P}^3_K$ with parametrization*

   $$w = u^r, x = u^{r-p}v^p, y = u^{r-q}v^q, z = v^r.$$

   Then $\overline{V}(p, q, r)$ in $\mathbb{P}^3$ is a set-theoretic complete intersection for $r \gg 0$.
4. *Let $p, q_0, q_1, \ldots, q_{n-1}$ be positive integers. Let $\overline{V}(p, q_0, q_1, \ldots, q_{n-2}) \subset \mathbb{P}^n_K$ be the projective toric curve with parametrization*

$$\begin{aligned} w &= u^{q_{n-2}}, \\ x &= u^{q_{n-2}-p}v^p, \\ y &= u^{q_{n-2}-q_0}v^{q_0}, \\ z_1 &= u^{q_{n-2}-q_1}v^{q_1}, \\ &\cdots \\ z_{n-2} &= v^{q_{n-2}}. \end{aligned}$$

Let $\overline{V}_1(p, q_0, q_1, \ldots, q_{n-2}, q_{n-1}) \subset \mathbb{P}^{n+1}_K$ *be the projective curve defined by*

$$\begin{aligned} w &= u^{q_{n-1}}, \\ x &= u^{q_{n-1}-p} v^p, \\ y &= u^{q_{n-1}-q_0} v^{q_0}, \\ z_1 &= u^{q_{n-1}-q_1} v^{q_1}, \\ &\ldots \\ z_{n-2} &= u^{q_{n-1}-q_{n-2}} v^{q_{n-2}}, \\ z_{n-1} &= v^{q_{n-1}}. \end{aligned}$$

Let $\gcd(p, q_{n-2}) = l$, $p' = p/l, q' = q_{n-2}/l$. *Assume that* $q_{n-1} \geq p'q'(q' - 1) + q'l$. *If* $\overline{V}(p, q_0, q_1, \ldots, q_{n-2})$ *is a set-theoretic complete intersection, then so is* $\overline{V}_1(p, q_0, q_1, \ldots, q_{n-2}, q_{n-1})$.

Moreover, the proofs presented here are constructive. It should be mentioned that there is no general way to study set-theoretically generation of ideals. This is not surprising because one can not give an answer to this most famous problem on this subject, which deals a very simple case of projective curve in $\mathbb{P}^3_K$:

Question 4.1.3 Assume that K is a field of characteristic 0. Let $\overline{V}(1, 3, 4)$ be the projective toric curve with parametrization

$$w = u^4, x = u^3v^1, y = u^1v^3, z = v^4.$$

Is $\overline{V}(1, 3, 4)$ a set-theoretic complete intersection?

4.2 Definition of Toric Varieties by Parametrization, Semigroups or Lattices

There are several ways to introduce a toric variety, which is associated with a set of n vectors $\mathbf{a_i} = (a_{i,1}, \ldots, a_{i,m}) \in \mathbb{Z}^m, i = 1, \ldots, n$.

1. **Parametrization.** A *toric variety* $V \subset K^n$ is a variety having a following parametrization of the form

$$\begin{aligned} x_1 &= \underline{u}^{\mathbf{a_1}}, \\ &\vdots \\ x_n &= \underline{u}^{\mathbf{a_n}}, \end{aligned}$$

where $\underline{u}^{\mathbf{a_i}} = u_1^{a_{i,1}} \cdots u_m^{a_{i,m}}$, $i = 1, \ldots, n$, are monomials in a polynomial ring $K[u_1, \ldots, u_m]$. Sometimes we simply say that V is parametrized by $\underline{u}^{\mathbf{a_1}}, \ldots, \underline{u}^{\mathbf{a_n}}$.

2. **Semigroups.** The coordinate ring of the above toric variety is isomorphic to the subring $K[\underline{u}^{\alpha}, \alpha \in \Sigma_A] \subset K[u_1, \ldots, u_m]$. This subring can be considered as the semigroup ring $K[\Sigma_A]$ of the semigroup

$$\Sigma_A = \mathbb{N}\mathbf{a_1} + \ldots + \mathbb{N}\mathbf{a_n} \subset \mathbb{Z}^m.$$

Note that $K[\Sigma_A]$ is a domain and that $\dim K[\Sigma_A] = \operatorname{rank} A$, where A is the $m \times n$ matrix whose i-th column vector is $\mathbf{a_i}$.

There is a canonical surjective map $\Psi : S = K[X_1, \ldots, X_n] \to K[\Sigma_A]$. Let $I_A = \ker \Psi$. Then I_A is the defining ideal of the toric variety in S. One calls I_A a *toric ideal*.

We give now a short proof of the fact that I_A is generated by binomials. Observe that

- For any non zero monomial $M \in S$ its image $\Psi(M)$ is non zero.
- For any monomials $M_1, M_2 \in S$, if $\Psi(M_1) = \Psi(M_2)$ then $M_1 - M_2 \in I_A$.
- For any non zero monomials $M_1, M_2 \in S$, if $\Psi(M_1) \neq \Psi(M_2)$ then any linear combination $\alpha\Psi(M_1) + \beta\Psi(M_2)$ with $(\alpha, \beta) \in (K^2)^*$ is non zero.

Let $F = \sum_{i=1}^{t} \alpha_i M_i \in I_A$, where $\alpha_i \in K^*$ and M_i is a nonzero monomial, $i = 1, \ldots, t$. By the observation above we may assume that $\Psi(M_i) = \Psi(M_j)$ for any $i, j = 1, \ldots, t$. It is clear that this implies $\sum_{i=1}^{t} \alpha_i = 0$ and consequently $\alpha_1 = -\sum_{i=2}^{t} \alpha_i$. Hence $F = \sum_{i=2}^{t} \alpha_i (M_i - M_1)$. That shows that the toric ideal I_A is generated by binomials of the type $M - N$, where M, N are monomials with coefficients 1 without common divisor.

3. **Lattice of relations.** Note that any vector $\alpha \in \mathbb{Z}^n$ can be uniquely written as $\alpha = \alpha_+ - \alpha_-$, with $\alpha_+, \alpha_- \in \mathbb{N}^n$ such as $(\alpha_+)_i(\alpha_-)_i = 0$ for all $i = 1, \ldots, n$. Let

$$L_A := \{\alpha \in \mathbb{Z}^n \mid X^{\alpha_+} - X^{\alpha_-} \in I_A\}.$$

Then $L_A \subset \mathbb{Z}^n$ is a subgroup of finite rank. We call it the *lattice of relations* associated to I_A. It is easy to see that $L_A \subset \mathbb{Z}^n$ is the set of integer solutions of the linear system $AX = 0$.

In general, given a subgroup of finite rank (lattice) $L \subset \mathbb{Z}^n$, we can define the ideal $I_L \subset S$ generated by the binomials $X^{\alpha_+} - X^{\alpha_-}$, $\alpha \in L$. It is called the *lattice ideal* associated to L. We call L saturated if $dv \in L$ for some $d \in \mathbb{Z} \setminus \{0\}$, $v \in \mathbb{Z}^n$, implies $v \in L$.

Remark The lattice of relations of a toric ideal I_A is saturated and has the property

$$I_{L_A} = I_A.$$

For any vector $\alpha \in \mathbb{Z}^n$, we set $F_\alpha := X^{\alpha_+} - X^{\alpha_-}$. Note that F_α is a reduced binomial, that is it can't be factored by a monomial.

Lemma 4.2.1 *Let I_A be a toric ideal and $\mathbf{v_1}, \dots, \mathbf{v_r}$ a basis of L_A. Let $F_{\mathbf{v_i}} \in I_A$ be the binomial associated to $\mathbf{v_i}$. Then*

$$Z(F_{\mathbf{v_1}}, \dots, F_{\mathbf{v_r}}) \cap (K^*)^n = V(I_A) \cap (K^*)^n.$$

Proof We have only to prove the inclusion $Z(F_{\mathbf{v_1}}, \dots, F_{\mathbf{v_r}}) \cap (K^*)^n \subset V(I_A)$.

Let $P \in Z(F_{\mathbf{v_1}}, \dots, F_{\mathbf{v_r}}) \cap (K^*)^n$. Then $F_{\mathbf{v_1}}(P) = 0, \dots, F_{\mathbf{v_r}}(P) = 0$. Let $F \in I_A$ be any reduced binomial then there exist $\mathbf{v} \in L_A$ such that $F = X^{\mathbf{v}_+} - X^{\mathbf{v}_-}$. Since $\mathbf{v_1}, \dots, \mathbf{v_r}$ is a basis of L_A, we can write $\mathbf{v} = \alpha_1\mathbf{v_1} + \cdots + \alpha_r\mathbf{v_r}$ for some integers α_i. Let $P = (x_1, \dots, x_n) \in (K^*)^n$. We have $\underline{x}^{\mathbf{v_i}_+} - \underline{x}^{\mathbf{v_i}_-} = 0$ for $i = 1, \dots, r$. Since $P = (x_1, \dots, x_n) \in (K^*)^n$, this is equivalent to $\underline{x}^{\mathbf{v_i}} = 1$ for $i = 1, \dots, r$, which implies $\underline{x}^{\alpha_i\mathbf{v_i}} = 1$ for $i = 1, \dots, r$. Hence $1 = \underline{x}^{\sum_{i=1}^r \alpha_i\mathbf{v_i}} = \underline{x}^{\mathbf{v}}$, and so $F(P) = 0$.

The following result [7, Corollary 2.6] gives an exact relationship between binomial ideals and toric ideals.

Theorem 4.2.2 *Let K be an algebraically closed field. A binomial ideal is toric if and only if it is prime.*

For simplicity, we say that a binomial ideal is a set-theoretic complete intersection of binomials if $bar(I) = ht(I)$. We have the following theorem from [16].

Theorem 4.2.3 *Let K be a field of characteristic zero. A toric ideal is a set-theoretic complete intersection of binomials if and only if it is a complete intersection.*

By virtue of this theorem, we always assume that our toric ideal is not a complete intersection in the rest of the lecture notes.

4.3 Simplicial Toric Varieties Which Are Set-Theoretic Complete Intersections

Most of the results on set-theoretic complete intersections in this lecture notes concern the following class of toric varieties.

Let $\mathbf{e_1}, \dots, \mathbf{e_n}$ denote the elements of the canonical basis of $\mathbb{Z}^n$. Let $\mathbf{a_i} = (a_{i,1}, \dots, a_{i,n})$, $i = 1, \dots, r$, be non zero vectors in $\mathbb{N}^n$.

Definition 4.3.1 Let A be a matrix with column vectors $d_1\mathbf{e_1}, \dots, d_n\mathbf{e_n}, \mathbf{a_1}, \dots, \mathbf{a_r}$, where $d_1 \dots, d_n \in \mathbb{N}^*$, that is

$$A = \begin{pmatrix} d_1 & 0 & \dots & 0 & a_{1,1} & \dots & a_{r,1} \\ 0 & d_2 & \dots & 0 & a_{1,2} & \dots & a_{r,2} \\ \dots & \dots & \dots & \dots & \dots & \dots & \dots \\ 0 & 0 & \dots & d_n & a_{1,n} & \dots & a_{r,n} \end{pmatrix}.$$

Then I_A is called the *simplicial toric ideal* associated to A and its affine variety $V_A = V(I_A)$ in K^{n+r} is called an *affine simplicial toric variety.*

In this case, the dimension of the affine semigroup ring $K[\Sigma_A]$ is n. Note that V_A has codimension $r \geq 2$ in K^{n+r} and has the following parametrization:

$$
\begin{aligned}
x_1 &= u_1^{d_1},\\
&\vdots\\
x_n &= u_n^{d_n},\\
y_1 &= u_1^{a_{1,1}} \cdots u_n^{a_{1,n}},\\
&\vdots\\
y_r &= u_1^{a_{r,1}} \cdots u_n^{a_{r,n}},
\end{aligned}
$$

One can define a *projective simplicial toric variety* similarly as above. For that we need to assume that $d_1 = \cdots = d_r = \deg u^{\mathbf{a_i}}$ for all $i = 1, \ldots, r$.

For any vector $\mathbf{v} \in \mathbb{Z}^m$, we set $\operatorname{supp}(\mathbf{v}) = \{j \in \{1, \ldots, m\} \mid v_j \neq 0\}$ and call it the support of $\mathbf{v}$.

Definition 4.3.2 We say that the parametrization of V_A is *full* if $\operatorname{supp} \mathbf{a_i} = \operatorname{supp} \mathbf{a_j}$ for $i, j = 1, \ldots, r$. The parametrization of V_A is *almost full* if $\operatorname{supp} \mathbf{a_1} \subset \operatorname{supp} \mathbf{a_2} \subset \cdots \subset \operatorname{supp} \mathbf{a_r}$.

Note that when working with full or almost full parametrization we may always assume that $\operatorname{supp} \mathbf{a_r} = \{1, \ldots, m\}$.

In this section we extend the results on simplicial varieties with full parametrization of [2] to those with almost full parametrization. Namely, we will prove the following results.

1. *In characteristic $p > 0$, any simplicial toric affine or projective variety with almost full parametrization is a set-theoretic complete intersection* (see Theorem 4.3.8).
2. *In any characteristic, any simplicial toric affine or projective variety with full parametrization is an almost set-theoretic complete intersection* (see Theorem 4.3.11).

4.3.1 Lattice of Relations of Simplicial Toric Varieties

As we said above for toric ideals I_A the lattice L_A is the set of integer solutions of the linear system $AX = 0$. That is the problem of finding binomials in I_A is equivalent to finding solutions of $AX = 0$ or more generally of $AX = \mathbf{b}$. For any matrix with integer coefficients A, we set $\mid A \mid$ to be the greatest common divisor of all its maximal minors. We say that the matrix A, has *full rank* if at least one of its maximal minors is non null. Suppose that A has full rank. If there exists one column vector for which some integer multiple belongs to the lattice generated by the other column vectors, we can delete this column vector preserving our search of solutions for the equation $AX = \mathbf{b}$. That means that we can assume that all the maximal minors are non zero.

We have the following basic lemma in Number Theory (see [11], or for a modern presentation, [23, p. 51]):

Lemma 4.3.3 *Assume that* $|A| \neq 0$. *The linear Diophantine system* $AX = \mathbf{b}$ *has an integer solution if and only if* $|A| \neq 0$ *and* $|A| = |A\mathbf{b}|$, *where* $A\mathbf{b}$ *is the augmented matrix.*

Another important ingredient is given by the chapter IV of [5] about basis of Lattices. We learn in this chapter that we can find triangular basis of a lattice that we will describe thanks to Lemma 4.3.3 in the case of simplicial toric varieties.

With the notations of Definition 4.3.1, for all $i = 0 \ldots, r$, let A_i be the matrix:

$$A_i = \begin{pmatrix} d_1 & 0 & \ldots & 0 & a_{1,1} & \ldots & a_{i,1} \\ 0 & d_2 & \ldots & 0 & a_{1,2} & \ldots & a_{i,2} \\ \ldots & \ldots & \ldots & \ldots & \ldots & \ldots & \ldots \\ 0 & 0 & \ldots & d_n & a_{1,n} & \ldots & a_{i,n} \end{pmatrix}.$$

We denote by $\mathbf{d_i}$ the ith column vector of A for all $i = 1, \ldots, n$, and by $\mathbf{a_i}$ the $(n+i)$th column vector of A for all $i = 1, \ldots, r$. We set $D[j_1, \ldots, j_n]$ the determinant of the $n \times n$ submatrix consisting of the columns of A with the indices $j_1, \ldots, j_n$, where $\{j_1, \ldots, j_n\}$ is an n-subset of $\{1, 2, \ldots, n+r\}$. For all $i = 0, \ldots, r$ let $|A_i| := \gcd\{D[j_1, \ldots, j_n] : 1 \leq j_1 < j_2 < \cdots < j_n \leq n+i\}$; for the sake of simplicity we set $g_i = |A_i|$. Moreover, let $\zeta_i = g_{i-1}/g_i$, for all $i = 1, \ldots, r$.

Let us remark that any integer solution α of the linear system $A_i Z = 0$ gives rise to a binomial, more precisely, let write $\alpha = \beta + \gamma$, with $\operatorname{supp}\beta \subset \{1, 2, \ldots, n\}$, $\operatorname{supp}\gamma \subset \{n+1, n+2, \ldots, n+i\}$, then the binomial $F_{\alpha+\beta} = \underline{x}^{\beta_+} y^{\gamma_+} - x^{\beta_-} y^{\gamma_-}$ in the variables $x_1, \ldots, x_n, y_1, \ldots, y_i$ belongs to I_A.

In our situation we have the following corollary of Lemma 4.3.3 which can be seen as a generalization of [15, Remark 2.1.2]:

Theorem 4.3.4 *Keep the above notations. Then*

1. *For any* $i = 1, \ldots, r$, *the linear Diophantine system* $A_{i-1} Z = \theta \mathbf{a_i}$ *has an integer solution if and only if* $\theta \in \zeta_i \mathbb{Z}$.
2. *The lattice* $L_A \subset \mathbb{Z}^{n+r}$ *of rank* r *has a triangular basis:*

$$\{(\mathbf{w_1}, s_{(1,1)}, 0, \ldots, 0), \ (\mathbf{w_2}, s_{(2,1)}, s_{(2,2)}, 0, \ldots, 0), \ \ldots, \\ (\mathbf{w_r}, s_{(r,1)}, s_{(r,2)}, \ldots, s_{(r,r)})\},$$

where $\mathbf{w_1}, \ldots, \mathbf{w_r} \in \mathbb{Z}^n$ *and* $s_{(i,i)} = \zeta_i$.

3. *Let* $\mathbf{s_1} = (s_{(1,1)}, 0, \ldots, 0), \mathbf{s_1} = (s_{(2,1)}, s_{(2,2)}, 0, \ldots, 0), \ldots, \mathbf{s_r} = (s_{(r,1)}, s_{(r,2)}, \ldots, s_{(r,r)})$. *For* $i \in \{1, \ldots, r\}$ *we have the reduced binomials*

$$F_{\mathbf{w_i}+\mathbf{s_i}} := M_i - N_i y_i^{\zeta_i} \in I_A,$$

where M_i, N_i *are monomials in* $K[x_1, \ldots, x_n, y_1, \ldots, y_{i-1}]$.
4. $Z(F_{\mathbf{w_1}+\mathbf{s_1}}, \ldots, F_{\mathbf{w_r}+\mathbf{s_r}}) \cap (K^*)^{n+r} \subset V_A$.

Proof

1. We have $g_0 = d_0 d_1 \ldots d_n$ and for all $1 \le i \le r$, the numbers g_{i-1} are non null. On the other hand it holds:

$$g_i = \gcd\{g_{i-1}, D[j_1, \ldots, j_{n-1}, n+i] : 1 \le j_1 < j_2 < \cdots < j_n \le n+i-1\}, \tag{4.1}$$

which yields

$$1 = \gcd\{\frac{g_{i-1}}{g_i}, \frac{D[j_1, \ldots, j_{n-1}, n+i]}{g_i} : 1 \le j_1 < j_2 < \cdots < j_n \le n+i-1\}, \tag{4.2}$$

$$\begin{aligned} |A_{i-1}, \theta \mathbf{a_i}| &= \gcd\{g_{i-1}, \theta D[j_1, \ldots, j_{n-1}, n+i] : 1 \le j_k \le n+i-1\} \\ &= \gcd\{(\frac{g_{i-1}}{g_i})g_i, \theta D[j_1, \ldots, j_{n-1}, n+i] : 1 \le j_k \le n+i-1\} \\ &= g_i \gcd\{(\frac{g_{i-1}}{g_i}), \theta \frac{D[j_1, \ldots, j_{n-1}, n+i]}{g_i} : 1 \le j_k \le n+i-1\} \end{aligned}$$

Hence $|A_{i-1}, \theta \mathbf{a_i}| g_{i-1} = |A_{i-1}|$ if and only if

$$g_i \gcd\{(\frac{g_{i-1}}{g_i}), \theta \frac{D[j_1, \ldots, j_{n-1}, n+i]}{g_i} : 1 \le j_k \le n+i-1\} = g_{i-1},$$

or equivalently

$$\gcd\{\zeta_i, \theta \frac{D[j_1, \ldots, j_{n-1}, n+i]}{g_i} : 1 \le j_k \le n+i-1\} = \zeta_i.$$

Using (2) it implies that $|A_{i-1}, \theta \mathbf{a_i}| g_{i-1} = |A_{i-1}|$ if and only if $\theta \in \zeta_i \mathbb{Z}$.
2. By the first part, for every $i \in \{1, \ldots, r\}$ the Diophantine system $A_{i-1}\mathbf{x} = \zeta_i \mathbf{a_i}$ always has a solution. This means that the vector $\zeta_i \mathbf{a_i}$ can be expressed as a linear combination of the vectors $\mathbf{d_1}, \ldots, \mathbf{d_n}, \mathbf{a_1}, \ldots, \mathbf{a_{i-1}}$ with integer coefficients, i.e., one has

$$\zeta_i \mathbf{a_i} = w_{(i,1)}\mathbf{d_1} + \cdots + w_{(i,n)}\mathbf{d_n} + s_{(i,1)}\mathbf{a_1} + \cdots + s_{(i-1,i-1)}\mathbf{a_{i-1}}, \tag{4.3}$$

for some integers $w_{(i,j)}, \dots, s_{(i,j)}$. Setting for every $i \in \{1, \dots, r\}$ $\mathbf{w_i} = (w_{(i,1)}, \dots, w_{(i,n)})$, we have that

$$\{(\mathbf{w_1}, s_{(1,1)}, 0, \dots, 0), \ (\mathbf{w_2}, s_{(2,1)}, s_{(2,2)}, 0, \dots, 0), \ \dots, \\ (\mathbf{w_r}, s_{(r,1)}, s_{(r,2)}, \dots, s_{(r,r)})\},$$

is a triangular basis of L_A.
3. The expression (3) gives us monomials M_i, N_i in $K[x_1, \dots, x_n, y_1, \dots, y_{i-1}]$ such that $F_{\mathbf{w_i}+\mathbf{s_i}} := M_i - N_i y_i^{\zeta_i}$.
4. Follows from the above items and Lemma 4.2.1.

Triangular basis will give us some particular binomials which will play an important role in our proofs.

Remark For the sake of simplicity we shall set $\mathbf{s} = (s_1, \dots, s_{r-1})$, $\underline{y} = (y_1, \dots, y_{r-1})$. In particular, if $(\mathbf{w}, \mathbf{s}, t) \in L_A$, then $t \in \zeta_r \mathbb{Z}$ and, conversely, for all multiples t of ζ_r there is $\mathbf{s} \in \mathbb{Z}^{r-1}$, $\mathbf{w} \in \mathbb{Z}^n$ such that $(\mathbf{w}, \mathbf{s}, t) \in L_A$.

For all $\mathbf{s} \in \mathbb{Z}^{r-1}$, we can write $\mathbf{s} = \mathbf{s}_+ - \mathbf{s}_-$. Fix an element $(\mathbf{w}, \mathbf{s}, s_r) \in L_A$. Let $\mathbf{w} = \mathbf{w}_+ - \mathbf{w}_-$. Then the binomial corresponding to $(\mathbf{w}, \mathbf{s}, s_r) \in L_A$ is

$$\underline{y}^{\mathbf{s}_+} \underline{x}^{\mathbf{w}_+} - y_r^{-s_r} \underline{y}^{\mathbf{s}_-} \underline{x}^{\mathbf{w}_-},$$

provided $s_r \le 0$; otherwise it is

$$\underline{y}^{\mathbf{s}_+} y_r^{s_r} \underline{x}^{\mathbf{w}_+} - \underline{y}^{\mathbf{s}_-} \underline{x}^{\mathbf{w}_-}.$$

Remark Let

$$J = I_A \cap K[x_1, \dots, x_n, y_1, \dots, y_{r-1}].$$

Then J is the defining ideal of the simplicial toric variety of codimension $r - 1$ having the following parametrization:

$$\begin{aligned} x_1 &= u_1^{d_1}, \\ &\vdots \\ x_n &= u_n^{d_n}, \\ y_1 &= u_1^{a_{1,1}} \cdots u_n^{a_{1,n}}, \\ &\vdots \\ y_{r-1} &= u_1^{a_{r-1,1}} \cdots u_n^{a_{r-1,n}}. \end{aligned}$$

Note that if the parametrization of the variety defined by I_A is full (resp. almost full), then the parametrization of the variety defined by J satisfies the same property.

4.3.2 Simplicial Toric Varieties in Characteristic $p > 0$

We introduce one more piece of notation. Let M_1, M_2 be monomials, and let $h = M_1 - M_2$. For all positive integers q we set

$$h^{(q)} = M_1^q - M_2^q.$$

Lemma 4.3.5 *Let* $J = I_A \cap K[x_1, \ldots, x_n, y_1, \ldots, y_{r-1}]$, *and* $\delta > 0$ *an integer for which there is a binomial*

$$f_r = y_r^{\zeta_r \delta} - \underline{y}^{\mathbf{s}_\delta} x_1^{l_1} \cdots x_n^{l_n} \in I_A.$$

Then for any binomial h *in* I_A *we have*

$$h^{(\delta)} \in (J, f_r).$$

Proof Let $h \in I_A$ be a binomial. Since I_A is a prime ideal, we may assume that

$$h = y_r^{\zeta_r \rho} g_1 - g_2$$

for some monomials $g_1, g_2 \in K[x_1, \ldots, x_n, y_1, \ldots, y_{r-1}]$. Then

$$\begin{aligned} h^{(\delta)} &= y_r^{\zeta_r \rho \delta} g_1^\delta - g_2^\delta \\ &= (f_r^{(\rho)} + (\underline{y}^{\mathbf{s}_\delta} x_1^{l_1} \cdots x_n^{l_n})^\rho) g_1^\delta - g_2^\delta \\ &\in (J, f_r). \end{aligned}$$

Lemma 4.3.6 *Suppose that* $\operatorname{supp} \mathbf{a_r} = \{1, \ldots, m\}$. *For all sufficiently large integers* $\delta > 0$ *there is a binomial*

$$f_r = y_r^{\zeta_r \delta} - \underline{y}^{\mathbf{s}} x_1^{l_1} \cdots x_n^{l_n} \in I_A.$$

Proof Let $\delta > 0$. There is $\mathbf{s}'$ such that $(\mathbf{s}', -\zeta_r) \in Ker\Phi$. Hence there are integers $r'_1, \ldots, r'_n$ such that for all i

$$\sum_{j=1}^{r-1} s'_j a_{j,i} - \zeta_r a_{r,i} = r'_i d_i$$

for all i. Multiplying this relation by $\delta > 0$ we get

$$\sum_{j=1}^{r-1} \delta s'_j a_{j,i} - \zeta_r \delta a_{r,i} = \delta r'_i d_i$$

for all i. Let $d = \operatorname{lcm}\{d_1, \ldots, d_n\}$. Replacing $\delta s'_j$ by its residue s_j modulo d, we get a relation

$$\sum_{j}^{r-1} s_j a_{j,i} - \zeta_r \delta a_{r,i} = r_i d_i.$$

Thus, if δ is sufficiently large, we will have $r_i < 0$ for all i. Then $f_r = y_r^{\zeta_r \delta} - \underline{y}^{\mathbf{s}} x_1^{-r_1} \cdots x_n^{-r_n} \in I_A$ as required.

As an immediate consequence we have:

Corollary 4.3.7 *Suppose that* $\operatorname{supp} \mathbf{a_r} = \{1, \ldots, m\}$. *Let p be a prime number. For any sufficiently large integer m there is a binomial*

$$f_r = y_r^{\zeta_r p^m} - \underline{y}^{\mathbf{s}} x_1^{l_1} \cdots x_n^{l_n} \in I_A.$$

The next theorem improves [2, Theorem 1], where the case of full parametrization was considered.

Theorem 4.3.8 *Suppose that char $K = p > 0$. Then every simplicial toric variety having an almost full parametrization is a set-theoretic complete intersection.*

Proof We proceed by induction on $r \geq 1$. Since the polynomial ring $K[x_1, \ldots, x_n, y_1]$ is an UFD the claim is true for $r = 1$.

Suppose that $r \geq 2$ and the claim is true in codimension $r - 1$. Let $h \in I_A$ be a binomial, then by Corollary 4.3.7 and Lemma 4.3.6, for m sufficiently large we get

$$h^{p^m} = h^{(p^m)} \in (f_r, J).$$

By the induction hypothesis the ideal J is set-theoretically generated by $r - 1$ binomials $f_1, \ldots, f_{r-1}$. Hence some power of h lies in $(f_1, \ldots, f_r)$.

Remark Note that the proof of the preceding result yields an effective and recursive construction of the defining equations of a simplicial toric variety having almost full parametrization over any field K of characteristic $p > 0$.

Exercise 4.3.9 Assume that K is a field of characteristic p. Let $\overline{V}(1, 3, 4)$ be the projective toric curve in $\mathbb{P}^3$ with parametrization

$$w = u^4, x = u^3 v^1, y = u^1 v^3, z = v^4.$$

1. Write the matrix A corresponding to $\overline{V}(1, 3, 4)$.
2. Use Theorem 4.3.4 to find a triangular basis of the Lattice L_A.
3. Give f_1, f_2 such that $\operatorname{rad}(f_1, f_2) = I(\overline{V}(1, 3, 4))$ showing that in characteristic p, $\overline{V}(1, 3, 4)$ is a set-theoretic complete intersection.

4.3.3 Almost Set-Theoretic Complete Intersections

We have studied the case where the field K is of characteristic $p > 0$, so now we assume that the field K is algebraically closed of characteristic 0, since we will use the Hilbert's Nullstellensatz.

In this section we show that simplicial toric varieties having a full parametrization are almost set-theoretic complete intersections.

If the parametrization of V_A is full we will improve the triangular basis of I_A founded in Theorem 4.3.4.

Lemma 4.3.10 *Let V_A be a simplicial toric variety. If the parametrization of V_A is full, then for every $i = 2, \ldots, r$ there exists a binomial*

$$F_i = y_{i-1}^{\mu_i} - x_1^{\nu_{i,1}} \cdots x_n^{\nu_{i,n}} y_1^{\mu_{i,1}} \cdots y_{i-2}^{\mu_{i,i-2}} y_i^{\zeta_i} \in I_A,$$

and there also exists a binomial

$$F_1 = y_1^{\zeta_1} - x_1^{\nu_{1,1}} \cdots x_n^{\nu_{1,n}} \in I_A,$$

for some strictly positive integers μ_i, $\mu_{i,j}$ and $\nu_{i,j}$.

Proof In this proof, for all $i = 1, \ldots, n$, $\mathbf{d_i}$ will denote the ith column vector of A and for all $i = 1, \ldots, r$, $\mathbf{a_i}$ will denote the $(n+i)$th column vector of A.

Set $\mu = \gcd(d_1, \ldots, d_n)$ and $q_i = \gcd(\mu, a_{i,1}, \ldots, a_{i,n})$ for all $i = 1, \ldots, r$. For all $i = 1, \ldots, r$ and all $j = 1, \ldots, n$ let $\rho_{i,j} = a_{i,j}\mu/d_j q_j$. Then, for all $i = 1, \ldots, r$, one has that

$$G_i = y_i^{\mu/q_i} - x_1^{\rho_{i,1}} \cdots x_n^{\rho_{i,n}} \in I_A.$$

It is easy to see that $\zeta_1 = \mu/q_1$, then for $i = 1$ the preceding formula yields the required binomial F_1.

As we have seen in Theorem 4.3.4, for all $i = 1, \ldots, r$ the vector $\zeta_i \mathbf{a_i}$ can be expressed as a linear combination of the vectors $\mathbf{d_1}, \ldots, \mathbf{d_n}, \mathbf{a_1}, \ldots, \mathbf{a_{i-1}}$ with integer coefficients, i.e., one has

$$\zeta_i \mathbf{a_i} = w_{(i,1)}\mathbf{d_1} + \cdots + w_{(i,n)}\mathbf{d_n} + s_{(i,1)}\mathbf{a_1} + \cdots + s_{(i-1,i-1)}\mathbf{a_{i-1}}, \tag{4.2}$$

for some integers $w_{(i,j)}, \ldots, s_{(i,j)}$ and this expression gives us monomials M_i, N_i in $K[x_1, \ldots, x_n, y_1, \ldots, y_{i-1}]$ such that $M_i - N_i y_i^{\zeta_i} \in I_A$.

Now suppose that the parametrization of V_A is full. From the binomial G_j we see that for each $\mathbf{a_j}$ there exist positive integers $\rho_j = \mu/q_i, \rho_{j,1}, \ldots, \rho_{j,n}$ such that $\rho_j \mathbf{a_j} = \rho_{j,1}\mathbf{d_1} + \cdots + \rho_{j,n}\mathbf{d_n}$. Furthermore, for all $1 \le j \le i-2$ there exists a positive integer ν_j such that, after adding all the zero vectors $\nu_j(\rho_{j,1}\mathbf{d_1} + \cdots + \rho_{j,n}\mathbf{d_n} - \rho_j \mathbf{a_j})$ to the right-hand side of (2), the new coefficient $-\mu_{i,j}$ of $\mathbf{a_j}$ is negative for all $j = 1, \ldots, i-2$. There also exists a large positive integer ν_{i-1}

such that after adding the zero vector $\nu_{i-1}(\rho_{i-1}\mathbf{a_{i-1}} - (\rho_{i-1,1}\mathbf{d_1} + \cdots + \rho_{i-1,n}\mathbf{d_n}))$ on the right-hand side of the new equation, for all $j = 1, \ldots, n$ the new coefficient $-\nu_{i,j}$ of $\mathbf{d_j}$ is negative and the new coefficient μ_i of $\mathbf{a_{i-1}}$ is positive. It follows that for all $i = 2\ldots, r$ we have a binomial

$$F_i = y_{i-1}^{\mu_i} - x_1^{\nu_{i,1}} \cdots x_n^{\nu_{i,n}} y_1^{\mu_{i,1}} \ldots y_{i-2}^{\mu_{i,i-2}} y_i^{\zeta_i} \in I_A.$$

Theorem 4.3.11 *Assume that K is algebraically closed field of characteristic 0. Let V_A be a simplicial toric variety having a full parametrization. Then $r \leq bar(I_A)) \leq r+1$. In fact $bar(I_A) = r+1$ unless I_A is a complete intersection.*

Proof Consider the r binomials $F_1, F_2, \ldots, F_r$ which were defined in Lemma 3 and let F_{r+1} be any binomial monic in y_r, for example G_r. We claim that $I_A = \text{rad}\,(F_1, \ldots, F_{r+1})$.

By virtue of Hilbert Nullstellensatz the claim is proved once it has been shown that every point $\mathbf{x} = (x_1, \ldots, x_n, y_1, \ldots, y_r)$ which is a common zero of $F_1, \ldots, F_{r+1}$ in K^{n+r} is also a point of V_A. First of all note that if $x_k = 0$ for some index k, then $y_j = 0$ for all indices j. It is then easy to find $u_1, \ldots, u_n \in K$ which allow us to write $\mathbf{x}$ as a point of V_A. Now suppose that $x_k \neq 0$ for all indices k, $F_1(\mathbf{x}) = 0, \ldots, F_{r+1}(\mathbf{x}) = 0$, we have inductively that $y_1 \neq 0, \ldots, y_r \neq 0$. So we can assume that all the coordinates of $\mathbf{x}$ are non zero. Note that the vectors in L_A corresponding to $F_1, F_2, \ldots, F_r$ form a triangular basis of L_A, hence by applying Theorem 4.3.4 we have that $\mathbf{x}$ is a point of V_A.

Exercise 4.3.12 Assume that K is an algebraically closed field of characteristic 0. Let $V(1,3,4)$ be the projective toric curve in $\mathbb{P}^3$ with parametrization

$$w = u^4, x = u^3v^1, y = u^1v^3, z = v^4.$$

Use Exercise 4.3.9 and the above section to give F_1, F_2, F_3 binomials such that $\text{rad}\,(F_1, F_2, F_3) = I(V(1,3,4))$.

4.4 Equations in Codimension 2

This section is an English shorten version of the results in [15].

In this section we suppose that $r = 2$, i.e., V_A is a simplicial toric variety of codimension 2 in K^{n+2}. The parametrization of V_A now is:

$$x_1 = u_1^{d_1},$$

$$\vdots$$

$$x_n = u_n^{d_n},$$

$$y_1 = u_1^{a_{1,1}} \cdots u_n^{a_{1,n}},$$
$$y_2 = u_1^{a_{2,1}} \cdots u_n^{a_{2,n}},$$

where the vectors $\mathbf{a_1}, \mathbf{a_2}$ may have zero components.

4.4.1 The Lattice Associated in Codimension Two

In this section, we introduce the reduced lattice associated to V_A, which determines the associated lattice L_A, in this particular case.

Consider the morphism of groups:

$$\Phi : \mathbb{Z}^2 \longrightarrow \mathbb{Z}/d_1\mathbb{Z} \times \cdots \times \mathbb{Z}/d_n\mathbb{Z} \quad (s, p) \mapsto (sb_1 - pc_1, \ldots, sb_n - pc_n)$$

Definition 4.4.1 The reduced lattice associated to V_A is

$$Ker(\Phi) := \{(s, p) \in \mathbb{Z}^2 \mid sb_i - pc_i \equiv 0 \, mod \, d_i, \forall i = 1, \ldots, n\}.$$

Remark $Ker(\Phi)$ is not the lattice of V_A in the sense given in Sect. 4.3, but it determines the lattice of V_A. For any $i = 1, \ldots, n$ there exists integers numbers l_i such that $sb_i - pc_i = l_i d_i$. To the vector $(s, p) \in Ker(\Phi)$ corresponds the vectors $(-l_1, \ldots, -l_n, s, -p)$ in the Lattice L_A. As a consequence, we associate to the vector $(s, p) \in Ker(\Phi)$ with $s \geq 0$ a binomial $F_{(-l_1,\ldots,-l_n,s,-p)} \in I_A$ and we call it the binomial associated to (s, p). Reciprocally, any vector $(\mathbf{w}, s, -p) \in L_A$, with $s \geq 0$, determines a unique $(s, p) \in Ker(\Phi)$.

Proposition 4.4.2 *We will define a fan decomposition of* $Ker(\Phi)$ *in* $\mathbb{R}^2_+$, *i.e. we will determine a family of vectors* $\epsilon_{-1}, \epsilon_0, \ldots, \epsilon_{m+1} \in Ker(\Phi) \cap \mathbf{Z}^2_+$ *such that* $\epsilon_i, \epsilon_{i+1}$ *is a base of* $Ker(\Phi)$, *with* $det(\epsilon_i, \epsilon_{i+1}) > 0$.

Proof We use the notion of base adapted to a lattice used in [5] p. 67. This allows us to determine a base $\epsilon_{-1}, \epsilon_0$ of $Ker(\Phi)$. Precisely $\epsilon_{-1} = (s_{-1}, 0)$, $\epsilon_0 = (s_0, p_0)$ where s_{-1} is the smallest positive integer $s \neq 0$ such that $(s, 0) \in Ker(\Phi)$ and p_0 is the smallest positive integer $p \neq 0$ such that there is a vector $(s, p) \in Ker(\Phi)$, s_0 is unique defined such that $s_0 < s_{-1}$.

Consider Euclide's algorithm, with negative rest, for the computation of the greatest common divisor, gcd (s_{-1}, s_0):

$$s_{-1} = q_1 s_0 - s_1$$
$$s_0 = q_2 s_1 - s_2$$
$$\ldots$$
$$s_{m-1} = q_{m+1} s_m$$

$$s_{m+1} = 0$$

$$\forall i \; q_i \geq 2 \; , \; s_i \geq 0.$$

Let us define the sequence: $p_i \; (-1 \leq i \leq m+1)$, by $p_{-1} = 0$ and:

$$p_{i+1} = p_i q_{i+1} - p_{i-1} \; , (0 \leq i \leq m).$$

We set $\epsilon_i = (s_i, p_i)$. By induction it is easy to check that $s_i p_{i+1} - s_{i+1} p_i = p_0 s_{-1}$ for all $-1 \leq i \leq m+1$, completing the proof.

In particular we have defined two sequences $\{s_i\}$, $\{p_i\}$.

Example 4.4.3 Let consider the projective monomial curve with parametrization:

$$X = s^{10}, Y = s^7 t^3 Z = s^3 t^7, W = t^{10}.$$

The lattice $Ker(\Phi)$ is given by the vectors (s, p) such that (r, r', s, p) is an integer solution of the system:

$$7s - 3p = 10r$$

$$s - 7p = 10r'$$

Note that the Lattice L_A is given by the vectors $(-r, -r', s, -p)$ such that (s, p, r, r') is an integer solution of the above system.

We have the following table

i	s_i	p_i	r_i	r'_i	q_i
−1	10	0	7	3	0
0	9	1	6	2	0
1	8	2	5	1	2
2	7	3	4	0	2
3	6	4	3	−1	2
4	5	5	2	−2	2
5	4	6	1	−3	2
6	3	7	0	−4	2
7	2	8	−1	−5	2
8	1	9	−2	−6	2
9	0	10	−3	−7	2

Corollary 4.4.4 *For* $i = -1, \ldots, m+1$ *we set* $\epsilon_i = (s_i, p_i)$. *With the above notations, the vectors*

$$\epsilon_{m+1}, \ldots, \epsilon_0, \epsilon_{-1}, \epsilon_{-1} - \epsilon_0, \ldots, \epsilon_{-1} - (q_1 - 1)\epsilon_0 = \epsilon_0 - \epsilon_1, \ldots, \epsilon_0 - (q_2 - 1)\epsilon_1$$
$$= \epsilon_1 - \epsilon_2, \ldots,$$
$$\epsilon_{m-2} - (q_{m-1} - 1)\epsilon_{m-1} = \epsilon_{m-1} - \epsilon_m, \ldots, \epsilon_{m-1} - (q_m - 1)\epsilon_m$$
$$= \epsilon_m - \epsilon_{m+1}, -\epsilon_{m+1}$$

are a fan decomposition of $\mathbb{R}_+ \times \mathbb{R}$. *The determinant of two consecutive vectors is* $-p_0 s_{-1}$.

Proof The conclusion is a consequence of the above Proposition, since :

$$\det(\epsilon_{i-1} - j\epsilon_i, \epsilon_{i-1} - (j+1)\epsilon_i) = -\det(\epsilon_{i-1}, \epsilon_i).$$

The fan decomposition of $\mathbb{R}_+ \times \mathbb{R}$ is represented in Fig. 4.1:

Corollary 4.4.5 *The set of binomials associated to the vectors*

$$\epsilon_{m+1}, \ldots, \epsilon_0, \epsilon_{-1}, \epsilon_{-1} - \epsilon_0, \ldots, \epsilon_{-1} - (q_1 - 1)\epsilon_0$$
$$= \epsilon_0 - \epsilon_1, \ldots, \epsilon_0 - (q_2 - 1)\epsilon_1 = \epsilon_1 - \epsilon_2, \ldots,$$
$$\epsilon_{m-2} - (q_{m-1} - 1)\epsilon_{m-1} = \epsilon_{m-1} - \epsilon_m, \ldots, \epsilon_{m-1} - (q_m - 1)\epsilon_m = \epsilon_m - \epsilon_{m+1}, -\epsilon_{m+1}$$

is a Universal Grobner Basis of I_A.

Fig. 4.1 Fan decomposition

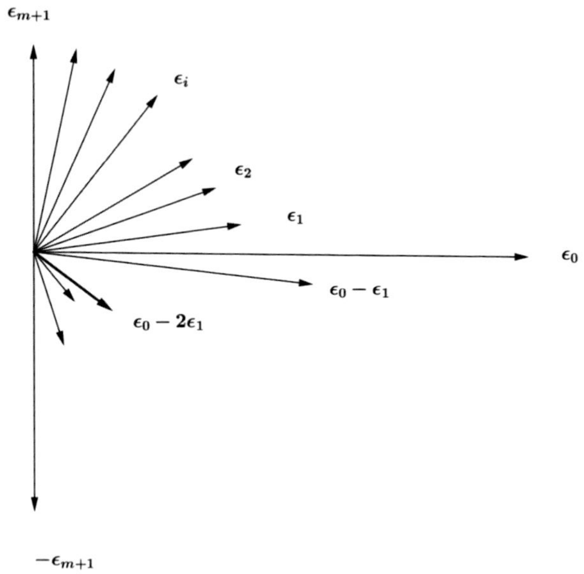

4.4.2 Effective Computation of the Fan Associated to the Universal Grobner Basis of I_A

We can assume that d_i, b_i, c_i are coprime.

Lemma 4.4.6 *For any i let* $\delta_i = \gcd(d_i, b_i)$, *and*

$$\Phi_i : \mathbb{Z}^2 \longrightarrow \mathbb{Z}/d_i\mathbb{Z} \quad (s, p) \mapsto (sb_i - pc_i)$$

Then $Ker(\Phi_i)$ *is a* $\mathbb{Z}$*–free submodule of* $\mathbb{Z}^2$ *generated by the vectors* $(d_i/\delta_i, 0)$, $(\tilde{s}_{i,0}, \delta_i)$ *where* $\tilde{s}_{i,0}$ *is the unique integer such that* $\tilde{s}_{i,0}b_i - (\delta_i)c_i \equiv 0 \; mod \; d_i$ *and* $0 \leq \tilde{s}_{i,0} < d_i/\delta_i$.

The proof is elementary. We have the following consequence:

Lemma 4.4.7 *Let*

$$\begin{aligned} \rho_i &= \gcd(d_1/\delta_1, d_i/\delta_i)\;, \;\; \chi_i = \gcd(\delta_1, \delta_i), \\ \kappa_i &= \gcd((\delta_1 s_{i,0})/(\chi_i) - (\delta_i s_{1,0})/(\chi_i), \rho_i), \\ s_{-1} &= \text{lcm}(d_1/\delta_1, \ldots, d_n/\delta_n), \quad p_{-1} = 0, \quad \text{and} \\ p_0 &= \text{lcm}_{2 \leq i \leq n}((\rho_i/\kappa_i)\text{lcm}(\delta_1, \delta_i)). \end{aligned}$$

Then $Ker\Phi$ *is a subgroup of* $\mathbb{Z}^2$ *generated by the vectors:* (s_{-1}, p_{-1}) (s_0, p_0) *where* s_0 *is the unique integer such that:*

$$0 \leq s_0 < \text{lcm}(d_1/\delta_1, \ldots, d_n/\delta_n) \quad \text{and}$$

$$\forall i \in \{1, \ldots, n\} \;\; s_0 \equiv s_{i,0} p_0/\delta_i \; mod \; d_i/\delta_i.$$

For the proof we refer to [15].

Definition 4.4.8 We define the sequences of integers $\{s_i\}$, $\{p_i\}$ as in Proposition 4.4.2. That is $\{s_i\}$ is defined by Euclid algorithm and $\{p_i\}$ by $p_{-1} = 0$ and:

$$p_{i+1} = p_i q_{i+1} - p_{i-1}\;, (0 \leq i \leq m).$$

For all $j \in \{1, \ldots, n\}$ we define the sequences $\{r_{j,i}\}$ by

$$r_{j,i} = (s_i b_j - p_i c_j)/d_j \quad -1 \leq i \leq m+1, \; 1 \leq j \leq n.$$

Lemma 4.4.9

1) The sequences $\{s_i\}$, $\{p_i\}, \{r_{j,i}\}$, $1 \leq j \leq n$ satisfy the following recurrent relations:
$v_{i+2} = q_{i+2}v_{i+1} - v_i$ *for* $-1 \leq i \leq m-1$.
2) $r_{j,-1} = s_{-1}b_i/d_i.\forall 1 \leq j \leq n$
3) For any index i such that $-1 \leq i \leq m$, *we have:*

$$i)\ s_i p_{i+1} - s_{i+1} p_i = s_{-1} p_0$$

$$ii)\ s_{i+1} r_{j,i} - s_i r_{j,i+1} = s_{-1} p_0 c_i / d_i$$

$$iii)\ p_{i+1} r_{j,i} - p_i r_{j,i+1} = s_{-1} p_0 b_i / d_i$$

Lemma 4.4.10 *For all j the sequences* $\{s_i\}, \{r_{j,i}\}$ *are strictly decreasing, and the sequence* $\{p_i\}$ *is strictly increasing.*

Definition 4.4.11

1) Let D_j be the line with equation $sb_j - pc_j = 0$. By changing if necessary the order of the variables x_j we can assume that the slopes of the lines D_j are in increasing order.
2) Let ν (resp. μ) the unique integer such that $r_{1,\nu} \geq 0 > r_{1,\nu+1}$, (resp. $r_{n,\mu} > 0 \geq r_{n,\mu+1}$).
3) Suppose that $\mu \neq \nu$. For $1 \leq i \leq \mu - \nu$ let k_i be the smallest integer $j \leq n-1$ such that $r_{j,\nu+i} < 0$. We set $k_{\mu-\nu+1} = n$.

Lemma 4.4.12 *We have:*

i) $-1 \leq \nu \leq \mu \leq m$,
ii) let $1 \leq i \leq \mu - \nu$. *If* $l \leq k_i$ *then* $r_{l,\nu+i} < 0$ *and if* $l > k_i$ *then* $r_{l,\nu+i} \geq 0$,
iii) if $r_{j,\mu+1} = 0$ *then* $r_{n,\mu+1} = 0$, *and*
iv) $\mu = \nu$ *if and only if* $r_{j,u} \leq 0$ *for all* $j \in \{1, \ldots, n\}$ *and* $u \geq \nu + 1$.

Theorem 4.4.13 ([15]) *Let* V_A *be a simplicial toric variety of codimension 2.* V_A *is arithmetically Cohen-Macaulay if and only if* $\mu = \nu$. *If* V_A *is not arithmetically Cohen-Macaulay the ideal* I_A *is minimally generated by the binomials associated to the vectors*

$$\begin{gathered} \epsilon_\nu\ ,\ \epsilon_{\nu+1}\ ,\ \epsilon_\nu - \epsilon_{\nu+1}, \\ \epsilon_\nu - 2\epsilon_{\nu+1}\ ,\ \ldots\quad ,\ \epsilon_\nu - q_{\nu+2}\epsilon_{\nu+1}\ ,\ \epsilon_{\nu+2}, \\ \ldots \\ \epsilon_{\mu-1} - 2\epsilon_\mu\ ,\ \ldots\quad ,\ \epsilon_{\mu-1} - q_{\mu+1}\epsilon_\mu\ ,\ \epsilon_{\mu+1}. \end{gathered}$$

The proof consist to check that the mentioned binomials are a Grobner basis of I_A.

Example 4.4.14 We consider again the Toric variety of Example 4.4.3, with parametrization

$$X = s^{10}, Y = s^7t^3 Z = s^3t^7, W = t^{10}.$$

Its defining ideal is generated by the polynomials:

$$\begin{aligned} F_1 &= Z^7 - Y^3W^4, \\ F_2 &= YZ - XW, \\ F_3 &= Y^4W^3 - XZ^6, \\ F_4 &= Y^5W^2 - X^2Z^5, \\ F_5 &= Y^6W - X^3Z^4, \\ F_6 &= Y^7 - X^4Z^3. \end{aligned}$$

Theorem 4.4.15 *[15] Let V_A be a simplicial toric variety of codimension 2. Assume that V_A is arithmetically Cohen-Macaulay that is $\mu = \nu$. The ideal I_A is generated by three binomials $F_{\epsilon_\nu}, F_{\epsilon_{\nu+1}}, F_{\epsilon_\nu - \epsilon_{\nu+1}}$ associated to the vectors*

$$\epsilon_\nu, \epsilon_{\nu+1}, \epsilon_\nu - \epsilon_{\nu+1}.$$

That is

$$\begin{aligned} F_{\epsilon_\nu} &= z^{s_\nu} - y^{p_\nu} x_1^{r_{1,\nu}} \ldots x_n^{r_{n,\nu}}, \\ F_{\epsilon_{\nu+1}} &= y^{p_{\nu+1}} - z^{s_{\nu+1}} x_1^{-r_{1,\nu+1}} \ldots x_n^{-r_{n,\nu+1}}, \\ F_{\epsilon_\nu - \epsilon_{\nu+1}} &= z^{s_\nu - s_{\nu+1}} y^{p_{\nu+1} - p_\nu} - x_1^{r_{1,\nu} - r_{1,\nu+1}} \ldots x_n^{r_{n,\nu} - r_{n,\nu+1}}. \end{aligned}$$

In fact $F_{\epsilon_\nu}, F_{\epsilon_{\nu+1}}, F_{\epsilon_\nu - \epsilon_{\nu+1}}$ are the 2×2 minors of the matrix

$$M = \begin{pmatrix} x_1^{r_{1,\nu}} \ldots x_n^{r_{n,\nu}} & y^{p_\nu} & z^{s_\nu - s_{\nu+1}} \\ y^{p_{\nu+1} - p_\nu} & z^{s_{\nu+1}} & x_1^{-r_{1,\nu+1}} \ldots x_n^{-r_{n,\nu+1}} \end{pmatrix}.$$

Moreover I_A is a complete intersection if and only if either $p_\nu = 0$ or $s_{\nu+1} = 0$.

Exercise 4.4.16 Let K be any field. Let $V(1, 3, 4)$ be the projective toric curve in $\mathbb{P}^3$ with parametrization

$$w = u^4, x = u^3v^1, y = u^1v^3, z = v^4.$$

1. Draw the fan decomposition of $V(1, 3, 4)$.
2. Use Theorem 4.4.13 to find a minimal generating set F_1, F_2, F_3, F_4 of the ideal I_A.
3. Use the fact that we have an explicit formulation of $Ker(\Phi)$, and so of L_A, together with the fan decomposition to prove directly Theorem 4.4.13 for this example. (Hint. Binomials are represented by plane vectors.)

The material developed in this section help to understand not only generators but also syzygies for codimension two simplicial toric ideals. See for example [6].

4.5 Almost-Complete Intersections and Set-Theoretic Complete Intersections

From now on, we assume that the field K is algebraically closed of characteristic 0, since we will use the Hilbert's Nullstellensatz.

4.5.1 Almost-Complete Intersections: The General Case

Lemma 4.5.1 *Assume that we have r binomials in* $K[x_1, \ldots, x_n, y_1, \ldots, y_r]$*:*

$$\begin{aligned} F_1 &= y_1^{\rho_1} - y_2^{\beta_{1,2}} \cdots y_r^{\beta_{1,r}} h_1(\underline{x}), \\ F_2 &= y_2^{\rho_2} - y_1^{\beta_{2,1}} y_3^{\beta_{2,3}} \cdots y_r^{\beta_{2,r}} h_2(\underline{x}), \\ F_3 &= y_3^{\rho_3} - y_1^{\beta_{3,1}} y_4^{\beta_{3,4}} \cdots y_r^{\beta_{3,r}} h_3(\underline{x}), \\ &\cdots \\ F_{r-1} &= y_{r-1}^{\rho_{r-1}} - y_1^{\beta_{r-1,1}} y_r^{\beta_{r-1,r}} h_{r-1}(\underline{x}), \\ F_r &= y_r^{\rho_r} - y_1^{\beta_{r,1}} h_r(\underline{x}), \end{aligned}$$

where $h_1(\underline{x}), \ldots, h_r(\underline{x})$ *are monomials in* $x_1, \ldots, x_n$, $\rho_1 > \sum_{k=2}^{r} \beta_{k,1}$, *and for* $j = 2, \ldots, r$, $\rho_j \geq \sum_{k=1}^{j-1} \beta_{k,j}$. *Let* $\sigma = \rho_2 \cdots \rho_r$. *Then we have*

$$F_1^{\sigma} = y_1^{\sum_{j=2}^{r} \alpha_{j,\sigma} \beta_{j,1}} \widetilde{F_1^{\sigma}}, \quad \bmod (F_2, \ldots, F_r),$$

with

$$\widetilde{F_1^{\sigma}} = \sum_{k=0}^{\sigma} (-1)^k \binom{\sigma}{k} y_1^{\gamma_{k,1}} y_2^{\delta_{2,k}} y_3^{\delta_{3,k}} \cdots y_r^{\delta_{r,k}} h_1^{\alpha_{1,k}} h_2^{\alpha_{2,k}} \cdots h_2^{\alpha_{r,k}},$$

where all exponents are non negative integer numbers such that $0 \leq \delta_{j,k} < \rho_j$, $\alpha_{j,0} = \delta_{j,0} = 0$, $\delta_{j,\sigma} = 0$ *for* $j = 2, \ldots, r$, $k = 0, \ldots, \sigma$, *and* $\gamma_{0,1} > \gamma_{1,1} > \cdots > \gamma_{\sigma,1} = 0$.

Proof We have

$$(y_1^{\rho_1} - y_2^{\beta_{1,2}} \cdots y_r^{\beta_{1,r}} h_1(\underline{x}))^\sigma = \sum_{k=0}^{\sigma} (-1)^k \binom{\sigma}{k} y_1^{(\sigma-k)\rho_1} y_2^{k\beta_{1,2}} \cdots y_r^{k\beta_{1,r}} h_1^k(\underline{x}).$$

Let $\alpha_{1,k} = k$, we define $\alpha_{2,k}, \delta_{2,k}$ by the relation

$$\alpha_{1,k}\beta_{1,2} = \alpha_{2,k}\rho_2 + \delta_{2,k},\ \alpha_{2,k} \geq 0,\ 0 \leq \delta_{2,k} < \rho_2.$$

Note that $\alpha_{1,0} = 0$, hence $\alpha_{2,0} = \delta_{2,0} = 0$, and $\alpha_{1,\sigma} = \sigma$, hence $\alpha_{2,\sigma} = (\sigma/\rho_2)\beta_{1,2}, \delta_{2,\sigma} = 0$. By using F_2 we get:

$$F_1^\sigma = \sum_{k=0}^{\sigma} (-1)^k \binom{\sigma}{k} y_1^{(\sigma-k)\rho_1 + \alpha_{2,k}\beta_{2,1}} y_2^{\delta_{2,k}} \cdots y_3^{k\beta_{1,3}+\alpha_{2,k}\beta_{2,3}} + \cdots$$
$$+ y_r^{k\beta_{1,r}+\alpha_{2,k}\beta_{2,r}} h_1^{\alpha_{1,k}}(\underline{x}) \mod F_2.$$

We define $\alpha_{3,k}, \delta_{3,k}$ by the relation:

$$\alpha_{1,k}\beta_{1,3} + \alpha_{2,k}\beta_{2,3} = \alpha_{3,k}\rho_3 + \delta_{3,k},\ \alpha_{3,k} \geq 0,\ 0 \leq \delta_{3,k} < \rho_3.$$

Note that $\alpha_{3,0} = \delta_{3,0} = 0$, and $\delta_{3,\sigma} = 0$. By using F_3 we get:

$$F_1^\sigma = \sum_{k=0}^{\sigma} (-1)^k \binom{\sigma}{k} y_1^{(\sigma-k)\rho_1 + \alpha_{2,k}\beta_{2,1} + \alpha_{3,k}\beta_{3,1}} y_2^{\delta_{2,k}} y_3^{\delta_{3,k}} \cdots y_r^{k\beta_{1,r}+\alpha_{2,k}\beta_{2,r}+\alpha_{3,k}\beta_{3,r}}$$
$$\times h_1^{\alpha_{1,k}} h_2^{\alpha_{2,k}}$$

modulo the ideal (F_2, F_3). We can inductively define the numbers $\alpha_{j,k}, \delta_{j,k}$ by the relation:

$$\alpha_{1,k}\beta_{1,j} + \alpha_{2,k}\beta_{2,j} + \cdots + \alpha_{j-1,k}\beta_{j-1,j} = \alpha_{j,k}\rho_j + \delta_{j,k},\ \alpha_{j,k} \geq 0,\ 0 \leq \delta_{j,k} < \rho_j.$$

Note that $\alpha_{j,0} = \delta_{j,0} = 0$, and $\delta_{j,\sigma} = 0$. Hence

$$F_1^\sigma = \sum_{k=0}^{\sigma} (-1)^k \binom{\sigma}{k} y_1^{(\sigma-k)\rho_1 + \sum_{j=2}^{r} \alpha_{j,k}\beta_{j,1}} y_2^{\delta_{2,k}} y_3^{\delta_{3,k}} \cdots y_r^{\delta_{r,k}} h_1^{\alpha_{1,k}} h_2^{\alpha_{2,k}} \cdots$$
$$\times h_2^{\alpha_{r,k}} \mod (F_2, \ldots, F_r).$$

It is easy to prove by induction that

$$\forall k = 0, \ldots, \sigma - 1, \ 0 \leq \alpha_{1,k+1} - \alpha_{1,k} \leq 1.$$

Hence

$$(\sigma - k)\rho_1 + \sum_{j=2}^{r} \alpha_{j,k}\beta_{j,1} - (\sigma - k - 1)\rho_1 + \sum_{j=2}^{r} \alpha_{j,k+1}\beta_{j,1}$$

$$= \rho_1 + \sum_{j=2}^{r} (\alpha_{j,k} - \alpha_{j,k+1})\beta_{j,1}$$

$$> \sum_{j=2}^{r} \beta_{j,1} + \sum_{j=2}^{r} (\alpha_{j,k} - \alpha_{j,k+1})\beta_{j,1} = \sum_{j=2}^{r} (1 + \alpha_{j,k} - \alpha_{j,k+1})\beta_{j,1} \geq 0$$

We can factor by $y_1^{\sum_{j=2}^{r} \alpha_{j,\sigma}\beta_{j,1}}$ and finally get

$$F_1^{\sigma} = y_1^{\sum_{j=2}^{r} \alpha_{j,\sigma}\beta_{j,1}} (\sum_{k=0}^{\sigma} (-1)^k \binom{\sigma}{k} y_1^{\gamma_{k,1}} y_2^{\delta_{2,k}} y_3^{\delta_{3,k}} \cdots y_r^{\delta_{r,k}} h_1^{\alpha_{1,k}} h_2^{\alpha_{2,k}} \cdots h_2^{\alpha_{r,k}})$$
$$\mod (F_2, \ldots, F_r).$$

with $\gamma_{k,1} > \gamma_{k+1,1}$.

Theorem 4.5.2 *Let V_A be a simplicial toric variety. Let*

$$\begin{aligned}
F_1 &= y_1^{\rho_1} - y_2^{\beta_{1,2}} \cdots y_r^{\beta_{1,r}} h_1(\underline{x}), \\
F_2 &= y_2^{\rho_2} - y_1^{\beta_{2,1}} y_3^{\beta_{2,3}} \cdots y_r^{\beta_{2,r}} h_2(\underline{x}), \\
F_3 &= y_3^{\rho_3} - y_1^{\beta_{3,1}} y_4^{\beta_{3,4}} \cdots y_r^{\beta_{3,r}} h_3(\underline{x}), \\
&\cdots \\
F_{r-1} &= y_{r-1}^{\rho_{r-1}} - y_1^{\beta_{r-1,1}} y_r^{\beta_{r-1,r}} h_{r-1}(\underline{x}), \\
F_r &= y_r^{\rho_r} - y_1^{\beta_{r,1}} h_r(\underline{x}), \\
F_{r+1} &= y_1^{\rho_1 - \sum_{k=2}^{r} \beta_{k,1}} y_2^{\rho_2 - \beta_{1,2}} y_3^{\rho_3 - \sum_{k=1}^{2} \beta_{k,3}} \cdots y_r^{\rho_r - \sum_{k=1}^{r-1} \beta_{k,r}} - h_1(\underline{x}) \cdots h_r(\underline{x}),
\end{aligned}$$

be $r + 1$ binomials in $I_A \subset K[x_1, \ldots, x_n, y_1, \ldots, y_r]$, where $h_1(\underline{x}), \ldots, h_r(\underline{x})$ are monomials in $x_1, \ldots, x_n$, $\rho_1 > \sum_{k=2}^{r} \beta_{k,1}$, and for $j = 2, \ldots, r$, $\rho_j \geq \sum_{k=1}^{j-1} \beta_{k,j}$. Note that if for $i = 1, \ldots, r$, F_i corresponds to the vector v_i in the lattice L_A, then F_{r+1} corresponds to the vector $v_1 + \cdots + v_r$. Suppose that $I_A = J + (F_1, \ldots, F_{r+1})$

and $J \subset \mathrm{rad}\,(F_1, \ldots, F_{r+1})$. *Then,* $I_A = \mathrm{rad}\,(F_2, \ldots, F_r, \widetilde{F_1^\sigma})$; *in particular* V_A *is a set-theoretic complete intersection.*

Proof Since

$$F_1^\sigma = y_1^{\sum_{j=2}^{r} \alpha_{j,\sigma}\beta_{j,1}} \widetilde{F_1^\sigma} \mod (F_2, \ldots, F_r),$$

and I_A is a prime ideal, we have that $\widetilde{F_1^\sigma} \in I_A$. So we only need to prove that if $P = (x_1, \ldots, x_n, y_1, \ldots, y_r)$ is a zero of $F_2, \ldots, F_r, \widetilde{F_1^\sigma}$, then P is also a zero of I_A. We note that $F_1^\sigma(P) = 0$. Since $J \subset \mathrm{rad}\,(F_1, \ldots, F_{r+1})$, we have $H(P) = 0$ for any $H \in J$. So we only have to check that $F_{r+1}(P) = 0$.

Note that for $i = 1, \ldots, r$, $\rho_i \neq 0$. Let examine the terms of $F_{r+1}(P)$. We have four cases:

1. Suppose that $h_i(P) = 0$ for some $i = 1, \ldots, r$. Since $F_i(P) = 0$, we have $y_i = 0$. If $\rho_i - \sum_{k=1}^{i-1} \beta_{k,i} > 0$, we have $F_{r+1}(P) = 0$. If $\rho_i - \sum_{k=1}^{i-1} \beta_{k,i} = 0$, let $1 \leq k_1 \leq i-1$ be the smallest integer such that $\beta_{k_1,i} \neq 0$. Since $F_{k_1}(P) = 0$ we have $y_{k_1} = 0$. If $\rho_{k_1} - \sum_{k=1}^{k_1-1} \beta_{k,k_1} > 0$, we have $F_{r+1}(P) = 0$. If $\rho_{k_1} - \sum_{k=1}^{k_1-1} \beta_{k,k_1} = 0$, there exists $1 \leq k_2 \leq k_1 - 1$ such that $\beta_{k_2,k_1} \neq 0$, a contradiction.
2. If $y_1 = 0$, then $\widetilde{F_1^\sigma}(P) = 0$ implies $h_i(P) = 0$ for some i, so we are done.
3. If $y_j = 0$ for some $j > 1$, let $i > 1$ be the biggest one such that $y_i = 0$. Then from $F_i(P) = 0$ we have either $h_i(P) = 0$, or $y_1 = 0$. We are done.
4. If for all $i = 1, \ldots, r$, $h_i(P) \neq 0$ and $y_i \neq 0$. For $i = 1, \ldots, r$, assume that F_i corresponds to the vector v_i in the lattice L_A, then F_{r+1} corresponds to the vector $v_1 + \ldots + v_r$. Since $F_i(P) = 0$ for $i = 1, \ldots, r$, the assertion follows trivially.

The following examples are applications of the above Theorem 4.5.2.

Example 4.5.3 Let V be the projective toric curve in $\mathbb{P}^4$ with parametrization

$$w = t^7, x = s^7, y = t^3 s^4, z = t^4 s^3, a = t^2 s^5.$$

Then $I(V)$ is generated by

$$F_{v_1} = a^2 - xz, F_{v_2} = y^2 - az, F_{v_3} = z^3 - yaw, F_{v_1+v_2+v_3} = yz - xw.$$

Example 4.5.4 Let V be the toric surface in K^4 with parametrization

$$w = t^9, x = s^9, y = ts^5, z = t^2 s^7, a = ts^8.$$

Then $I(V)$ is generated by

$$F_{v_1} = y^3 - az, F_{v_2} = a^2 - xz, F_{v_3} = z^5 - x^3 aw, F_{v_1+v_2+v_3} = y^3 z^3 - x^4 w.$$

Example 4.5.5 Let V be the projective toric surface in $\mathbb{P}^5$ with parametrization

$$x = s^9, w = t^9, v = u^9, y = t^4 s^4 u, z = t^5 s^2 u^2, a = t^3 s^6.$$

Then $I(V)$ is generated by

$$F_{v_1} = y^2 - az, F_{v_2} = a^3 - wx^2, F_{v_3} = z^5 - vw^2 ya, F_{v_1+v_3} = yz^4 - vw^2 a^2.$$

Example 4.5.6 Let V be the projective toric curve in $\mathbb{P}^4$ with parametrization

$$\begin{aligned} x &= u^{11}, \\ w &= v^{11}, \\ y &= u^6 v^5, \\ z &= u^7 v^4, \\ a &= u^3 v^8. \end{aligned}$$

Then the ideal $I(V)$ is generated by:

$$\begin{aligned} F_{v_1} &= y^3 - wxz, \\ F_{v_2} &= -wa + z^2, \\ F_{v_3} &= -xy + a^2, \\ F_{v_1+v_2+v_3} &= -w^2x^2 + y^2 az. \end{aligned}$$

$I(V)$ is a set-theoretic complete intersection.

We can compute $F_{v_1}^4$ modulo F_{v_2}, F_{v_3}, and we get:

$$F_{v_1}^4 = y(y^{11} - 4y^8 wxz + 6y^5 w^3 x^2 z - 4y^2 w^4 x^3 za + w^6 x^5) \text{ modulo } (F_{v_2}, F_{v_3}).$$

Let $F := y^{11} - 4y^8 wxz + 6y^5 w^3 x^2 z - 4y^2 w^4 x^3 za + w^6 x^5$. Our theorem says that $I(V) = \text{rad}\,(F_{v_2}, F_{v_3}, F)$.

Example 4.5.7 Let the projective surface $\overline{V}$ with parametrization

$$x = s^{15}, w = t^{15}, v = u^{15}, y = t^4 s^2 u^9, z = t^6 s^3 u^6, a = t^{10} s^5.$$

We have

$$I(V) = (y^2 a - z^3, y^3 - vz^2, w^2 x - a^3, -va + yz).$$

Note that if $y^2 a - z^3$ corresponds to a vector v_1, $y^3 - vz^2$ corresponds to a vector v_2, then $-va + yz$ corresponds to the vector $v_2 - v_1$. So V is a stci.

Example 4.5.8 Let V be the projective toric curve in $\mathbb{P}^4$ with parametrization

$$x = s^5, w = t^5, y = t^4 s, z = t^3 s^2, a = t^2 s^3.$$

The ideal $I(V)$ is generated by

$$xy - a^2, -wx + az, -ya + z^2, -wa + yz, y^2 - wz.$$

It is a Gorenstein projective curve in $\mathbb{P}^4$. We prove now that $I(V)$ is a set-theoretic complete intersection. We follow the ideas of Brezinsky [3]:

First note that $z(-wa + yz) = y(-ya + z^2) + a(y^2 - wz)$ implies $-wa + yz \in \text{rad}\,(xy - a^2, -wx + az, -ya + z^2, y^2 - wz)$. Next if $a^2 - xy$ corresponds to a vector v_1, $z^2 - ya$ corresponds to a vector v_2, $y^2 - wz$ corresponds to a vector v_3, then $az - wx$ corresponds to the vector $v_1 + v_2 + v_3$, so by our Theorem 4.5.2, $\text{rad}\,(xy - a^2, -wx + az, -ya + z^2, y^2 - wz) = \text{rad}\,(xy - a^2, -ya + z^2, y^2 - wz)$.

Now let $\alpha_1, \ldots, \alpha_5$ be any positive numbers, $\alpha := \alpha_1 + \cdots + \alpha_5 > 0$. Let us consider the variety W:

$$\begin{aligned} x &= s^{5\alpha}, \\ w &= t^{5\alpha}, \\ y &= t^{4\alpha} s^{\alpha}, \\ z &= t^{3\alpha} s^{2\alpha}, \\ a &= t^{2\alpha} s^{3\alpha}, \\ b &= t^{5\alpha_2 + 4\alpha_3 + 3\alpha_4 + 2\alpha_5} s^{5\alpha_1 + \alpha_3 + 2\alpha_4 + 3\alpha_5}. \end{aligned}$$

Then W is a set-theoretic complete intersection. Note that the ideal $I(\overline{W})$ is generated by: $xy - a^2, -wx + az, -ya + z^2, -wa + yz, y^2 - wz, b^{\alpha} - x^{\alpha_1} w^{\alpha_2} y^{\alpha_3} z^{\alpha_4} a^{\alpha_5}$. $I(\overline{W}) = \text{rad}\,(xy - a^2, -ya + z^2, y^2 - wz, b^{\alpha} - x^{\alpha_1} w^{\alpha_2} y^{\alpha_3} z^{\alpha_4} a^{\alpha_5})$.

Example 4.5.9 Let V be the projective toric curve in $\mathbb{P}^3$, with parametrization

$$w = t^9, x = s^9, y = t^8 s, z = t^4 s^5.$$

V_A is arithmetically Cohen-Macaulay. Let $\overline{V}$ be the projective toric curve in $\mathbb{P}^4$, with parametrization

$$w = t^9, x = s^9, y = t^8 s, z = t^4 s^5, a = t^6 s^3.$$

Its ideal is generated by five elements: $-y^3 + w^2 a$, $-y^2 a + w^2 z$, $-w^2 x + yaz$, $a^2 - yz$, $-xy + z^2$ but is not Gorenstein, However we can still apply the method used by Brezinsky [3]. First note that $a(-y^2 a + w^2 z) = y^2(yz - a^2) + z(-y^3 + w^2 a)$ implies by studying both cases when $a = 0$ or when $a \neq 0$ that $-y^2 a + w^2 z \in$

rad $(xy - a^2, -wx + az, -ya + z^2, y^2 - wz)$. Secondly if $-y^3 + w^2a$ corresponds to a vector v_1, $a^2 - yz$ corresponds to a vector v_2, $-xy + z^2$ corresponds to a vector v_3, then $-w^2x + yaz$ corresponds to the vector $v_1 + v_2 + v_3$, so by our Theorem 4.5.2, rad $(-y^3 + w^2a, -y^2a + w^2z, -w^2x + yaz, a^2 - yz, -xy + z^2) =$ rad $((-y^3 + w^2a)^4, a^2 - yz, -xy + z^2)$.

We have the following open question:

Question 4.5.10 : Let V be the toric variety with parametrization

$$w = t^d, x = s^d, y = \underline{s}^{\mathbf{a_1}}, z = \underline{s}^{\mathbf{a_2}}$$

and let V_1 be the toric variety with parametrization

$$w = t^d, x = s^d, y = \underline{s}^{\mathbf{a_1}}, z = \underline{s}^{\mathbf{a_2}}, a = \underline{s}^{\frac{\mathbf{a_1}+\mathbf{a_2}}{2}},$$

where we assume that $\frac{\mathbf{a_1}+\mathbf{a_2}}{2}$ has integer coordinates. We know by Theorem 4.5.12, that if V is arithmetically Cohen-Macaulay then it is a set-theoretic complete intersection. Can we say if $I(V_1)$ is a set-theoretic complete intersection?

We can answer to this question in Theorem 4.6.2 if one of the components of $\mathbf{a_1} + \mathbf{a_2}$ is odd.

Example 4.5.11 Let the projective curve with parametrization

$$w = t^5, x = s^5, y = t^3s^2, z = t^1s^4,$$

it is arithmetically Cohen-Macaulay. The projective curve with parametrization

$$w = t^5, x = s^5, y = t^3s^2, z = t^1s^4, a = t^2s^3$$

is Gorenstein and generated by five elements.

4.5.2 Almost-Complete Intersections, The Codimension Two Case

In this subsection we apply Theorem 4.5.2 in the case of simplicial monomial varieties of codimension two which are arithmetically Cohen-Macaulay:

Theorem 4.5.12 *Let V_A be a simplicial toric variety of codimension 2, such that is arithmetically Cohen-Macaulay. Then V_A is a set-theoretic complete intersection.*

Proof By Theorem 4.4.15, the defining ideal of a simplicial monomial variety of codimension two arithmetically Cohen-Macaulay, is generated by three elements

$$F = z^{s_\mu} - y^{p_\mu} x_1^{r_{1,\mu}} \cdots x_n^{r_{n,\mu}},$$
$$G = y^{p_{\mu+1}} - z^{s_{\mu+1}} x_1^{-r_{1,\mu+1}} \cdots x_n^{-r_{n,\mu+1}},$$
$$H = z^{s_\mu - s_{\mu+1}} y^{p_{\mu+1}-p_\mu} - x_1^{r_{1,\mu}-r_{1,\mu+1}} \cdots x_n^{r_{n,\mu}-r_{n,\mu+1}},$$

for some positive integer exponents with $s_\mu > s_{\mu+1}$, $p_{\mu+1} > p_\mu$.

It is clear that we can apply the Theorem 4.5.2. Indeed let F_1 be the polynomial obtained from $(z^{s_\mu} - y^{p_\mu}\underline{x}^{r_\mu})^{p_{\mu+1}}$ by reduction modulo G. That is $F^{p_{\mu+1}} = AG + z^{p_\mu(s_{\mu+1})}F_1$. Then $I = \text{rad}\,(G, F_1)$.

Example 4.5.13 Let V_A be the projective toric surface in $\mathbb{P}^5$ with parametrization

$$\begin{aligned} v &= u^{10}, \\ x &= s^{10}, \\ w &= t^{10}, \\ y &= t^5 s^5, \\ z &= t^4 s^2 u^4, \\ a &= t^2 s^6 u^2. \end{aligned}$$

The ideal I_A is generated by:

$$\begin{aligned} F_{v1} &= z^3 - vwa, \\ F_{v2} &= a^2 - xz, \\ F_{v3} &= -y^2 + wx, \\ F_{v1+v2+v3} &= vy^2 - az^2. \end{aligned}$$

Then I_A is a set-theoretic complete intersection. In fact we can compute F_{v1}^4 modulo F_{v2}, F_{v3}, and we get:

$$F_{v1}^4 = z^2(v^4w^4x^2 - 4v^3w^3axz^2 + 6v^2w^2xz^5 - 4vwaz^7 + z^{11}) \mod (F_{v2}, F_{v3}).$$

Let $F := v^4w^4x^2 - 4v^3w^3axz^2 + 6v^2w^2xz^5 - 4vwaz^7 + z^{11}$. By Theorem 4.5.2 we have $I(V) = \text{rad}\,(F_{v2}, F_{v3}, F)$.

Another proof: Let us consider the variety W:

$$\begin{aligned} v &= u^5, \\ x &= s^5, \end{aligned}$$

$$w = t^5,$$
$$z = t^2 s u^2,$$
$$a = t s^3 u.$$

By the trick developed in Sect. 4.6.1, $I_A = (I(W) + (y^2 - xw))$. $I(W)$ has codimension 2 and is arithmetically Cohen-Macaulay. Hence $I(W)$ is a set-theoretic complete intersection and so is I_A.

Remark For an arithmetically Cohen-Macaulay projective curve, the shape of the equations and the above theorem was known, by Stuckrad and Vogel [22] and by Robbiano and Valla [19]. For an arithmetically Cohen-Macaulay simplicial toric variety of codimension two, in [15] it was proved that its equations are given by the 2×2 minors of a 2×3 matrix, so the above theorem can be also proved by using [22], or the next theorem. Our proof is simpler, it gives us the ideal I up to radical in one step, while the next theorem needs several steps.

Theorem 4.5.14 ([19]) *Let R be a commutative ring with identity, let m, n be non negative integers, and let J be the ideal generated by the 2×2 minors of the matrix $M = \begin{pmatrix} a & b^m & c \\ b^n & d & e \end{pmatrix}$, with entries in R. Then we can construct two elements $f, g \in J$, such that*

$$\operatorname{rad}(J) = \operatorname{rad}(f, g).$$

4.6 Some Set-Theoretic Complete Intersection Toric Varieties

4.6.1 Tricks on Toric Varieties

The following theorem was originally stated and proved in [14], in the case of numerical semigroups, but it can be extended in general and the proofs are unchanged.

Theorem 4.6.1 *Let H be the semigroup of $\mathbb{N}^m$ generated by $\mathbf{a_1}, \dots, \mathbf{a_n}$. Let $I_H \subset K[x_1, \dots, x_n]$ be the toric ideal associated to H.*

1. *Let $l \in \mathbb{N}^*$, and $H^{(l)}$ be the semigroup generated by $l\mathbf{a_1}, \dots, l\mathbf{a_{n-1}}, \mathbf{a_n}$. Then the ideal $I_{H^{(l)}}$ is generated by $\widetilde{f}(x_1, \dots, x_n) := f(x_1, \dots, x_{n-1}, x_n^l)$, where f runs over all the generators of I_H.*
2. *Let $l_1, \dots, l_n \in \mathbb{N}, l = l_1 + \cdots + l_n > 0$, let $\overline{H}^{(l_1,\dots,l_n)}$ be the semigroup generated by $l\mathbf{a_1}, \dots, l\mathbf{a_{n-1}}, l\mathbf{a_n}, \mathbf{a_{n+1}} := l_1\mathbf{a_1} + \cdots + l_n\mathbf{a_n}$. If l is relatively prime to a component of $\mathbf{a_{n+1}}$ then $I_{\overline{H}^{(l_1,\dots,l_n)}} = I_H + (x_{n+1}^l - x_1^{l_1} \cdots x_n^{l_n}) \subset K[x_1, \dots, x_{n+1}]$.*

The following theorem follows from [14], Lemmas 1.3, 1.4, and 1.5. See also [21] Corollary 2.5 and [16] Theorem 2.6.

Theorem 4.6.2

1. *If I_H is Cohen-Macaulay, Gorenstein, complete intersection or set-theoretic complete intersection then the same property holds for $I_{H^{(l)}}$.*
2. *If I_H is Cohen-Macaulay, Gorenstein, complete intersection or set-theoretic complete intersection and l is relatively prime to a component of $\mathbf{a_{n+1}}$ then the same property holds for $I_{\overline{H}^{(l_1,\ldots,l_n)}}$.*

We deduce a positive answer to Question 4.5.10 if one of the components of $\mathbf{a_1}+\mathbf{a_2}$ is odd. The following example shows that the hypothesis l is relatively prime to a component of $\mathbf{a_{n+1}}$ is necessary. We thank Mesut Sahin to pointed us this problem.

Example 4.6.3 Consider the projective surface with parametrization

$$x = s^9,\ w = t^9,\ v = u^9,\ z = t^5s^2u^2,\ a = t^3s^4u^2.$$

It is a complete intersection but the projective surface with parametrization

$$x = s^9,\ w = t^9,\ v = u^9,\ y = t^4s^3u^2,\ z = t^5s^2u^2,\ a = t^3s^4u^2$$

is not arithmetically Cohen-Macaulay. Its defining ideal is generated by six elements.

These tricks can be applied to the projective case using the following

Theorem 4.6.4 *Let H be the semigroup of $\mathbb{N}^m$ generated by $\mathbf{a_1},\ldots,\mathbf{a_n}$, which are not necessarily homogeneous with respect to the standard graduation. Suppose that $I_H = \mathrm{rad}\,(F_1,\ldots,F_r)$. Let $d = max\{\deg\mathbf{a_1},\ldots,\deg\mathbf{a_n}\}$, where $\deg\mathbf{a_i}$ is the sum of its components. Let H_1 be the semigroup in $\mathbb{Z}^{m+1}$ generated by $\mathbf{b_1},\ldots,\mathbf{b_{n+1}}$, where for $i = 1,\ldots,n$, $\mathbf{b_i} = \mathbf{a_i} + (d - \deg\mathbf{a_i})\mathbf{e_{m+1}}$ and $\mathbf{b_{n+1}} = d\mathbf{e_{m+1}}$. Let x_{n+1} be a new variable and let $F_1^h,\ldots,F_r^h$ be the homogenization of $F_1,\ldots,F_r$ with respect to x_{n+1}.*

Let $P = (x_1,\ldots,x_n,x_{n+1})$ be a zero of $F_1^h,\ldots,F_r^h$. If $x_{n+1} = 0$ implies that $F(P) = 0$ for all $F \in I_H^h$, then $I_{H_1} = \mathrm{rad}\,(F_1^h,\ldots,F_r^h)$.

Proof For projective closure and parametrization of toric varieties we refer to [4]. Let V_A be the zero set of I_H, then the projective closure $\overline{V}$ is the zero set of I_H^h and since both ideals I_H^h, I_{H_1}, are prime of the same height they coincide. This implies that $\mathrm{rad}\,(F_1^h,\ldots,F_r^h) \subset I_{H_1}$.

Let $P = (x_1,\ldots,x_n,x_{m+1})$ be a zero of $F_1^h,\ldots,F_r^h$. By hypothesis, if $x_{m+1} = 0$ then $P \in \overline{V}$. If $x_{m+1} \neq 0$ then $P \in \overline{V}$ since $\overline{V} \cap (x_{m+1} = 1) = \{(Q,1) \mid Q \in V\}$, by general arguments on the projective closure.

Example 4.6.5 Let V_A be the affine surface with parametrization

$$b = t^7, x = s^7, y = t^3s^2, z = t^4s^3, a = t^2s^5,$$

$\overline{V}$ be the projective surface with parametrization

$$b = t^7, x = s^7, w = u^7, y = t^3s^2u^2, z = t^4s^3, a = t^2s^5.$$

Then

$$I(V) = (-a^2 + xz, z^4 - xab^2, -az^3 + x^2b^2, y^7 - x^2b^3),$$

and

$$I(\overline{V}) = (-a^2 + xz, z^4 - xab^2, -az^3 + x^2b^2, y^7 - w^2x^2b^3).$$

Applying the proof of Theorem 4.5.2, we have that $\overline{V}$ is a set-theoretic complete intersection. Indeed, let $F_{\mathbf{v_1}} = z^4 - xab^2$, $F_{\mathbf{v_2}} = a^2 - xz$ then $F_{\mathbf{v_1}+\mathbf{v_2}} = az^3 - x^2b^2$ and $F_{\mathbf{v_1}}^2 = z(z^7 - 2z^3ab^2x + b^4x^3) \mod F_{\mathbf{v_2}}$. Hence $I(\overline{V}) = \mathrm{rad}\,(-a^2 + xz, z^7 - 2z^3ab^2x + b^4x^3, y^7 - x^2b^3)$.

4.6.2 *Toric Curves in* $\mathbb{P}^3$

In this section we consider curves, that is V_A is a simplicial toric variety of dimension 1 in K^3. The parametrization of V_A is:

$$x = v^p,$$
$$y = v^q,$$
$$z = v^r,$$

where $p < q < r$ are positive integers. We simply denote this curve by V or $V(p, q, r)$. Let $\overline{V}$ be the projective toric curve in $\mathbb{P}^3$, with parametrization

$$w = u^r,$$
$$x = u^{r-p}v^p,$$
$$y = u^{r-q}v^q,$$
$$z = v^r.$$

We simply denote this curve by $\overline{V}(p, q, r)$.

Theorem 4.6.6 ([20]) *Let a, b, p, q, r be natural integer numbers such that $r = ap + bq$. If $b \geq a(q - p - 1) + 1$, then $\overline{V}(p, q, r)$ is a set-theoretic complete intersection. Moreover $\overline{V}(p, q, r)$ is the zero set of the polynomials $F_1 := x^q - y^p w^{q-p}$, $F_2 = ((z - x^a y^b)^q)^h_{y^p = x^q}$, where $(H)_{y^p = x^q}$ means substitution when possible x^q by y^p, and H^h is the homogenization of H with respect to w.*

Proof This proof is more or less the proof given by Sahin [20].

Let us consider

$$(z - x^a y^b)^q = z^q + \sum_{k=1}^{q-1} (-1)^k \binom{q}{k} z^{q-k} x^{ka} y^{kb} + x^{qa} y^{qb}.$$

By setting $ka = s_k q + r_k$,with $0 \leq s_k, 0 \leq r_k < q$, we can write

$$((z - x^a y^b)^q)_{y^p = x^q} = z^q + \sum_{k=1}^{q-1} (-1)^k \binom{q}{k} z^{q-k} x^{r_k} y^{s_k p + kb} + y^{pa+qb}.$$

Note that for $k = 1, \ldots, q-1$, $q - k + r_k + s_k p + kb < q - k + ka + kb$, so it is enough to check the condition $q - k + ka + kb \leq pa + qb$ for $k = 0, \ldots, q-1$. This is equivalent to $q - k + ka - pa \leq qb + kb = q - k + (q - k)(b - 1)$, i.e., equivalent to $(k - p)a \leq (q - k)(b - 1)$ for $k = 0, \ldots, q - 1$. This last condition is equivalent to $(k - p)a \leq (q - k)(b - 1)$ for $k = p + 1, \ldots, q - 1$. We remark that if $b - 1 \geq a(q - p - 1)$, then $(k - p)a \leq a(q - p - 1) \leq b - 1 \leq (q - k)(b - 1)$. We can write:

$$((z - x^a y^b)^q)_{y^p = x^q} = \sum_{k=0}^{q-1} (-1)^k \binom{q}{k} z^{q-k} x^{r_k} y^{s_k + kb} w^{r - (q - k + r_k + s_k + kb)} + y^r.$$

By the preceding discussion $q - k + r_k + s_k p + kb < q - k + ka + kb \leq r$ if $b - 1 \geq a(q - p - 1)$. In conclusion the exponent of w in the monomial $z^{q-k} x^{r_k} y^{s_k + kb} w^{r - (q - k + r_k + s_k + kb)}$ is strictly positive.

Let $P = (w : x : y : z) \in Z(F_1, F_2)$. If $w = 0$, then $F_2(P) = 0$ implies $y = 0$, and $F_1(P) = 0$ implies $x = 0$. Hence $P = (0 : 0 : 0 : 1)$ belongs to $\overline{V}$. If $w \neq 0$, we can assume that $w = 1$. Hence $F_1(P) = 0$ implies that there exists $v \in K$ such that $x = v^p$, $y = v^q$ and $F_2(P) = 0$ implies $(z - v^r)^q = 0$, which finally implies $z = v^r$; that is $P = (1 : v^p : v^q : v^r)$ belongs to $\overline{V}$.

The next theorem uses a trick that improves Sahin's theorem in some cases:

Theorem 4.6.7 ([17])

1. *Let p, q, r be natural integer numbers and $\overline{V}(p, q, r)$ be the projective toric curve in $\mathbb{P}^3$, with parametrization $(u^r, u^{r-p} v^p, u^{r-q} v^q, v^r)$. Suppose that $r = ap + bq$, with $a, b \in \mathbb{N}$,*

a. if $p = 1$ *and* $0 \le a \le q-1, b \ge q-a$ *, or*
b. if $p > 1$ *and* $0 \le a \le q-1, b \ge (q-a-1)p$,

then $\overline{V}(p, q, r)$ *is a set-theoretic complete intersection. Moreover* $\overline{V}(p, q, r)$ *is the zero set of the polynomials* $F_1^h := x^q - y^p w^{q-p}$, F_2^h, *where* F_2^h *is obtained from* $((z - x^a y^b)^q)_{x^q - y^p}$, *by a trick, explained in the proof.*

2. *Let* l *be a natural number, let* $\overline{V}(lp, lq, r)$ *be the projective toric curve in* $\mathbb{P}^3$. *Suppose that* $r = ap + bq$, *with* $a, b \in \mathbb{N}$,

a. if $p = 1$ *and* $0 \le a \le q-1, b \ge q-a-1+l$, *or*
b. if $p > 1$ *and* $0 \le a \le q-1, b \ge q-a-p+l, b \ge (q-a-1)p$,

then $\overline{V}(lp, lq, r)$ *is a set-theoretic complete intersection. Moreover* $\overline{V}(lp, lq, r)$ *is the zero set of the polynomials* $F_1^h := x^q - y^p w^{q-p}, \overline{F}_2$ *where* $\overline{F}_2(w, x, y, z) = (F_2(x, y, z^l))^h$, *by the trick developed in Sect. 4.6.1.*

Proof We prove only the first claim, the second claim follows from the proof of the first and the trick developed in Sect. 4.6.1. The proof is more or less the one given in [17].

Let us consider

$$(z - x^a y^b)^q = z^q + \sum_{k=1}^{q-1} (-1)^{q-k} \binom{q}{k} z^k x^{(q-k)a} y^{(q-k)b} + x^{qa} y^{qb}.$$

By setting $(q-k)a = k(q-a) + q(a-k)$ and by using $y^p = x^q$, we get the polynomial

$$F_2 := z^q + \sum_{k=1}^{q-1} (-1)^{q-k} \binom{q}{k} z^k x^{k(q-a)} y^{r-k(b+p)} + y^r,$$

For $k = 1, \ldots, q-1$, the exponent of x in F_2 is $x^{k(q-a)}$ which is strictly positive. For $k = 1, \ldots, q-1$, the exponent of y in F_2 is $y^{r-k(b+p)}$ which is positive if and only if $b \ge (q-a-1)p$. Finally $\deg F_2 = r$ if and only if $b \ge q-a-p+1$.

It is easy to show that these conditions are equivalent to

1. if $p = 1$ and $0 \le a \le q-1, b \ge q-a$, or
2. if $p > 1$ and $0 \le a \le q-1, b \ge (q-a-1)p$.

We also remark that the affine curve $V(p, q, r)$ is a complete intersection by the trick developed in Sect. 4.6.1, so is clear that $V(p, q, r)$ is a set-theoretic complete intersection defined by (F_1, F_2).

Now we can prove that $\overline{V}(p,q,r)$ is a set-theoretic complete intersection defined by (F_1^h, F_2^h), where

$$F_1^h := x^q - y^p w^{q-p},$$

$$F_2^h := z^q w^{r-q} + \sum_{k=1}^{q-1}(-1)^{q-k}\binom{q}{k} z^k x^{k(q-a)} y^{r-k(b+p)} w^{k(b+p+a-q-1} + y^r.$$

Let $P = (w : x : y : z) \in Z(F_1, F_2)$. If $w = 0$ then $F_1(P) = 0$ implies $x = 0$, and $F_2(P) = 0$ implies $y = 0$. Hence $P = (0 : 0 : 0 : 1)$ and it is clear that it belongs to $\overline{V}$. If $w \neq 0$, we can assume that $w = 1$, the claim follows from the fact that $V(p,q,r)$ is a set-theoretic complete intersection defined by (F_1, F_2).

Remark We can compare the bounds on b given in Theorems 4.6.6 and 4.6.7. We assume that $0 \leq a \leq q-1$

1. If $p = 1$ then the bound given by Theorem 4.6.7 is better, that is $b \geq q - a$.
2. If $p > 1$ and $p \leq a$ then the bound given by Theorem 4.6.7 is better, that is $b \geq (q-a-1)p$,
3. If $p > 1$ and $p > a$ then the bound given by Theorem 4.6.6 is better, that is $b \geq a(q-p-1)+1$.

Proof We need a proof.

1. If $p = 1$ $q - a \leq a(q-1-1)+1 \Leftrightarrow (a-1)(q-1) \geq 0$.
2. If $p > 1$ $(q-a-1)p \geq a(q-p-1)+1 \Leftrightarrow (p-a)(q-1) \geq 1$.

Note also that the bound given in Theorem 4.6.7 is the best one given by the methods used, but the bound given by Theorem 4.6.6 is not the best obtained by the methods used. We sometimes can get better bounds by applying the proof of Theorem 4.6.6.

Theorem 4.6.8 *Suppose that* $\gcd(p,q) = l$. *We set* $p' = p/l, q' = q/l$. *If* $r \geq p'q'(q'-1)+q'l$, *then* $\overline{V}(p,q,r)$ *is a set-theoretic complete intersection.*

In particular given two positive numbers p,q *there is only a finite number of positive integers* r *for which we don't know if the projective toric curve* $\overline{V}(p,q,r)$ *in* $\mathbb{P}^3$ *is a set-theoretic complete intersection.*

Proof The Frobenius number for the semigroup generated by p', q' is $(p'-1)(q'-1)$, since $r \geq p'q'(q'-1)+q'l \geq (p'-1)(q'-1)$, we have that r belongs to the semigroup generated by p', q', and we can find a, b integers such that $r = ap' + bq', 0 \leq a \leq q-1, b \geq 1$. We will check the conditions for b in Theorem 4.6.7.

1. Suppose that $b < q' - a - p' - l$, then

$$r = ap' + bq' < r = ap' + q'(q'-a-p'-l) = q'(q'-p'-l) - a(q'-p') \leq q'(q'-p'-l),$$

and $q'(q'-p'-l) \leq p'q'(q'-1)+q'l$ is equivalent to $q' \leq p'q'$, so we get a contradiction.

2. Suppose that $b < (q' - a - 1)p'$, then

$$\begin{aligned} r = ap' + bq' < r = ap' + q'((q' - a - 1)p') &= q'((q' - 1)p') - a(q' - p') \\ &\leq p'q'(q' - 1) \leq p'q'(q' - 1) + q'l, \end{aligned}$$

we get again a contradiction.

We conclude that the conditions for b in Theorem 4.6.7 are satisfied, hence $\overline{V}(p, q, r)$ is a set-theoretic complete intersection.

Example 4.6.9 Let $\overline{V}(1, 2, r)$ be the projective toric curve in $\mathbb{P}^3$. Then $\overline{V}$ is a set-theoretic complete intersection for all integers $r \geq 3$, by applying the proof of the above theorem.

Example 4.6.10 Let $\overline{V}(1, 3, r)$ be the projective toric curve in $\mathbb{P}^3$. Then $\overline{V}$ is a set-theoretic complete intersection for all integers $r \geq 5$, by applying the proof of the above theorem.

Remark that in this case the only unsolved example is the famous projective quartic $\overline{V}(1, 3, 4)$.

Example 4.6.11 Let $\overline{V}(1, 4, r)$ be the projective toric curve in $\mathbb{P}^3$. Then $\overline{V}$ is a set-theoretic complete intersection for all integers $r \in \{7, 8, 10, \ldots\}$. By applying the proof of the above theorem, we get that $r \in \{7, 10, 11, 13, \ldots\}$. Now by a direct computation using [15], we get that $\overline{V}$ is a complete intersection for $r = 8, 12$.

The unsolved cases are $\overline{V}(1, 4, 5)$, $\overline{V}(1, 4, 6)$ and $\overline{V}(1, 4, 9)$.

Example 4.6.12 Let $\overline{V}(2, 3, r)$ be the projective toric curve in $\mathbb{P}^3$. Then $\overline{V}$ is a set-theoretic complete intersection for all integers $r \geq 4$. By applying the proof of the above theorem, we get that $r \in \{4, 7, 8, 10, 11, 12, \ldots\}$. Now by direct computation using [15], we get that $\overline{V}$ is a arithmetically Cohen-Macaulay for $r = 5$ and a complete intersection for $r = 6, 9$.

4.6.3 Toric Curves in $\mathbb{P}^n$

Let K be an algebraically closed field. In this subsection we consider curves in K^n, that is $V(p, q_0, q_1, \ldots, q_{n-2})$ is an affine simplicial toric variety of dimension 1. The parametrization of $V := V(p, q_0, q_1, \ldots, q_{n-2})$ is:

$$\begin{aligned} x &= v^p, \\ y &= v^{q_0}, \\ z_1 &= v^{q_1}, \\ &\cdots \\ z_{n-2} &= v^{q_{n-2}}. \end{aligned}$$

Theorem 4.6.13 *Let $p, q_0, q_1, \ldots, q_{n-2}$ be positive integers. Let $\overline{V}(p, q_0, q_1, \ldots, q_{n-2})$ be the projective toric curve in $\mathbb{P}^n$ with parametrization*

$$\begin{aligned} w &= u^{q_{n-2}}, \\ x &= u^{q_{n-2}-p}v^p, \\ y &= u^{q_{n-2}-q_0}v^{q_0}, \\ z_1 &= u^{q_{n-2}-q_1}v^{q_1}, \\ &\ldots \\ z_{n-2} &= v^{q_{n-2}}. \end{aligned}$$

Suppose that $\overline{V}(p, q_0, q_1, \ldots, q_{n-2})$ is a set-theoretic complete intersection, defined by $F_1, \ldots, F_{n-1}$. Let $q_{n-1} \in \mathbb{N}$, and $\overline{V}_1$ the projective curve defined by

$$\begin{aligned} w &= u^{q_{n-1}}, \\ x &= u^{q_{n-1}-p}v^p, \\ y &= u^{q_{n-1}-q_0}v^{q_0}, \\ z_1 &= u^{q_{n-1}-q_1}v^{q_1}, \\ &\ldots \\ z_{n-2} &= u^{q_{n-1}-q_{n-2}}v^{q_{n-2}}, \\ z_{n-1} &= v^{q_{n-1}}. \end{aligned}$$

If $q_{n-1} = ap + bq_{n-2}$, with $0 \leq a \leq q_{n-2} - 1$, $b \geq q_{n-2} - a$ when $p = 1$, or $0 \leq a \leq q_{n-2} - 1$, $b \geq (q_{n-2} - a - 1)p$ when $p > 1$, then $\overline{V}_1(p, q_0, q_1, \ldots, q_{n-2}, q_{n-1})$ is a set-theoretic complete intersection.

In particular, let $\gcd(p, q_{n-2}) = l$. We set $p' = p/l, q' = q_{n-2}/l$. If $q_{n-1} \geq p'q'(q' - 1) + q'l$, then $\overline{V}_1(p, q_0, q_1, \ldots, q_{n-2}, q_{n-1})$ is a set-theoretic complete intersection.

Proof By the hypothesis $\overline{V}$ is a set-theoretic complete intersection, defined by $F_1, \ldots, F_{n-1}$. We will prove that $\overline{V}_1$ is a set-theoretic complete intersection, defined by $F_1, \ldots, F_{n-1}, F_n$, where F_n is the polynomial

$$z_{n-1}^{q_{n-2}} w^{q_{n-1}-q_{n-2}} + \sum_{k=1}^{q_{n-2}-1} (-1)^{q_{n-2}-k} \binom{q_{n-2}}{k} z_{n-1}^k x^{k(q_{n-2}-a)} z_{n-2}^{q_{n-1}-k(b+p)}$$
$$\times\, w^{k(b+p+a-q_{n-2}-1)} + z_{n-2}^{q_{n-1}},$$

obtained from $((z_{n-1} - x^a z_{n-2}^b)^{q_{n-2}})_{z_{n-2}^p = x^{q_{n-2}}}$ by the trick used in the proof of the Theorem 4.6.7. Note that also by Theorem 4.6.7, all exponents are positive with our hypothesis.

First note that $F_n \in I(\overline{V}_1)$. Let $P = (w, x, y, z_1, \ldots, z_{n-1}) \in \overline{V}_1$. If $w = 0$ then from the parametrization we get $x = y = z = \cdots = z_{n-2} = 0$, hence $F_n(P) = 0$. If $w \neq 0$, we can assume that $w = 1$, there exists $v \in K$ such that

$$x = v^p, y = v^{q_0}, z_1 = v^{q_1}, \ldots, z_{n-1} = v^{q_{n-1}}.$$

If $v = 0$ then $x = y = z = \cdots = z_{n-1} = 0$, and $F_n(P) = 0$. If $v \neq 0$, we can perform the trick used in the proof of the Theorem 4.6.7, and we get that $F_n(P) = (z_{n-1} - x^a z_{n-2}^b)^{q_{n-2}} = 0$.

Secondly we prove that $F_1, \ldots, F_{n-1} \in I(\overline{V})_1$. For $i = 1, \ldots, n-1$, $F_i \in I(\overline{V})$. This implies $F_i^{deh} \in I(V)$, where F_i^{deh} is the dehomogenized polynomial, that is setting $w = 1$ in F_i, hence $F_i(1, v^p, v^q, v^{q_1}, \ldots, v^{q_{n-2}}) = 0$, so $F_i^{deh} \in I(V_1)$ and finally $F_i \in I(\overline{V}_1)$. As a conclusion, the zero set of $F_1, \ldots, F_{n-1}, F_n$, is included in $\overline{V}_1$.

Third, we have to prove that if $P = (w, x, y, z_1, \ldots, z_{n-1})$ is a zero of $F_1, \ldots, F_{n-1}, F_n$, then $P \in \overline{V}_1$. Let $P' = (w, x, y, z_1, \ldots, z_{n-2})$, since $F_1(P') = \cdots = F_{n-1}(P') = 0$, there exist $u, v \in K$ such that

$$w = u^{q_{n-2}}, x = u^{q_{n-2}-p} v^p, y = u^{q_{n-2}-q} v^{q_0},$$

$$z_1 = u^{q_{n-2}-q_1} v^{q_1}, \ldots, z_{n-2} = v^{q_{n-2}}.$$

Suppose that $w = 0$, then $x = y = z = \cdots = z_{n-3} = 0$. Hence $F_n(P) = 0$ implies $z_{n-2} = 0$, that is $P = (0, \ldots, 0, 1)$, which is a point of $\overline{V}_1$. Suppose that $w \neq 0$, we can assume that $w = 1$, hence there exists $v \in K$ such that

$$x = v^p, y = v^{q_0}, z_1 = v^{q_1}, \ldots, z_{n-2} = v^{q_{n-2}}.$$

In particular $x^{q_{n-2}} = (v^p)^{q_{n-2}} = z_{n-2}^p$. From $F_n(P) = 0$, we get $(z_{n-1} - x^a z_{n-2}^b)^{q_{n-2}} = 0$, that is $z_{n-1} = x^a z_{n-2}^b = v^{q_{n-1}}$.

Example 4.6.14 Consider the projective curve $\overline{V}(p, q_0, q_1, \ldots, q_{n-2})$. Let $q_{n-1} = bq_{n-2}$ for a natural number $b \geq 2$. Then $\overline{V}_1(p, q_0, q_1, \ldots, q_{n-2}, q_{n-1})$ is the zero set of $I(\overline{V}(p, q_0, q_1, \ldots, q_{n-2}))$ and $F^h := z_{n-1} w^{b-1} - z_{n-2}^b$. In particular if $\overline{V}(p, q_0, q_1, \ldots, q_{n-2})$ is a set-theoretic complete intersection, then $\overline{V}_1(p, q_0, q_1, \ldots, q_{n-2}, q_{n-1})$ is a set-theoretic complete intersection.

Example 4.6.15 Let $\overline{V}(1, 2, 3, r)$ be the projective toric curve in $\mathbb{P}^4$ with parametrization

$$w = u^r, x = u^{r-1} v^1, y = u^{r-2} v^2, z_1 = u^{r-3} v^3, z_2 = v^r.$$

Then $\overline{V}$ is a set-theoretic complete intersection for all integers $r \geq 4$. By the above theorem we have that $\overline{V}(1, 2, 3, r)$ is a set-theoretic complete intersection for $r \geq 5$. The case $r = 4$ was done in [19]. Note that the case $r = 5$ follows also from [8]. This example was independently studied in [18].

Example 4.6.16 Let $\overline{V}(1, 3, 5, r)$ be the projective toric curve in $\mathbb{P}^4$ with parametrization

$$w = u^r, x = u^{r-1}v^1, y = u^{r-3}v^3, z_1 = u^{r-5}v^5, z_2 = v^r.$$

Then by using the Theorem 4.6.7 $\overline{V}$ is a set-theoretic complete intersection for all integers $r \in \{9, 13, 14, 17, 18, 19, 21, 22, \ldots\}$, and by Example 4.6.14, for all $r = 5b, b \geq 2$.

The trick used above can be improved. Let us consider the following example. Let $\overline{V}(1, 3, 5, 11)$ be the projective toric curve in $\mathbb{P}^4$, then $\overline{V}(1, 3, 5, 11)$ is a set-theoretic complete intersection on $I(\overline{V}(1, 3, 5))$ and F, where F is obtained from $(z_2 - y^2 z_1)^5 = 0$ working modulo $y^5 - z_1^3$.

In conclusion the only unknown cases are for $r = 6, 7, 8, 12$.

Example 4.6.17 Let $\overline{V}(2, 3, 5, r)$ be the projective toric curve in $\mathbb{P}^4$. We have seen in Example 4.6.14, that $\overline{V}(2, 3, 5, r)$ is a set-theoretic complete intersection for $r = 5b, b \geq 2$. By using the method in Theorem 4.6.6, we can see that $\overline{V}(2, 3, 5, r)$ is a set-theoretic complete intersection for $r = 12 + 5b, 14 + 5b$, and by using the methods in Theorem 4.6.7, that $\overline{V}(2, 3, 5, r)$ is a set-theoretic complete intersection for $r = 8 + 5b, 16 + 5b$. In conclusion $\overline{V}(2, 3, 5, r)$ is a set-theoretic complete intersection for all positive integers, except possibly for $r \in \{6, 7, 11\}$. Note that the case $\overline{V}(2, 3, 5, 9)$ was solved in [24].

Theorem 4.6.18 *Let $p, q_0, q_1, \ldots, q_{n-2}$ be positive integers. Let $\overline{V}$ be the projective toric curve in $\mathbb{P}^n$, with parametrization*

$$\begin{aligned} w &= u^{q_{n-2}}, \\ x &= u^{q_{n-2}-p}v^p, \\ y &= u^{q_{n-2}-q_0}v^{q_0}, \\ z_1 &= u^{q_{n-2}-q_1}v^{q_1}, \\ &\ldots \\ z_{n-2} &= v^{q_{n-2}}. \end{aligned}$$

For $i = 0, \ldots, q_{n-3}$ let $\gcd(p, q_i) = l_i$. *We set $p' = p/l, q_i' = q_i/l_i$. Suppose that for $i = 1, \ldots, n-2$, $q_i \geq q_{i-1}'(q_{i-1}' - 1)(q_{i-1}' - p' - 1) + q_{i-1}' l_i$. Then $\overline{V}$ is a set-theoretic complete intersection.*

Proof The proof is by induction, the case $n = 3$ is Theorem 4.6.8. The case $n - 1$ implies n follows from Theorem 4.6.13. In the case where $l_i = 1$ for all i we have that for $i = 1, \ldots, n-2$, there exist positive integers a_i, b_i such that $q_i = a_i p' + b_i q'_{i-1}$, $0 \leq a_i \leq q'_{i-1} - 1$. $\overline{V}$ is the zero set of the polynomials

$$F_1 := x^{q_0} - y^p w^{q_0 - p}, F_2, \ldots, F_{n-1},$$

where F_{i-1} is obtained, by applying the trick used in the proof of Theorem 4.6.7, from

$$((z_i - x^{a_i} z_{i-1}^{b_i})^{q_{i-1}})^h_{z_{i-1}^p = x^{q_{i-1}}},$$

where $(H)_{y^p = x^{q_0}}$ means substitution when possible x^{q_0} by y^p, and H^h is the homogenization of H with respect to w.

Acknowledgements The author thanks Vietnam Institute for Advanced Study in Mathematics (VIASM) and the Institute of Mathematics of Hanoi for their support and hospitality. Sincere thanks are due to Le Tuan Hoa, Ngo Viet Trung, Mesut Sahin and the anonymous referees for many useful suggestions which has improved the presentation of the lecture notes.

References

1. M. Barile, M. Morales, On the equations defining projective monomial curves. Commun. Algebra, **26**, 1907–1912 (1998)
2. M. Barile, M. Morales, A. Thoma, On simplicial toric varieties which are set-theoretic complete intersections. J. Algebra **226**(2), 880–892 (2000)
3. H. Bresinsky, Monomial Gorenstein curves in A^4 as set-theoretic complete intersections. Manuscr. Math. **27**, 353–358 (1979)
4. H.A. Casillas, M. Morales, Sums of toric ideals (2012). arXiv:1211.5386
5. H. Cohn, *Advanced Number Theory*, vol. XI (Dover, New York, 1980), 276 p.
6. L. Coudurier, M. Morales, Classification of courbes toriques dans l'espace projectif, module de Rao et liaison. J. Algebra **211**(2), 524–548 (1999)
7. D. Eisenbud, B. Sturmfels, Binomial ideals. Duke Math. J. **84**(1), 1–45 (1996)
8. K. Eto, Defining ideals of semigroup rings which are Gorenstein. Commun. Algebra **24**(12), 3969–3978 (1996)
9. K. Fischer, W. Morris, J. Shapiro, Affine semigroup rings that are complete intersections. Proc. Am. Math. Soc. **125**, 3137–3145 (1997)
10. R. Hartshorne, Complete intersections in characteristic $p > 0$. Am. J. Math. **101**, 380–383 (1979)
11. J. Heger, Denkschriften. Kais. Akad. Wissensch. Mathem. Naturw. Klasse **14**, II (1858)
12. E. Kunz, *Introduction to Commutative Algebra and Algebraic Geometry* (Birkhüuser, Boston, 1985), 238 p.
13. T.T. Moh, Set-theoretic complete intersections. Proc. Am. Math. Soc. **94**, 217–220 (1985)
14. M. Morales, Syzygies of monomial curves and a linear Diophantine problem of Frobenius. Max Planck Institut für Mathematik, Bonn-RFA (1987). Preprint
15. M. Morales, Equations des Variétés Monomiales en codimension deux. J. Algebra **175**, 1082–1095 (1995)
16. M. Morales, A. Thoma, Complete intersection lattices ideals. J. Algebra **284**(2), 755–770 (2005)

17. T.H.N. Nhan, On set-theoretic complete intersection monomial curves in $\mathbf{P}^3$. Lobachevskii J. Math. **34**(2), 133–136 (2013)
18. T.H.N. Nhan, M. Sahin, Equations defining recursive extensions as set theoretic complete intersections. Tokyo J. Math. **38**, 273–282 (2015)
19. L. Robbiano, G. Valla, Some curves in $\mathbf{P}^3$ are set-theoretic complete intersections, in *Algebraic Geometry - Open Problems. Proceedings of Ravello 1982*, ed. by C. Ciliberto, F. Ghione, F. Orecchia. Lecture Notes in Mathematics, vol. 997 (Springer, Berlin, 1983), pp. 391–399
20. M. Sahin, Producing set-theoretic complete intersection monomial curves in $\mathbf{P}^n$. Proc. Am. Math. Soc. **137**(4), 1223–1233 (2009)
21. M. Sahin, Extensions of toric varieties. Electron. J. Comb. **18**(1), 93, 1–10 (2011)
22. T. Schmitt, W. Vogel, Note on set-theoretic intersections of subvarieties of projective space. Math. Ann. **245**, 247–253 (1979)
23. A. Schrijver, *Theory of Linear and Integer Programming* (Wiley, Chichester, 1998), xi, 471 p.
24. A. Thoma, On the binomial arithmetical rank. Arch. Math. **74**(1), 22–25 (2000)

Editorial Policy

1. Lecture Notes aim to report new developments in all areas of mathematics and their applications – quickly, informally and at a high level. Mathematical texts analysing new developments in modelling and numerical simulation are welcome.

   Manuscripts should be reasonably self-contained and rounded off. Thus they may, and often will, present not only results of the author but also related work by other people. They may be based on specialised lecture courses. Furthermore, the manuscripts should provide sufficient motivation, examples and applications. This clearly distinguishes Lecture Notes from journal articles or technical reports which normally are very concise. Articles intended for a journal but too long to be accepted by most journals, usually do not have this "lecture notes" character. For similar reasons it is unusual for doctoral theses to be accepted for the Lecture Notes series, though habilitation theses may be appropriate.

2. Besides monographs, multi-author manuscripts resulting from SUMMER SCHOOLS or similar INTENSIVE COURSES are welcome, provided their objective was held to present an active mathematical topic to an audience at the beginning or intermediate graduate level (a list of participants should be provided).

   The resulting manuscript should not be just a collection of course notes, but should require advance planning and coordination among the main lecturers. The subject matter should dictate the structure of the book. This structure should be motivated and explained in a scientific introduction, and the notation, references, index and formulation of results should be, if possible, unified by the editors. Each contribution should have an abstract and an introduction referring to the other contributions. In other words, more preparatory work must go into a multi-authored volume than simply assembling a disparate collection of papers, communicated at the event.

3. Manuscripts should be submitted either online at www.editorialmanager.com/lnm to Springer's mathematics editorial in Heidelberg, or electronically to one of the series editors. Authors should be aware that incomplete or insufficiently close-to-final manuscripts almost always result in longer refereeing times and nevertheless unclear referees' recommendations, making further refereeing of a final draft necessary. The strict minimum amount of material that will be considered should include a detailed outline describing the planned contents of each chapter, a bibliography and several sample chapters. Parallel submission of a manuscript to another publisher while under consideration for LNM is not acceptable and can lead to rejection.

4. In general, **monographs** will be sent out to at least 2 external referees for evaluation.

   A final decision to publish can be made only on the basis of the complete manuscript, however a refereeing process leading to a preliminary decision can be based on a pre-final or incomplete manuscript.

   Volume Editors of **multi-author works** are expected to arrange for the refereeing, to the usual scientific standards, of the individual contributions. If the resulting reports can be forwarded to the LNM Editorial Board, this is very helpful. If no reports are forwarded or if other questions remain unclear in respect of homogeneity etc, the series editors may wish to consult external referees for an overall evaluation of the volume.

5. Manuscripts should in general be submitted in English. Final manuscripts should contain at least 100 pages of mathematical text and should always include

   - a table of contents;
   - an informative introduction, with adequate motivation and perhaps some historical remarks: it should be accessible to a reader not intimately familiar with the topic treated;
   - a subject index: as a rule this is genuinely helpful for the reader.
   - For evaluation purposes, manuscripts should be submitted as pdf files.

6. Careful preparation of the manuscripts will help keep production time short besides ensuring satisfactory appearance of the finished book in print and online. After acceptance of the manuscript authors will be asked to prepare the final LaTeX source files (see LaTeX templates online: https://www.springer.com/gb/authors-editors/book-authors-editors/manuscriptpreparation/5636) plus the corresponding pdf- or zipped ps-file. The LaTeX source files are essential for producing the full-text online version of the book, see http://link.springer.com/bookseries/304 for the existing online volumes of LNM). The technical production of a Lecture Notes volume takes approximately 12 weeks. Additional instructions, if necessary, are available on request from lnm@springer.com.

7. Authors receive a total of 30 free copies of their volume and free access to their book on SpringerLink, but no royalties. They are entitled to a discount of 33.3 % on the price of Springer books purchased for their personal use, if ordering directly from Springer.

8. Commitment to publish is made by a *Publishing Agreement*; contributing authors of multiauthor books are requested to sign a *Consent to Publish form*. Springer-Verlag registers the copyright for each volume. Authors are free to reuse material contained in their LNM volumes in later publications: a brief written (or e-mail) request for formal permission is sufficient.

Addresses:
Professor Jean-Michel Morel, CMLA, École Normale Supérieure de Cachan, France
E-mail: moreljeanmichel@gmail.com

Professor Bernard Teissier, Equipe Géométrie et Dynamique,
Institut de Mathématiques de Jussieu – Paris Rive Gauche, Paris, France
E-mail: bernard.teissier@imj-prg.fr

Springer: Ute McCrory, Mathematics, Heidelberg, Germany,
E-mail: lnm@springer.com

Printed in the United States
By Bookmasters